Smart Membranes and Sensors

Scrivener Publishing
100 Cummings Center, Suite 541J
Beverly, MA 01915-6106

Publishers at Scrivener
Martin Scrivener(martin@scrivenerpublishing.com)
Phillip Carmical (pcarmical@scrivenerpublishing.com)

Smart Membranes and Sensors

Synthesis, Characterization, and Applications

Edited by

Annarosa Gugliuzza

Scrivener Publishing

WILEY

Co-published by John Wiley & Sons, Inc. Hoboken, New Jersey, and Scrivener Publishing LLC, Salem, Massachusetts.
Published simultaneously in Canada.

For general information on our other products and services or for technical support, please contact our Customer Care Department within the United States at (800) 762-2974, outside the United States at (317) 572-3993 or fax (317) 572-4002.

Wiley also publishes its books in a variety of electronic formats. Some content that appears in print may not be available in electronic formats. For more information about Wiley products, visit our web site at www.wiley.com.

For more information about Scrivener products please visit www.scrivenerpublishing.com.

Cover design by Russell Richardson

Library of Congress Cataloging-in-Publication Data:

ISBN 978-1-118-42379-0

Printed in the United States of America

10 9 8 7 6 5 4 3 2 1

To my Lovely Family and Honey Billa

Contents

Preface

Unquestionably, the human being is the organism provided with the finest perception, because he is the most complex system of receptors of heat, cool, sounds, light and smells. In the human body, physical/chemical agents, in fact, pass through biological membranes reaching the receptors, while electrical signals are transmitted to the brain through nerve networks. The brain transduces marvelously each single response into a sensation.

With this awareness, many scientists have attempted to reproduce artificial sensing systems over the last years, trying to mimic natural structures and processes. Despite how arduous the accomplishment of such a task seemed, many efforts were made in this direction, moving from 'sense' to 'sense-to-react' systems. Today the major ambition is to go further. The desired target is the creation of ultra-smart systems, wherein the functions of sensing, acting and adapting are sequentially integrated. Within this frame work, the membranes can play a key role in the build up of complex arrays, where complementary smart functions can be allocated and integrated. Indeed, the molecular manipulation in membranes allows effectively to tailor desired properties on different length scales, supplying confined functional spaces and geometries for storage, release, separation, chemical reaction, energy/mass transfer, but also for shield/microclimate regulation, cleaning, fluid flows on molecular length scale, controlled cell growth, high-throughput screening for biological processes interrogation (and so on).

In this context, it is convenient to introduce briefly the concepts of membrane and sensor. The former is a semipermeable interface enabling the selective passage of molecular species, while blocking others. The latter signifies a device capable of detecting a physical, chemical or electrical response, which is converted by a transducer into a signal immediately perceptible by the human eye or measurable on an instrument. It so happens that, when a detection function is coupled with an adaptive transport, the membrane works as an ultra-smart system. In this way, the membrane adapts itself to surrounding environment, adjusting its own structure and chemistry in order to regulate mass/energy flow and/or transfer signals/information in response to external physical and/or chemical inputs.

In this perspective, adaptive membranes are expected to accelerate the passage from smart to ultra-smart systems, bringing large benefits to many technologically sophisticated areas such as telemedicine, microfluidics, drug delivery targeting, (bio)separation, textiles, clean power production, environment monitoring, agro-food safety, cosmetics, architecture, and automotive and so on. The use of membrane sensors becomes much more attractive if the modular scalability of the membrane technology is considered. A wider potential of smart membranes-based systems may be explored in the design of integrated industrial plants as well as in the creation of miniaturized devices, where molecular objects can sense and respond on one chip.

Numerous publications emerged in the literature dealing with sensing materials or membrane separations distinctly. Few of these contributions, however, are dedicated to dealing with sensor-like membranes. The intent of this book is to join these two concepts catalyzing the process of integration between complementary disciplines in order to share knowledge and expertise on this matter and construct a mutual language which can draw many and many researchers, investigators, graduated students and final users in the world of the smart science and technology.

This book contains insightful contributions from scientists working in the field of sensor materials and membranes. It covers various points of view, including the choice of materials and techniques for assembling responsive membranes and interfaces with ability to transport mass and energy on demand, along with the description of appropriate techniques for monitoring molecular scale events, which regulate the smartness of multifunctional objects needed to the accomplishment of developed applications.

Part I comprises three chapters, which deal with some sensors materials for membranes such as carbon nanotubes, ionic liquids, and light-responsive hydrogels, along with self-assembling lipids, polymers, and small molecules for the fabrication of perm-selective membranes and vesicular structures with ability to work as submicro-reactors, catalysts and drug delivery vehicles.

Part II is entirely dedicated to the description of molecular interactions, which cause the interfaces to self-adjust and restore morphology, chemistry and charge for preserving original properties against hostile external conditions, self-powering molecular diffusion and directing biomolecule recognition. Weak interactions that dominate the world of self-assembled materials and supra-molecular structures are discussed from a theoretical and experimental point of view.

In Part III, three chapters describe molecular recognition mechanisms directed to control drug release and bioseparation. Following an overview on self-assembled nanoporous membranes used as platforms for biosensors, an extensive discussion is dedicated to the fabrication of membranes bearing recognition sites and their use in bioseparation processes; the responsive activity of mesoporous silica nanoparticles, zeolites, molecularly imprinted membranes, biomemitic affinity membranes, and membranes containing cyclodextrins is examined.

In Part IV, four advanced applications of like-sensors membranes are presented: electrospun membranes for the construction of ultrasensitive sensors, which facilitate analyte adsorption, mass and electric charge transport; 3-D conductive scaffolds enabling one to monitor cell behavior, study chronic disease models, and repeat dose experiments; sensing particles prepared by membrane emulsification and with ability to transport active substances and/or convert chemical and biochemical signals into optical, electrical, thermal and mechanical signals; adaptive membranes for ultra-smart textiles, which can provide self-maintenance, adaptability, auto-adjustment and long-distance communication through heat storage and thermo-regulation, modular breathability, protection, self-cleaning, odor capture and drug delivery as well as electrical signal transmission.

I am very pleased to have edited this book and I am very grateful to each of the contributors for their dedication and cooperation. This book would not have been possible without their enthusiasm to share knowledge, passion and time. My hope is that everyone enjoys reading and using this cross-disciplinary discussion for bringing innovation to there own research.

Annarosa Gugliuzza
July 2014

Part 1

SENSING MATERIALS FOR SMART MEMBRANES

1

Interfaces Based on Carbon Nanotubes, Ionic Liquids and Polymer Matrices for Sensing and Membrane Separation Applications

María Belén Serrano-Santos*, Ana Corres Ortega and Thomas Schäfer

POLYMAT, University of the Basque Country, Donostia-San Sebastián, Spain

Abstract
The combination of carbon nanotubes with widely tunable materials such as ionic liquids and polymers theoretically provides a tremendous degree of freedom for designing thin films (membranes) for specific sensing and separation applications. Not surprisingly, a plethora of applications has been reported in literature primarily for sensing devices. This chapter discusses some selected case studies which illustrate, on one hand, the exciting new opportunities which a combination of these materials offer; on the other, it stresses the strong need for evaluating their potential in the context of existing devices such as to appreciate their true benefit.

Keywords: Ionic liquids, polymer membranes, carbon nanotube hybrids, thin film sensors

1.1 Introduction

Sensing and membrane separation applications seem, at first sight, two very distinct areas of applications of interfaces. However, both have in common that their very first step and at times overall performance is governed by a selective interface establishing a selective interaction with desired compounds (Figure 1.1). Subsequently, this interaction has in most cases be transduced into a signal in sensors, or result in a transport of compounds

**Corresponding author*: serrano.belen@gmail.com

Annarosa Gugliuzza (ed.) Smart Membranes and Sensors, (1–20) 2014 © Scrivener Publishing LLC

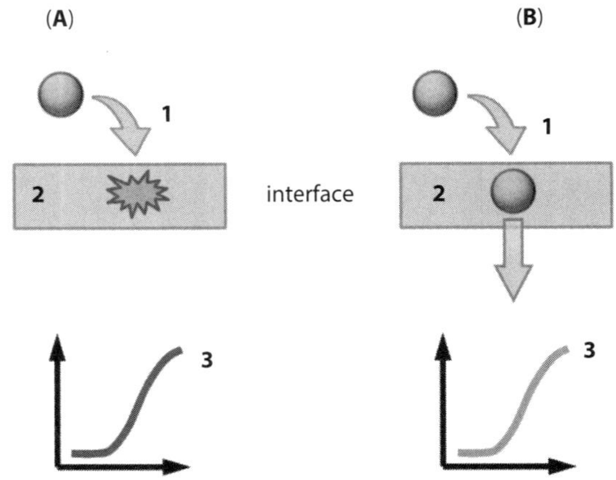

Figure 1.1 (*a*) In sensor applications, a selective interface warrants the interaction with a desired sample or target (1), resulting in an uptake, adsorption or recognition event, (2), which then is transduced into an output signal such as an electric current or an optical signal; (***b***) in membrane applications, the same interface can be used to preferentially concentrate or recognize desired compounds present upstream like in sensors (1), but these compounds then diffuse, convectively flow or are transported across the interface (2) and leading to an increase in the concentration downstream (3). Both concepts can also be combined, for example, to generate what may be considered "smart membranes".

across the interface in membranes, two phenomena which may require adaptation and possibly even compromising to some extent the pristine properties of the interface for the benefit of an overall improved performance. Therefore, a high intrinsic selectivity of interfaces and possibly a high degree of adaptation is indispensable. Sensor and membrane interfaces have, hence, in common that both require a maximum selectivity for target compounds. Membranes allow furthermore an optimization of the preferential transport of the target compounds across the interface. The complementarity of both sensors and membranes can therefore give rise to hybrid systems of enhanced overall performance [1].

Polymers are excellent matrices for such interfaces owing to their versatility, tremendous range of possible physico-chemical properties and their tunability. However, modifying the physico-chemical properties of polymers may go along with an undesired change in their mechanical properties. For example, polydimethylsiloxane (PDMS), widely known as silicone is a mechanically flexible polymer suitable for creating thin films as selective interfaces as much as self-standing thick films. Although the material is hydrophobic, its hydrophobicity and hence affinity for certain organic compounds can be further increased by gradually substituting short-chain methyl groups by longer-chain octyl groups. While the resulting polyoctylmethylsiloxane (POMS) yields thin films of significantly improved selectivity, its mechanical stability suffers slightly such as to not permit stable self-standing thick-films [2]. This drawback is of little relevance in practice where the focus is on supported thin films, such as is the case in sensor applications [3], but it illustrates that physico-chemical and mechanical properties are often interdependent.

As a consequence, hybrid materials or mixed-matrix materials are often conceived of that try to combine the best of several worlds into a single matrix. In this case, a support material providing the mechanical stability such as a polymer is doped with additives or carriers that further increase the selectivity of the interface. In such a modular system, one seeks to use as dopant another versatile material whose selectivity can be optimized without considering mechanical requirements, as these are taken care of by the support materials. In the following, examples of using ionic liquids and carbon nanotubes as materials for interfaces, individually or combined – also with other materials such as polymers - will be discussed. Rather than giving an exhaustive overview of the field which covers a tremendous amount of research articles published over the last 10–15 years, it will focus on individual case-studies in order to discuss in more detail key aspects of the respective applications.

1.2 Ionic Liquid-Carbon Nanotubes Composites for Sensing Interfaces

Carbon nanotubes (CNTs), cylindrically shaped sheets of graphene, can be arranged in thin films using either single-walled carbon nanotubes (SWCNT) or multi-walled carbon nanotubes (MWCNT). Based mainly on their outstanding electrochemical properties, as well as their overall optical transparency when used as thin films, there is a plethora of electrochemical, electronic and optical applications which can be found thoroughly reviewed elsewhere [4]. With regard to sensors, applications focus mainly on the electrochemical properties of CNTs given their high conductivity of up to 400 $kS{\cdot}cm^{-1}$ [4]. Similarly, ionic liquids possess attractive electrochemical properties given that their wide electrochemical window of up to 5.8 V [5].

Ionic liquids (ILs) are salts that remain liquid at or close to ambient temperature owing to the large size of their ions and their conformational flexibility, strong Coulomb and dispersion interactions, as this results in

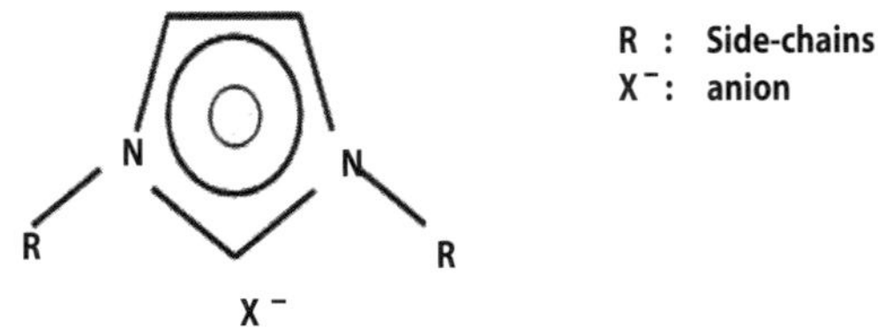

Figure 1.2 Basic structure of an imidazolium-based ionic liquid [RRIM][X] whose physico-chemical properties can be tailored through selecting an appropriate anion X^- as well as choosing adequate side-chains R. For example, 1-butyl-3-imidazolium hexafluorophosphate [BMIM][PF_6] is widely immiscible with water while 1-butyl-3-imidazolium tetrafluoroborate [BMIM][BF_4] is fully miscible; on the other hand, 1-ethyl-3-imidazolium tetrafluoroborate [EMIM][BF_4] possesses a dynamic viscosity η of 40 cP at room temperature while the dynamic viscosity of 1-decyl-3-imidazolium tetrafluoroborate [DMIM][BF_4] is about eight times higher, about 300 cP [10].

both a small lattice enthalpy as well as large entropy, favouring in this way thermodynamically the liquid rather than the solid state [6, 7]. By the time research on ILs grew exponentially, focus was mainly on their indeed unique property of possessing virtually no volatility given their electrolytic nature, and non-volatility was often than not used synonymous to non-toxicity. Both aspects have meanwhile been put into perspective [8, 9] and while ILs certainly are in the context of many applications virtually non-volatile and can possibly be non-toxic, these parameters need to be confirmed on a case to case basis.

Particularly regarding the toxicity of ILs, or their environmental benignity, also the synthetic route leading to the formation of ILs needs to be considered. In recent years, research in the field of ILs has therefore focused on other not less appealing assets of ionic liquids, such as the possibility to widely tune their physico-chemical properties depending on their structure (Figure 1.2). Introducing appropriate functional groups or side chains allows tuning to a certain extent their chemical affinity and/or miscibility with other solvents [10]. Choosing appropriate anions/cations can furthermore serve to modulate to a significant extent the viscosity of ILs [10], a highly important parameter to be taken into consideration for sensor and (membrane) separation applications. Because of their outstanding electrochemical properties, it therefore is understandable that ionic liquid/carbon nanotube composites have widely been studied as sensor interfaces. The group of Aida pioneered in 2003 what they called "bucky gels", dispersions of CNTs in imidazolium-based ionic liquids [11]. The principle breakthrough was in this case the use of a solvent which would allow a homogenous

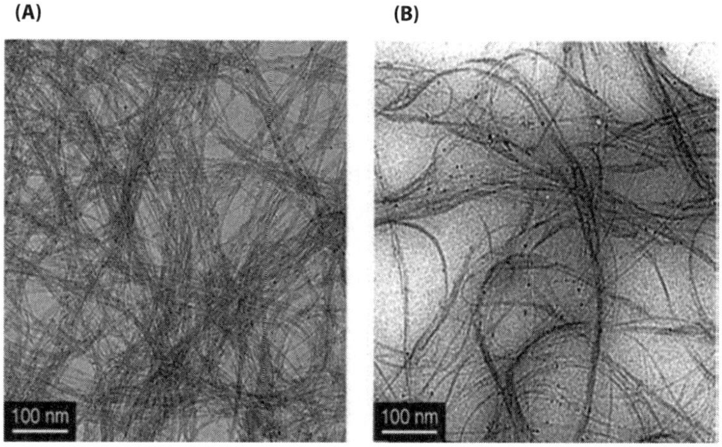

Figure 1.3 TEM micrographs of HighPeo SWNTs (A) as received from a commercial supplier and (B) obtained by mixing a bucky gel of [BMIM][BF_4] with deionized water (with permission from reference [13]).

dispersion of carbon nanotubes. It has later been demonstrated that this is due to a shielding effect the ionic liquid exerts on the π-π interactions between CNTs which otherwise is responsible for the low solubility of the CNTs in conventional solvents [12] as is illustrated in Figure 1.3.

"Bucky gels" have been employed as interfaces for a variety of analytical applications, ranging from biomolecules to gas compounds, which may appear surprising provided that the respective analytical challenge and experimental conditions are significantly distinct. Perspectives of these materials were discussed by Fukushima and Aida [13]; Tunckol *et al.* thoroughly reviewed recent developments and applications of these materials [14]. Some of the examples which reflect research efforts made in the area but also illustrate open questions and challenges to be tackled are discussed in the following.

DNA-hybridization was followed electrochemically by immobilizing first single-stranded DNA (ssDNA) on a composite membrane comprising 1-butyl-3-methylimidazolium hexafluorophosphate (BMIM PF_6), nano-sized CeO_2 and single-walled carbon nanotubes (SWNTs), and then expose this interface to target DNA [15]. The authors claim that the physico-chemical nature of the support allowed an enhanced loading of ssDNA and thus a higher sensitivity for the target DNA, based on voltammetric measurements. A very similar approach was taken with virtually the same interface concept by Li *et al.* [16] using thrombin-binding aptamer for the detection of lead cations (Pb^{2+}). In either case, no mention was made on the stability of the composite membrane, possible effects of dissolution of ionic liquid into the liquid environment and consequently interference on the DNA structure, but measurements involving DNA-targets with a single mismatch revealed a high specificity of the sensor surface suggesting that neither leaching took place, nor was the presence of ionic liquid apparently detrimental to the hybridization kinetics. In a conceptually similar work, electrochemical DNA biosensors were assembled by radiation-induced graft polymerization. Hereby poly(glycidyl methacrylate) was first grafted onto MWCNTs and then reacted with 1-butylimidazolium bromide. Owing to this reaction, the IL does in this case in fact lose its principal property, mainly being a liquid salt. Probe DNA was then physisorbed onto this surface and its binding to target DNA followed electrochemically [17]. The authors claim that the coverage of probe DNA onto the CNT-IL surface was significantly increased compared to other surfaces; however, direct comparisons with other surface coverage data as well as other immobilization techniques such as surface sensitive techniques (e.g., surface Plasmon resonance) were unfortunately not reported.

In general, it should be pointed out that the detection of biomolecules such as DNA can be very much prone to experimental artefacts as a high surface coverage of probe DNA requires a thorough shielding of the negative surface charge of DNA during immobilization, while the success of subsequent detection of target DNA can strongly depend on the accessibility of the probe DNA on the surface of the selective interface: depending on the surface chemistry, DNA may adsorb either in form of extended brushes, stretched flat over the surface, or both. In addition, while target DNA is expected to hybridize with probe DNA, it might just as well also adsorb non-specifically onto a sensor surface provided there is space enough. Hence, provided that electrochemical methods are indirect, results obtained on DNA hybridization would ideally be cross-checked with other, for example, optical methods in order to validate the analytical procedure.

A somewhat similar approach as the aforementioned was followed for the detection of purines using bucky gel surfaces. Simultaneous detection of adenine and guanine was reported using a PbO_2–CNT–IL composite film and voltammetric detection [18]. The simultaneous detection of adenine and guanine is challenging given their structural similarity but apparently the method proposed allowed the quantitative determination of adenine and guanine in DNA, although the respective results were not explored in detail. As in the previous case, also interaction mechanisms were not further elucidated such that it remains unclear why the bucky gels did indeed boost analytical specificity in comparison with other analytical methods.

The detection of xanthenes, and in particular theophylline, was reported using a system of platinum decorated MWCNTs with the ionic liquid 1-octyl-3-methylimidazolium hexafluorophosphate [OMIM][PF_6] serving as a binder between the Pt-nanoparticles and the CNTs [19]. The detection limit was indicated to be in the nanomolar range similar to the aforementioned. No explicit recognition or interaction mechanism was described which explains in more detail why purines were selectively detected by this system. An entirely different analytical challenged was posed in a work reporting on the detection of nitric oxide using gel electrodes consisting of 1-hexyl-3-methylimidazolium hexafluorophosphate ([HMIM][PF_6]) and SWCNTs, with the electrode furthermore protected by a Nafion-membrane [20]. The latter system was reported to have a good long-term stability even though nitric oxide reacts at the electrodes to nitrite going along with the formation of a proton. This is interesting as hexafluorophosphate-based ionic liquids are in general to be avoided for applications in aqueous environments as they react in time with protons causing not only disintegration of the IL but also leading to the formation of HF [21]. The use of Nafion membranes might even accelerate this effect through cation-exchange as Nafion possesses sulfonyl-groups which are neutralized either by sodium cations or, in aqueous solution, by protons [22].

CNT-IL composites have also been investigated as supports for enzymes, creating in this way interfaces that catalyze reactions leading to an electrochemically measurable signal. To name a few, glucose dehydrogenase (GDH), for example, was incorporated by drop coating onto a CNT-IL interface, again with the IL being [BMIM][PF_6] and with the aim of making use of the large surface area of CNTs and the supposedly biocompatible ionic liquid medium. Good analytical results were found for glucose detection but no further indications were given in how far the interface studied excelled existing and commercially available glucose-detection devices. Cytochrome C was reported to be immobilized in a chitosan-IL-CNT composite and the electrocatalytic activity of the interface studied via the reduction of H_2O_2 [23]. UV-vis absorption spectra revealed no significant change of the cytochrome C fingerprint when entrapped in the CNT-IL interface and both electrocatalytic activity and reproducibility of the measurements were found to be excellent. The IL involved in this study was 1-ethyl-3-methyl-imidazolium tetrafluoroborate [EMIM][BF_4]. In comparison, Baker *et al.* found that a similar ionic liquid, [BMIM][BF_4], could either stabilize or destabilize cytochrome C depending on its concentration [24]. [EMIM][BF_4] is entirely water miscible in contrast to the aforementioned [BMIM][PF_6]. While even [BMIM][PF_6] can be expected to leach from an interface when being in contact with water, as occurs in electrochemical applications, owing to its electrolytic nature, the stability of the chitosan-[EMIM][BF_4]-CNT interface in an aqueous environment is remarkable.

Although it can be repeatedly found stated in reports on electrochemical applications of ionic liquids involving biomolecules, ionic liquids cannot be considered solvents that are *per se* biocompatible. While there exist ILs that may heat-protect enzymes or warrant long-term stability of proteins during storage, others may detrimentally affect particularly biomolecules whose conformation determines activity and function, such as enzymes or oligonucleotides. This is why compatibility needs to be studied taking into account a variety of physico-chemical parameters that affect in particular biomolecule-IL interactions, requiring an evaluation on a case-to-case basis [25]. Detrimental interactions can be a reduced water activity through competition between biomolecules and ILs (ILs are electrolytes and, hence, inherently prone to favourably interact with water) similar to what is observed when precipitating DNA from ethanol-water solutions, or direct interference of ILs with the structural conformation of biomolecules through, e.g., hydrogen bonding. Choline-based Ils, for example, may change significantly the pH of aqueous solutions which in turn can negatively affect the activity or function of biomolecules, and hexafluorophosphate-based ionic liquids may react with protons setting free HF. Precisely with ILs offering a wide structural variety leading to manifold physico-chemical properties, it can easily be understood that a generalized perception of biocompatibility of ILs cannot be established.

In this section, examples of IL-CNT composites were discussed in which the electrochemical aspects of both materials were predominant while interaction with target compounds was of minor importance. IL-CNT composites, however, can also find applications in which specific interactions of either material with the target are not only beneficial, but strongly desired such as the detection or separation of gases and solvents.

At the early stage of what is now a booming research in ionic liquid technology, Brennecke *et al.* reported the combination of ILs with supercritical carbon dioxide (scCO_2) extraction as a possible benign hybrid process with manifold applications [26]. The particularity of the process was that ILs were found to take up significant amount of CO_2 without, however, mixing significantly with the CO_2-phase. As a consequence, ILs could be used as practically non-volatile extracting solvents in separation processes and scCO_2-extraction could be employed to recover without contamination the extracted compounds from the IL. With emerging concerns about greenhouse gases, in particular CO_2, focus shifted during the following years from the separation process aspect described by Brennecke *et al.* toward their observation that CO_2 dissolved in a high molar ratio (0.6 at 8MPa and 25°C) in the IL employed, [BMIM][PF_6], triggering a whole new research area on using ILs as matrices for capturing or separating gases, particularly from emissions.

1.3 Ionic Liquid Interfaces for Detection and Separation of Gases and Solvents

ILs have been reported to interact favourably with acid gases such as CO_2 and SO_2 which contribute significantly to air pollution and the so-called greenhouse effect. Making use of the tunability of ionic liquids, Wu *et al.* reported on the absorption capacity of SO_2, by tetramethylguanidinium ([TMG]) lactate [27]. At 1 bar and 40 °C, an equimolar absorption of SO_2 took place in the IL equalling a mass fraction of 0.305 g SO_2/g IL. This was significantly higher than absorption capacities for the common [BMIM][PF_6] and [BMIM][BF_4] which gave weight fractions orders of magnitude lower, namely 0.14 wt-% and 0.1 wt-%, respectively. The reason for the very high affinity of the [TMG]-based IL was due to a chemical reaction between the NH_2-group of the TMG-cation, which was claimed to be reversible under vacuum. The work of Wu *et al.* illustrates several aspects of interactions of ILs with gases.

First, it sets to a certain extent the level of absorption capacity that is expected from a material such that it may be considered a competitive candidate for creating selective interfaces. For example, metal-organic frameworks (MOFs) are currently studied as materials of exceptional high absorption capacities for gases, amongst which CO_2, and MOF-74 absorbs at 1.1 bar and 25°C about 0.214 g CO_2/g MOF-74 [28] which is about a third lower than what Wu *et al.* reported for [TMG] lactate, but as opposed to the latter, based on physisorption and in absence of any chemical reaction. Hence, the weight fractions of 0.14 wt-% of CO_2 absorbed under similar conditions in [BMIM][PF_6] seem far from being exceptional, although it can be found manifold stated otherwise in literature.

Second, the study of Wu *et al.* shows that experimental findings need to be put into a context. [BMIM][PF_6] possesses a 40 % higher CO_2 absorption capacity than [BMIM][BF_4] under their experimental conditions – but about 150–200 times less than [TMG] lactate and MOF-74, respectively. Certainly, literature data reveal that a high CO_2 solubility can be achieved in some ILs. In a recent work [29] it was shown that at 25 °C, 1-ethyl-3-methyl butylimidazolium trifluoroacetate, [EMIM][TFA], absorbs CO_2 up

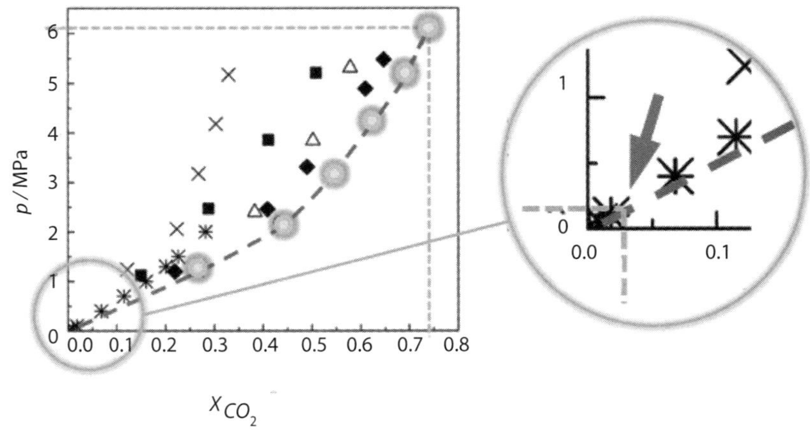

Figure 1.4 CO_2 solubility in ILs at 298 K, with ◆: 1-ethyl-3-methylimidazolium diethylphosphate [EMIM][DEP]; ▪: 1-ethyl-3-methylimidazolium methylsulfate [EMIM][$MeSO_4$]; x: 1-ethyl-3-methylimidazolium methane sulfonate [EMIM][$MeSO_3$]; Δ: 1-(2-hydroxyethyl)-3-methylimidazolium trifluoroacetate [OHEMIM][TFA]; □,: 1-ethyl-3-methylimidazolium trifluoroacetate [EMIM][TFA]; ○: 1-hexyl-3-methylimidazolium bis(trifluoromethylsulfonyl)imide [HMIM][Tf_2N], highlighted in orange (modified reprint with permission from [29]).

to a molar fraction of 0.8 (Figure 1.4) which can be converted into a mass fraction of 0.157 g CO_2/g IL – a value in the order of magnitude of the aforementioned exceptional MOF-74.

However, this high solubility is measured at a pressure of 6 MPa or 60 bars; a closer look at virtual standard conditions, 0.1 MPa or 1bar, reveals a solubility of only about 0.004 CO_2/g IL or 0.4 wt-% and thus 40 times less as well as far from exceptional. This observation might seem trivial, but is unfortunately more than justified considering wide-spread assumptions made in literature where high absorption capacities of ILs are considered synonymous with high affinity irrespective of the experimental conditions.

Third, both sensor and membrane systems aim at large-scale production of devices and therefore require a robust and simple design which allows maintaining production and operation costs low. Any new material should therefore ideally be seen in the context of the performance of existing devices and a pondering of gain in performance versus increase in production costs. For example, in the field of sensors, commercially available metal-oxide sensors which are well-characterized have been widely employed in manifold sensing applications. In particular SnO_2-sensors have been successfully used for the detection of carbon dioxide by optimizing their operation mode rather than modifying the selective metal-oxide interface [30, 31]. To the authors' best knowledge there unfortunately does not exist any comparative cost-benefit study between such commercial sensors operated under optimized conditions and the IL-based sensors under study.

Nevertheless, the aforementioned example also proved that ILs can indeed be to some extent tailor-made for improving their interaction with gases. Furthermore, being liquids the mobility/diffusivity of solutes and gases can be of 1–2 orders of magnitude higher in ILs than in solid matrices such as polymers [32]. This is of utmost importance when dealing with IL-interfaces for both sensor and membrane applications. As a consequence, in practice RTILs exhibit higher response and transport rates, both of which are crucial parameters in both sensor and separation systems. In the area of gas sensors, ILs have mainly be used as liquid depositions on piezoelectric devices such as quartz crystal microbalances (QCMs). The deposition is straightforwardly achieved by coating of the quartz electrode surface by by drop-coating [32] or spin-coating [33]. Different vapour absorption patterns could be found when contacting QCM-sensors coated with imidazolium-based ILs of varying anions and side-chain length [34].

Figure 1.5a depicts, for example, the response of [BMIM][PF_6] (the original figure erroneously states [BMIM][**B**F_6]) and 1-hexyl-3-methylimidazolium chloride [HMIM][Cl]. The anions of ILs are commonly regarded as crucial for interactions with other solutes. Following theoretical a priori considerations, both ionic liquids should exhibit a similar uptake pattern of the vapours tested since [PF_6] and [Cl] both establish mainly hydrogen bonds, with those bonds being stronger via [Cl]. Indeed, this was corroborated by the observation that [HMIM][Cl] absorbed more ethanol than [BMIM][PF_6] while the latter apparently responded most to acetone. Both ILs seemed to absorb high amounts of dichloromethane, however care has to be taken in interpreting these vapour uptake measurements as the vapours tested resulted in different vapour concentrations according to their saturated vapour pressure. A respective normalization and graphical representation visualizes clearly (Figure 1.5b) that while [HMIM][Cl] responded principally to ethanol and hardly to any other vapour, the responsiveness of [BMIM][PF_6] was more equilibrated but an order of magnitude lower. One may also conclude that [HMIM][Cl] is at best an ethanol sensitive IL owing to the halogen-anion while [BMIM][PF_6] is not sensitive to any of the vapours. The finding indicates a current bottleneck in the application of ILs, namely the apparent limitation of not making full use of the about 10^{14} potential variations of RTILs reported [10] from which those with the most desired physico-chemical properties can be chosen for a particular application. This limitation results from a still limited understanding of how ILs interact as bulk materials with gases and vapours. A priori calculations

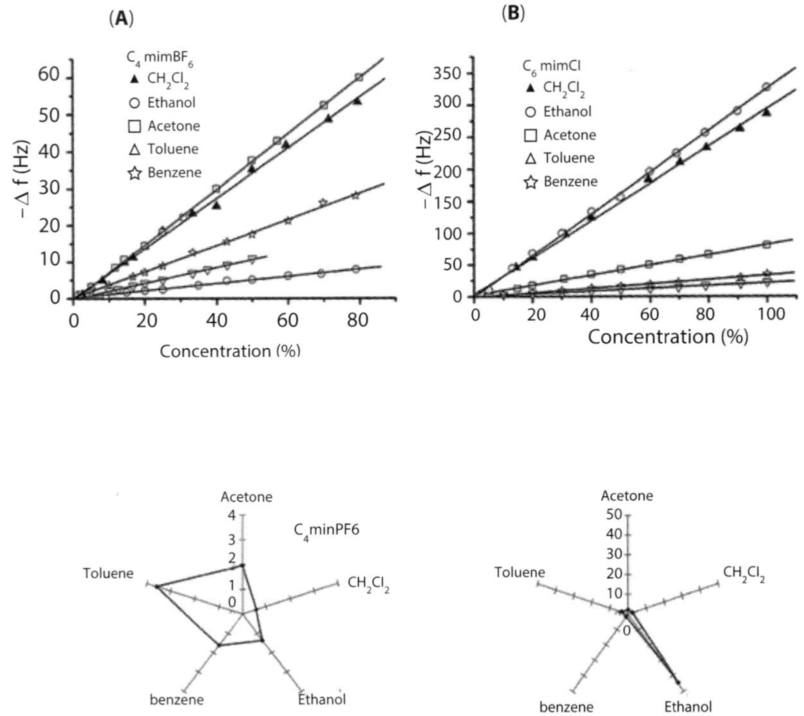

Figure 1.5 (***a***) Frequency shifts measured upon contacting a quartz-crystal microbalance on which an ionic liquid film was deposited with various organic vapours and (***b***) resulting radar plots using the slope of the linear frequency response of each vapour as the discriminating parameter; the ionic liquids used were (left) 1-butyl-3-methylimidazolium hexafluorophosphate [BMIM][PF_6] and (right) 1-hexyl-3-methylimidazolium bromide [HMIM][Cl] (reprinted with permission from [34]).

provide important insights into how ILs and other molecules interact on a molecular basis [7] but may not fully predict the behaviour of ILs as a bulk phase [35]. Consequently and in absence of a global theoretical interaction model, the quest for particularly selective ILs remains a trial and error approach rather than a systematic search.

Irrespective of the progress in this direction, IL sensing surfaces nonetheless find their application in the field of characterization of physical and mass transport phenomena occurring in ILs under gas/vapour uptake. The viscosity and viscoelastic behaviour of ILs strongly depends on the concentrations of dissolved volatile compounds or gases, and access to the respective data is crucial for process applications where ILs serve as a solvent or interface. While these data are widely known for the existing limited lot of conventional solvents, ILs as "designer-solvents" are under continuous development and every all new structure with a potential application will require characterization from the scratch. Therefore, and accounting for the somewhat trial and error approach with which ILs are still developed, one easily becomes aware of the need for rapid screening methods of application-relevant physicochemical parameters of ionic liquids. In this context, IL-modified sensors that not only allow quick characterization of gas/vapour uptake and selectivity but also simultaneously provide insights into mass transport (diffusivity), viscosity or viscoelastic behaviour, can make a significant contribution. Acoustic sensors seem to be particularly suitable for this purpose. Ouali *et al.* used a Love wave sensor in order to verify the Newtonian behaviour of ILs [36]. Love Waves are horizontally polarized shear waves. Hereby the IL was not used as a stationary phase but rather passed over the sensor and the acoustic response correlated with the viscosity-density product. QCMs have been used for the same purpose for decades but it is claimed that Love wave sensors could be miniaturized further and in this way reduce sample volume which would allow an in-line high-throughput characterization of ionic liquids with regard to their Newtonian behaviour particularly in combination with microfluidic devices. A complementary acoustic method is the use of QCMs in combination with dissipation monitoring (QCM-D). Serrano-Santos *et al.* showed that depositing thin 1-octyl-3-methyl imidazolium chloride ([OMIM][Cl]) films on the respective sensors, insights into the viscoelastic behaviour of ILs can be obtained upon exposure to vapour and gases [37, 38]. In the case of water, the response of the QCM-D to changing vapour concentrations proved to be fast, reaching a very stable plateau value within less than a minute (Figure 1.6a).

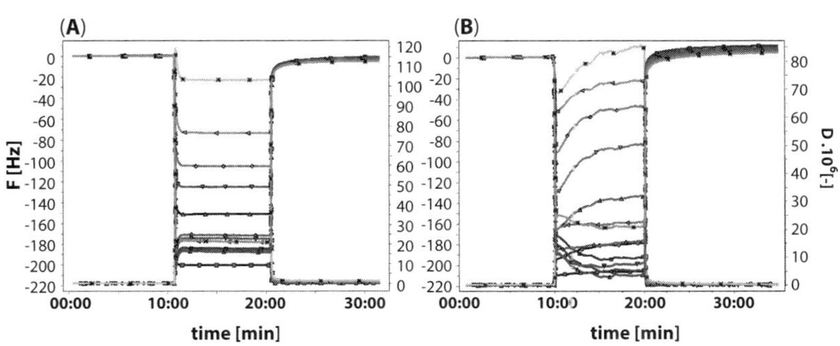

Figure 1.6 Frequency shifts (blue) and dissipation changes (red) as measured upon contacting a IL-modified quartz-crystal microbalance with dissipation monitoring (QCM-D) with ***(a)*** water and ***(b)*** toluene vapour. The ionic liquid employed was 1-octyl-3-methylimidazolium chloride [OMIM][Cl] and experiments were conducted at room temperature. Harmonics 3–13 are shown with darker colours indicating lower harmonics.

In contrast, measured frequency changes using toluene as a sample vapour started to level off significantly later (Figure 1.6b), which was attributed to both different diffusivities of the solutes as well as their interaction with the IL resulting in a rearrangement of the IL molecules within the film. Water as the smaller molecule possesses an intrinsically higher diffusivity than toluene; however, it could also be seen that even when the frequency shift of toluene had already reached a stable signal (at about 15 min, Fig. 1.6b), the respective dissipation values were still rising, indicating changes of the viscoelasticity properties resulting from an ongoing rearrangement of the ionic liquid phase which apparently was independent of diffusion and directly related to the interaction between toluene and [OMIM][Cl]. Figure 1.6 depicts the various harmonics of the base frequency of the sensor used, namely 5 MHz. Higher harmonics are considered to be more sensitive to phenomena occurring close to the sensor surface, while the lower harmonics protrude more into the liquid film. As a consequence, anisotropies that might occur across the film upon solute uptake can be qualitatively detected which can give hints on how the solute is embodied within the liquid matrix. It is noteworthy that in both water and toluene vapour uptake measurements the baseline was satisfactorily recovered, with a slight spreading of the frequencies being observed during desorption after contacting with toluene which might indicate incomplete desorption. The latter, however, appeared to be negligible in view of the magnitude of the response to solute vapour. The authors concluded that QCM-D is an outstanding tool for determining concurrently uptake of volatile solutes and associated changes of the viscoelastic properties.

1.4 Ionic Liquid-Polymer Interfaces for Membrane Separation Processes

The development of selectively capturing components of flue gases such as CO_2, H_2S or SO_2 from industrial gas mixtures as well as organic vapour emissions has become an important issue in recent years due to the increasing demand to reduce green house gas emission and air pollution in general. Membrane separation processes have proven to successfully tackle many gas and vapour related separation challenges. They suffer, however, from the drawback that when adjusting the affinity of the membrane polymer by introducing respective functional groups, other parameters such as mechanical stability might also change [2]. A possible way to overcome this handicap is the use of polymer blends or mixed-matrix membranes. Hereby, the affinity of a polymer support structure is enhanced by integrating a highly selective second phase whose physicochemical properties can to some degree be adjusted without sacrificing the overall membrane stability. Based on the aforementioned, ILs are an excellent candidate for serving as a second liquid phase in membranes [39]. Conventionally, membrane separations involving ionic liquids are based on the principle of supported ionic liquid membranes (SILMs) involving the immobilization of the ionic liquids in a porous membrane support structure. Although SILMs have been widely reported in literature owing to their facile preparation [39–42], important disadvantages have also been identified such as little stability during operation under pressure, limited control over actual ionic liquid content in the pores as well as instabilities during solute uptake, resulting from a decreased viscosity and surface tension of the ionic liquid.

Integrating ILs directly into a non-porous membrane polymer matrix can overcome these drawbacks. Rather than immobilizing the ILs in a porous membrane structure that by itself does not bear any selectivity, ILs can be added to polymers as a selective plasticizer for forming dense polymeric mixed-matrix membranes during casting [43, 44]. Owing to the high mobility of the ionic liquid phase incorporated, it can be anticipated that both membrane selectivity as well as membrane permeability can be increased. Corres *et al.* studied this concept employing polyether block amide (commercial trade name PEBAX®) as a base material owing to its widespread use in a variety of applications and its excellent mechanical stability [44]. ILs were chosen according to the supposed different degrees of interactions with the block co-polymer in order to elucidate how membrane performance could be correlated to molecular interactions occurring in the polymer blend. The ionic liquid-polymer blend membranes synthesized with 20 wt-% of imidazolium-based ILs incorporated possessed an excellent mechanical stability and transparency, the latter indicating that the IL was dispersed throughout the polymer and that no phase segregation had taken place (Figure 1.7). The membrane transport properties could be modulated in accordance with the affinity of the ionic liquids incorporated and the respective sample vapours, confirming that the incorporation of the ILs indeed allowed tuning the overall membrane performance. For example, PEBAX® blended with 1-butyl-3-methyl imidazolium acetate possessed a lower selectivity for toluene and ethyl acetate in comparison with the pristine membrane polymer. All ILs employed, however, resulted in membrane blends which were also more permeable to water and ethanol. This finding was in line with [34] and raises the doubt whether the interaction of polar compounds such as water and ethanol with ILs can at all be avoided given the electrolytic

Figure 1.7 PEBAX®-[OMIM][Cl] composite, non-porous membrane with 20 wt-% of 1-octyl-3-methylimidazolium chloride. The background text illustrates the optical transparency of the membrane with a thickness of about 60 μm.

nature of the latter. In principle, however, it could be shown that stable and functional membrane-IL blends can be obtained in which the selectivity can be modulated by simply adjusting the physico-chemical properties of the IL alone and without compromising the overall membrane stability.

1.5 Conclusions

Discussing specific applications involving ionic liquids, carbon nanotubes and polymers it could be seen that while these materials and their combination can offer exciting new opportunities for creating selective interfaces either for sensors or membranes, their performance must nevertheless be evaluated in the context of existing devices. This will warrant that the potential of these materials is being assessed thoroughly rather than focusing only on the most predominant benefits, as the latter might result in the worst case in an unnecessary investment of significant research efforts.

Acknowledgement

M.B.S.S. would like to acknowledge the Marie Curie Fellowship 236628-ILISENSE and T.S. the ERC Starting Grant 209842-MATRIX.

References

1. Schäfer, T.; Serrano-Santos, M.B.; Rocchi, S.; Fuoco, R. *Analytical and Bionanalytical Chemistry* 34 (2006) 860–866
2. Schäfer, T.; Heintz, A.; Crespo, J.G. *Journal of Membrane Science* 254 (2005) 259–265
3. Fort, A.; Rocchi, S.; Serrano-Santos, M.B.; Ulivieri, N.; Vignoli, V.; Pioggia, G.; Di Francesco, F. *Sensors and Actuators B: Chemical* 111–112 (2005) 193–199
4. Hu, L.; Hecht, D.S.; Grüner, G. *Chemical Reviews* 110 (2010) 5790–5844
5. Torimoto, T.; Tsuda, T.; Okazaki, K.; Kuwabata, S. *Advanced Materials* 22 (2010) 1196–1221
6. Krossing, I.; Slattery, J.M.; Daguenet, C.; Dyson, P.J.; Oleinikova, A.; Weingärtner, H. *Journal of the American Chemical Society* 128 (2006) 13427–13434
7. Rezabal, E.; Schäfer, T. *Journal of Physical Chemistry B* 117 (2013) 553–562
8. Earle, M.J.; Esperança, J.M.S.S.; Gilea, M.A.; Lopes, J.N.C; Rebelo, L.P.N.; Magee, J.W.; Seddon, K.R.; Widegren, J.A. *Nature* 439 (2006) 831–834
9. Pretti, C.; Chiappe, C.; Pieraccini, D.; Gregori, M.; Abramo, F.; Monni, G.; Intorre, L. *Green Chemistry* 8 (2006) 238–240
10. Zhang, S.; Sun, N.; He, X.; Lu, X.; Zhang, X. *Journal of Physical and Chemical Reference Data* 35 (2006) 1475–1517
11. Fukushima, T.; Kosaka, A.; Ishimura, Y.; Yamamoto, T.; Takigawa, T.; Ishii, N.; Aida, T. Science 300 (2003) 2072
12. Wang, J.; Chu, H.; Li, Y. *ACS Nano* 2 (2008) 2540-2546
13. Fukushima, T.; Aida, T. *Chemistry - A European Journal* 13 (2007) 5048-5058
14. Tunckol, M.; Durand, J.; Serp, P. *Carbon* 50 (2012) 4303–4334
15. Zhang, W.; Yang, T.; Zhuang, X.; Guo, Z.; Jiao, K. *Biosensors and Bioelectronics* 24 (2009) 2417–2422
16. Li, Y.; Liu, X.; Ning, X.; Huang, C.; Zheng, J. Zhang. J. *Journal of Pharmaceutical Analysis* 1 (2011) 258–263
17. Chung, D.; Whittaker, A.K.; Choi, S. *Journal of Applied Polymer Science* 126 (2012) 28–33
18. Liu, T.; Zhu, X.; Cui, L.; Ju, P.; Qu, X.; Ai, S. *Journal of Electroanalytical Chemistry* 651 (2011) 216–221
19. Liu, L.; Xiao, F.; Li, J.; Wu, W.; Zhao, F.; Zeng, B. *Electroanalysis* 20 (2008) 1194 - 1199
20. Li, C.M.; Zang, J.; Zhan, D.; Chen, W.; Sun, C.Q.; Teo, A.L.; Chua, Y.T.;Lee, V.S.; Moochhala, S.M. *Electroanalysis* 18 (2006) 713- 718

21. Swatloski, R.P.; Holbrey, J.D.; Rogers, R.D. *Green Chemistry* 5 (2003) 361–363
22. Schäfer, T.; Di Paolo, R.E.; Franco, R.; Crespo, J.G. *Chemical Communications* 20(2005) 2594–2596
23. Zhang, Y.; Zheng, J. *Electrochimica Acta* 54 (2008) 749–754
24. Baker, S.N.; Zhao, H.; Pandey, S.; Heller,W.T.; Bright, F.V.; Baker, G.A. *Physical Chemistry Chemical Physics* 13 (2011) 3642–3644
25. Zhao, H. *Journal of Chemical Technology and Biotechnology* 85 (2010) 891–907
26. Blanchard, L.A.; Hancu, D.; Beckman, E.; Brennecke, J.F. *Nature* 399 (1999) 28–29
27. Wu, W.;Han, B.; Gao, H.; Liu, Z.; Jiang, T.; Huang, J. *Angewandte Chemie International Edition* 43 (2004) 1415–1417
28. Millward, A.G.; Yaghi, O.M. *Journal of the American Chemical Society* 127 (2005) 17998–17999
29. Mejía, I.; Stanley, K.; Canales, R.; Brennecke, J.F. *Journal of Chemical Engineering Data* 58 (2013) 2642-2653
30. Fort, A.; Gregorkiewitz, M.; Machetti, N.; Rocchi, S.; Serrano, B.; Tondi, L.; Ulivieri, N.; Vignoli, V.; Faglia, G.; Comini, E. *Thin Solid Films* 418 (2002) 2–8
31. Fort, A.; Rocchi, S.; Serrano-Santos, M.B.; Mugnaini, M.; Vignoli, V.; Atrei, A.; Spinicci, R. *Sensors and Actuators B: Chemical* 116 (2006) 43–48
32. Schäfer, T.; Di Francesco, F.; Fouco, R. *Microchemical Journal* 85 (2007) 52–56
33. Serrano Santos, M.B; Llobet, E.; Schäfer, T. *Procedia Engineering* 25 (2011), 1169–1172
34. Xu, X.; Li, C.; Pei, K.; Zhao, K.; Zhao, Z.K.; Li, H. *Sensors and Actuators B: Chemical* 134 (2008) 258–265
35. Carvalho, P.J.; Coutinho, J.A.P. *Energy Fuel*s 24 (2010) 6662–6666
36. Ouali, F.F.; Doy, N. McHale, G.; Hardacre, C.; Ge, R.; Allen, R.W.K.; MacInnes, J.M.; Newton, M.I. *Analytical Chemistry* 83 (2011) 6717–6721
37. Serrano-Santos, M.B.; Llobet, E.; Schäfer, T. *Procedia Engineering* 25 (2011) 1169–1172
38. Serrano-Santos, M.B.; Schäfer, T. *submitted* (2014)
39. Schäfer, T.; Branco, L.C.; Fortunato, R.; Izák, P.; Rodrigues, C.M.; Afonso, C.A.M.; Crespo, J.G. *ACS Symposium Series* 902 (2005) 97–110
40. Noble, R.D.; Gin, D.L. *Journal of Membrane Science* 369 (2011) 1–4
41. Fortunato, R.; Afonso, C.A.M.; Reis, M.A.M.; Crespo, J.G. *Journal of Membrane Science* 242 (2004) 197–209
42. Lozano, L.J.; Godínez, C; De los Ríos, A.P.; Hernández-Fernández, F.J.; Sánchez-Segado, S.; Alguaci, F.J. Journal of Membrane Science 376 (2011) 1–14
43. Bernardo, P.; Jansen, J.C.; Bazzarellia, F.; Tassellia, F.; Fuoco, A.; Friess, K.; Izák, P.; Jarmarovác, V.; Kačírkovác, M.; Clarizia, G. *Separation and Purification Technology* 97 (2012) 73–97
44. Corres, A.; Schäfer, T. *submitted* (2014)

2

Photo-Responsive Hydrogels for Adaptive Membranes

David Díaz Díaz[1,2,*] and Jeremiah A. Johnson[3]

[1]*Institut für Organische Chemie, Fakultät für Chemie und Pharmazie, Universität Regensburg, Regensburg, Germany*
[2]*IQAC-CSIC, Barcelona, Spain*
[3]*Department of Chemistry, Massachusetts Institute of Technology, Cambridge, Mass, USA*

Abstract
Synthetic membranes that show a selective and reversible response to a specific environmental cue (e.g., a change in pH, temperature, ionic strength, electromagnetic field, irradiation) are platform materials for high-tech applications that demand switchable material properties. Mechanistically, stimuli-responsive membranes leverage reversible molecular properties of the membrane components, like conformation, polarity, and/or reactivity, to induce changes in the bulk pore structure. Among the many possible external stimuli, light is particularly attractive as it can be applied both spatially and temporally without diffusion limitations. This chapter describes the permeability properties of a selection of representative photo-responsive hydrogel membranes. The membranes are classified according to their chromophore structures.

Keywords: Hydrogels, polymer gels, photo-responsive, membranes

2.1 Introduction

A membrane is usually defined as an interphase between two bulk phases that acts as a selective and porous barrier for regulating mass transport between two compartments. Membranes are essential for life. For example, natural skin acts as a permeable multifunctional membrane that interacts

**Corresponding author*: david.diaz@chemie.uni-regensburg.de

Annarosa Gugliuzza (ed.) Smart Membranes and Sensors, (21–52) 2014 © Scrivener Publishing LLC

with the surrounding environment and responds to light, temperature, humidity, chemicals, and mechanical stress. Another example is the nuclear pore membrane, which regulates selective transport of biological molecules and polymers into and out of the nucleus of all eukaryotic cells. Inspired by the incredible material properties of biological membranes, extensive research efforts have focused on the design of artificial membranes that are able to respond, especially in a reversible manner (i.e., valve or gate function), to specific environmental cues. Such materials could have applications in a broad range of modern technologies that demand switchable material properties (e.g., sensors, delivery systems, optoelectronics, transducers, actuators, catalysis, purification). The broad potential of these materials is also reflected in the significant number of published reviews focused on the latest developments and applications of multifunctional responsive polymeric membranes [1–7].

From a mechanistic point of view, stimuli-responsive membranes utilize changes in the conformation, polarity, and/or reactivity of specific responsive functional groups in the membrane bulk or on the membrane surface to induce a change in pore structure and, hence, filtration properties. During the last three decades, functional groups that respond to changes in pH, temperature, ionic strength, specific chemical species, light, electric fields, magnetic fields, and mechanical stresses have all been applied to regulate membrane properties; pH, temperature, and light-responsive membranes have been the most studied [1–7, 8]. Light is a very attractive stimulus for switching material properties because its intensity and wavelength can be easily controlled in a spatial and temporal manner to achieve precise, selective, fast, energy-efficient functional group modification without diffusion limitations. Mother Nature leverages these unique properties of light in the context of biological membranes; photo-induced proton-coupled electron transfer reactions across cellular membranes are responsible for all life on Earth. Again, where Mother Nature leads the way scientists often follow: the incorporation of photo-responsive functional groups (i.e., chromophores) into polymeric membranes has received wide theoretical and experimental attention [3, 9–10]. In these systems, light drives actuation through reversible photo-induced isomerization or ionization reactions of the chromophores, or simply via photo-thermal effects. If the chromophore is attached to a suitable polymeric support, then light-induced changes at the chromophore level (e.g., dipole moment, charge, color, size) can manifest themselves as changes in macroscopic material properties (e.g., wettability, permeability, density, viscosity). Azobenzene, spiropyran, triphenylmethane leuco, diarylethene, and viologen derivatives have arguably been the most investigated chromophores in the context of photo-responsive membranes. These systems have been discussed in several excellent reviews [9, 11, 12].

Hydrogels are molecular networks that swell in water. Responsive hydrogels [7, 13–19] and tunable hydrogel membranes [4, 20–21] have attracted considerable interest in biomedicine due to their unique properties such as environmental-responsiveness to biological cues, reversible phase transitions at physiological temperatures, swellability in water, hydrophilicity, and biocompatibility [22]. The molecular composition of the hydrogel, the chemical methods used for its preparation, the cross-linking density, and the nature of the permeant strongly impact the functionality of hydrogel membranes [1, 23]. As is the case for all responsive membranes, incorporation of pendant stimuli-responsive units into hydrogel networks can factilitate control of the hydrogel membrane's bulk properties. Once again, pH- and temperature-sensitive materials obtained from synthetic and/or naturally derived hydrophilic building blocks are common platform components of hydrogel-based membranes [4–5, 18, 20, 24–26]. Despite the considerable number of photo-responsive hydrogel systems reported in the literature [27–28], their potential use for adaptive membranes began to receive more attention after the earlier studies carried out with their pH- and thermally-actuated analogues.

This chapter presents key examples of photo-responsive hydrogel membranes. The examples are classified by their chromophore, which is the key molecular specie that interacts with light to induce a bulk material response. In principle, any photo-responsive hydrogel could be fabricated into a planar membrane material; most phoro-responsive hydrogels have not been studied in this context. Therefore, this chapter focuses on examples where the permeability properties of the materials were characterized, i.e., the material was intended to be a membrane. A discussion of photo-thermally sensitive materials is also included. These materials represent a particular subclass of photo-responsive membranes wherein light is used to heat the material and induce a bulk thermal transition. Other photo-responsive polymer systems (i.e., photo-responsive polymer brushes [29]), or hydrogel membranes with responses to external stimuli other than light, are excluded from the scope of this chapter.

2.2 Photo-Responsive Hydrogel Membranes

2.2.1 Photo-Responsive Moiety: Cinnamylidene

Cinnamylidene (CA) is a versatile photo-crosslinkable group that undergoes an efficient [2+2] dimerization reaction in response to visible and long-wavelength UV light. The reaction is photo-reversible; retro [2+2] cycloaddition occurs upon exposure to short-wavelength UV light.

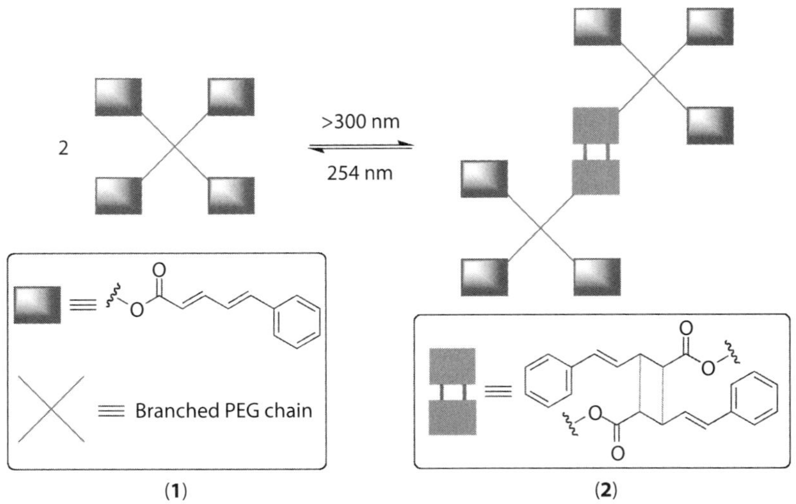

Figure 2.1 Reversible [2+2] photo-dimerization/photo-scission of b-PEG-CA **1** (λ = 313 nm; ε = 37390 L mol^{-1} cm^{-1}). Adapted with permission from Ref. [30]. Copyright (1996) American Chemical Society.

Almost two decades ago, Russell and coworkers described the preparation of poly(ethylene glycol) (PEG) hydrogels via photo-initiated end-linking of four-armed PEG star polymers bearing terminal cinnamylidene acetyl (CA) groups (Figure 2.1) [30]. As a proof of concept, tetrakis-CA macromer **1** was obtained in ca. 70% yield by simple esterification of CA acid chloride with hydroxyl terminated four-armed PEG (b-PEG, MW = 15,000 Da). A film of this polymer was cast onto a glass surface and irradiated with light (450 W medium pressure UV lamp, l > 300 nm) for 5 min to 3 h to produce hydrogels. Upon light exposure in solution, the parent *trans*-CA species undergoes rapid isomerization to the corresponding *cis*-isomer. Then, end-linking takes place via a photo-induced [2+2] cycloaddition reaction between an excited (*) CA group of one chain with a ground state CA group of another chain to yield cyclobutane products [30].

The intermolecular dimerization of CA groups from two different **1** molecules was necessary to form gels; no gels formed with either unmodified b-PEG (i.e., absence of CA groups) or an equimolar aqueous mixture of cinnamylidene acetic acid and b-PEG.

These hydrogels possess many unique features with important practical implications: (*1*) The gelation occurs without the need for potentially toxic photo-sensitizers and/or photo-initiators; (*2*) the junction density, and therefore the degree of swelling ($DS = (Ws\text{-}Wd)/Wd$, where Ws = weight of the gel in air after swelling and Wd = weight of the dry gel), of the hydrogels can be fine-tuned by varying the average number of CA moieties on macromer **1** and/or the irradiation time (*DS* decreases as the degree of substitution or the time of irradiation increase. For instance, $DS = 58 \pm 4$ for a degree of substitution of 25%, whereas $DS = 35 \pm 5$ for a degree of substitution of 64%); (*3*) The "degelation" via photo-induced retro-[2+2] cycloaddition of **2** occurs within minutes under 254 nm irradiation; and (*4*) the hydrogels display remarkable anti-thrombogenic properties, e.g., 99.6% reduction in platelet deposition onto poy(methyl methacrylate) (PMMA) coated surfaces, which is of fundamental importance for applications that require contact with blood (Figure 2.2).

Similar CA-based photo-responsive hydrogels comprised of a larger b-PEG (MW = 20,000, $Đ = 1.026$) was later used for the fabrication of membranes for protein transport [31]. In that study, the permeation of model proteins with varied molecular weight and hydrodynamic radius (equine myoglobin (Mb), bovine hemoglobin (Hb), and bovine lactate dehydrogenase-L (LDH)) through these membranes was evaluated using a horizontal side-by-side diffusion cell apparatus with defined compartment volume and diffusion cross-section area (Figure 2.3) [31]. A typical experiment consisted of three main steps: (*1*) A homogeneous membrane of uniform thickness (STDV $< 5 \times 10^{-4}$ mm) was prepared by photo-induced gelation of an aqueous solution of **1** within a quartz mold (mold dimensions: $2 \times 1.5 \times 0.08$ cm^3); (*2*) the hydrogel membrane was swollen for 12 h in a Tris-HCl buffer (pH 7.5), and then placed between the donor and receptor cells, which were filled with the protein solution and Tris-HCl, respectively; and (*3*) aliquots from the receptor cell were periodically collected and analyzed by UV-vis spectroscopy. Aliquots from the receptor cell were replaced with fresh buffer in order to maintain a concentration gradient between the two cells. The

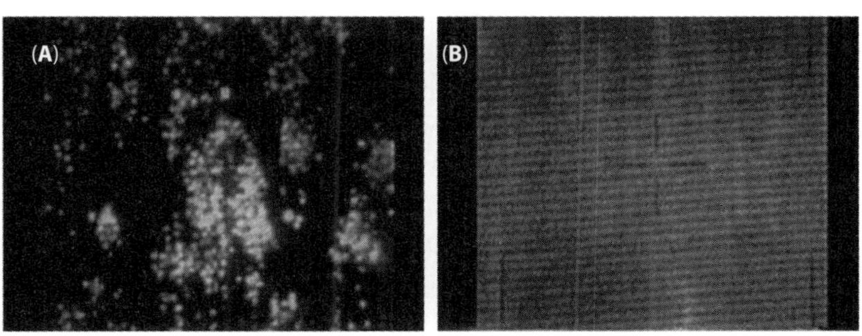

Figure 2.2 Images of blood flow over (A) PMMA and (B) PMMA coated with b-PEG-CA hydrogel. The photographs show that platelets adhere readily to the PMMA surface. In contrast, platelets flow over the b-PEG-CA-coated PMMA surface without depositing on it. Adapted with permission from Ref. [30]. Copyright (1996) American Chemical Society.

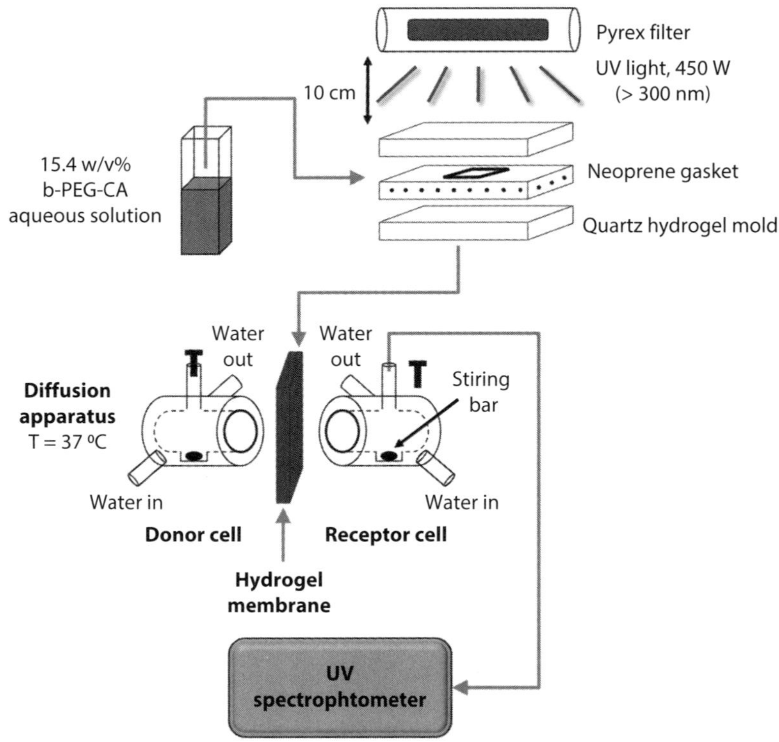

Figure 2.3 Representation of the procedure followed to prepare membranes based on b-PEG-CA, and the diffusion apparatus used for the permeation studies. Cell volume = 3 mL. Cell diffusion cross-section area, ID = 9 mm. Adapted with permission from Ref. [31]. Copyright (1998) Elsevier.

diffusion coefficient of the proteins through the hydrogel membrane was calculated from the equation $D = P \times l / K$, where P is the permeability coefficient, l is the thickness of the swollen membrane, and K is the partition coefficient of the proteins through the membrane measured spectrophotometrically by solute uptake experiments. P was estimated from Colton's equation: $\ln\{(C_o - 2Ct) / C_o\} = -2APt/V$, where C_o is the concentration of the protein in the donor cell (mg mL^{-1}), Ct the concentration of the protein (mg mL^{-1}) in the receptor cell at time t (min), A is the area of permeation (cm^2), V is the volume of each cell (mL), and t the time of permeation (min).

The authors predicted that "as mesh size and swellability decrease, the rate of protein diffusion would also decrease as long as the mesh size was close to that of the protein diameter." For Mb, the smallest protein studied, increased irradiation time (decreased mesh size) gave a relatively small decrease in diffusion coefficient (Table 2.1). In this case, the mesh size was

Table 2.1 The diffusion coefficients of Mb, Hb, and LDH are shown as a function of light exposure, and degree of functionality. Adapted with permission from Ref. [31]. Copyright (1998) Elsevier.

Proteins *a*	**Thickness of gel membrane (mm ± STDV)** *b*	**Time of light exposure (λ> 300 nm)**	**Average mesh size (Å)** *c*	**Diffusion coefficients ($cm^2\ min^{-1}$)**	**PEG-CA modification**
Myoglobin (MW = 17600 Da) (Æ = 44 Å)	1.62 ± 0.01	2	242	7.0e-5	85%
	1.58 ± 0.02	3	200	6.20e-5	85%
	1.52 ± 0.08	12	154	3.70e-5	85%
Hemoglobin (MW = 64500 Da) (Æ = 77 Å)	1.57 ± 0.01	6	181	8.0e-6	85%
	1.52 ± 0.08	12	154	2.76e-6	85%
	1.50 ± 0.01	21	171	1.05e-5	68%
	1.50 ± 0.01	21	142	9.70e-6	68%
	1.30 ± 0.02	21 + 1 h of 254 nm irradiation	155	1.12e-5	68%
Lactate dehydrogenase-L (MW =140000 Da) (Æ = 172 Å)	1.59 ± 0.05	12	165	2.25e-6	68%
	1.50 ± 0.01	12 + 45 min of 254 nm irradiation	216	4.62e-6	68%

a The concentrations of myoglobin and hemoglobin were evaluated from a corresponding protein calibration curve. An activity assay (Diagnostics Lactate Dehydrogenase reagent) based on the oxidation of lactate to pyruvate was used to measure the LDH activity in the receptor cell.

b The thickness of the gels was measured using an electron micrometer and an average between different experiments was taken. Each membrane was measured at three different places to assess its un iformity and an average was taken.

c The Flory-Huggins model was utilized to determine the mesh size and the average molecular weight between cross-links of the synthesized hydrogels.

consistently larger than the protein diameter; the permeation of Mb was minimally affected. As the protein size increased, the decrease in diffusion coefficient with increased irradiation time was much more pronounced; for Hb there was a 56% decrease in the diffusion coefficient in gels irradiated for 12 h compared to gels irradiated for 6 h.

An alternative strategy to modify the mesh size is to change the average degree of functionality of the monomer. For example, there was a 73% decrease in the diffusion coefficient of Hb through hydrogel membranes prepared from b-PEG-CM macromers with 85% CM functionality compared to similar gels from macromers with 68% functionality. Increasing the macromer functionality has the same effect as increasing the irradiation time; the 73% decrease in diffusion coefficient observed upon increasing the macromer functionality from 68% to 73% could also be achieved by irradiating samples of equivalent functionality for ca. 7.5 h longer.

Therefore, there is a direct relationship between the mesh size of the b-PEG-CA hydrogel membrane (which is controlled by the initial functional group concentration, the irradiation time, and the irradiation wavelength) and the diffusion coefficients of permeating solutes. The solute size also plays a critical role. In a visual experiment, UV irradiation was used to decrease the mesh size of the b-PEG-CA hydrogel membranes, and consequently initiate the release of fluorescein-isothiocyanate (FITC) dextran from the network into the external aqueous solution (Figure 2.4).

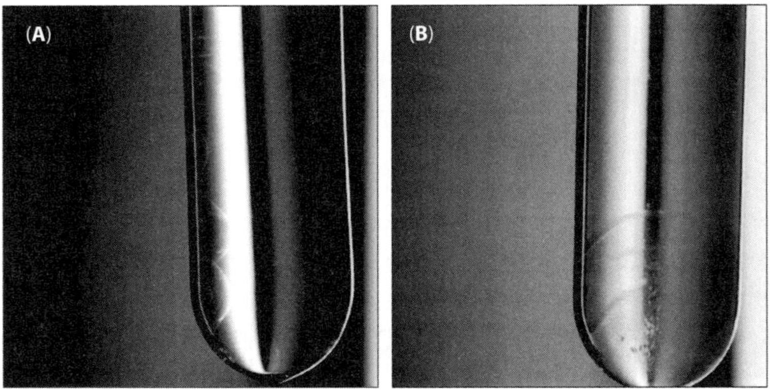

Figure 2.4 The release of FITC dextran (MW = 145,000 Da) from b-PEG-CA hydrogel (degree of modification = 70%) in (A) water, and (B) upon irradiation with a 254 nm light for 40 min. 0.02 g of dextran-FITC were dissolved in aqueous b-PEG-CA solution (25 w/v%) and the sample irradiated with light ($\lambda > 300$ nm) at the side of a quartz vial for 2 h. Following gelation the gel was equilibrated in water for 1 day. Reproduced with permission from Ref. [31]. Copyright (1998) Elsevier.

The hydrogel matrix was almost completely dissolved after exposure to 254 nm irradiation for 40 min [31].

2.2.2 Photo-Responsive Moiety: Triphenylmethane Leuco Derivatives

Triphenylmethane leuco derivatives are well-known photo-chromic molecules that undergo heterolytic S_N1-type bond cleavage under UV irradiation to expel an anionic species and generate intensely colored triphenylmethyl cations. In the absence of light, the reverse nucleophilic addition can occur to regenerate the triphenylmethane leuco derivative. Irie, Tanaka, and co-workers were the first to study the photo-induced reversible swelling of polyacrylamides (PAAm) that possessed pendant triphenylmethane leucohydroxide (TPMLH) or leucocyanide (TPMLC) groups [32–34] (compounds **3a-b**, Figure 2.5). The gels were prepared in polyethylene tubes by free radical copolymerization of acrylamide and diphenyl(4-vinylphenyl)-methane leucohydroxide or leucocyanide, respectively, in DMSO at 60 °C for 3–4 h. After this time, the gels were first removed from the tubes and soaked in DMSO and then in water to remove all residual monomers and initiatiors, and finally swollen to the equilibrium condition on standing in water overnight.

The so-obtained chromophoric hydrogels displayed remarkable swelling properties. When exposed to UV irradiation ($\lambda > 270$ nm) for 1 h at fixed temperature and pH the hydrogels underwent a large (up to 1700%) weight increase, which was the result of photo-induced ionization of the triphenylmethane leuco derivatives and a corresponding increase in the material's hydrophilicity [33]. After removal of the light source, the materials returned to their original size within hours due to ionic recombination

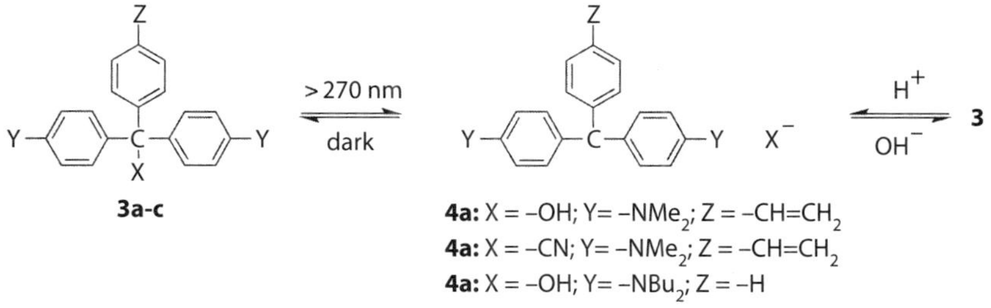

Figure 2.5 Chemical structures of triphenylmethane leuco derivatives and their response to UV irradiation and pH change. Adapted with permission from Ref. [32–33,35]. Copyrights (1986, 1990) American Chemical Society, (1988) The Chemical Society of Japan.

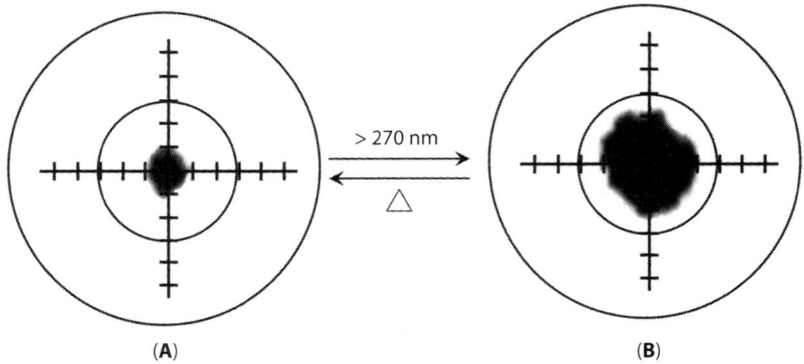

Figure 2.6 Photo-induced swelling of polyacrylamide hydrogel with pendant triphenylmethane leucocyanide groups (3.1 mol %): (A) Before photo-irradiation; (B) after irradiation with UV light ($\lambda > 270$ nm; pH 6.5). (B) Returned to the size of (A) in the dark. Upon irradiation, gel weight increases of as much as 18 times (1700%) were reported for TPMLC-based hydrogels. Adapted with permission from Ref. [32]. Copyright (1986) American Chemical Society.

and a corresponding decrease in hydrophilicity (Figure 2.6). Similar swelling phenomena were not observed for a PAAm gel that lacked TPMLH groups in the gel network. Furthermore, the swelling of this class of hydrogels could be suppressed by addition of external salts (e.g., NaCl, KBr) [32].

Interestingly, the swelling of the TPMLH-based hydrogels could also be induced in the dark simply by exposure to acidic buffer (pH < 4.0). In this case, swelling results from chemical (rather than photo-chemical) ionization of the leucohydroxide moieties. This phenomenon is similar to the swelling of polymer gels that possess carboxylic acid groups in an alkaline solution.

The TPMLC functional group is more hydrophobic and is less prone to chemical ionization; the degree of swelling of TPMLC-based hydrogels at pH 6.5 in the dark was less than that of TPMLH-based hydrogels. These results show that swelling behavior can be tuned via choice of triphenylmethane leuco derivative. The key role of the ionization of polymer networks in their phase transition of gels was previously studied [36].

Triphenylmethane leuco derivatives have potential applications for selective transport of ionic versus netural species. Osa and coworkers studied the **3c**-mediated transport of organic anions such as methyl orange (MO^-, Figure 2.7) and sodium picrate across a liquid membrane (Figure 2.5) [35]. In this system, a solution of **3c** in CH_2Cl_2 (Figure 2.7, phase B) was sequestered between two immiscible aqueous layers: a solution of MO^-Na^+ (Figure 2.7, phase A) and a solution of NaOH (Figure 2.7,

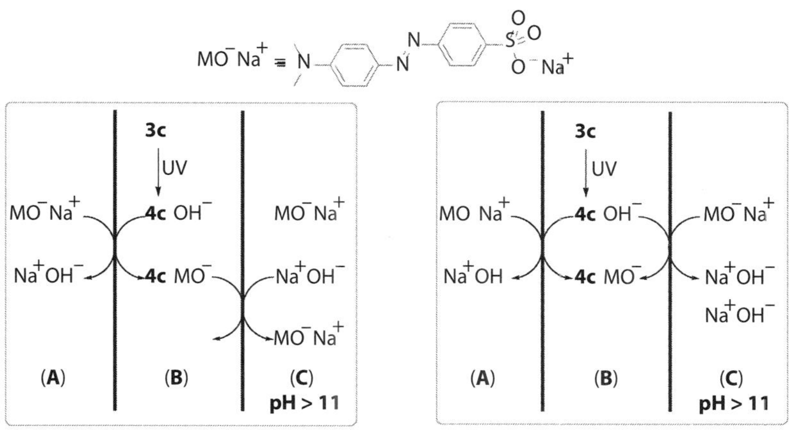

Figure 2.7 Schematic sketches of the photo-induced active transport of methyl orange anion (MO^-) at different pH values of phase C. Adapted with permission from Ref. [35]. Copyright (1988) The Chemical Society of Japan.

Table 2.2 Methyl orange concentrations in phases A and C after UV irradiation for 10 min. Adapted with permission from Ref. [35]. Copyright (1988) The Chemical Society of Japan. *a*

	Methyl orange concentration/10^{-5} mol dm^{-3}			
pH in phase C	**8 h after UV irradiation**		**72 h after UV irradiation**	
	Phase A	**Phase C**	**Phase A**	**Phase C**
13.0	0.39	1.40	0.01	1.96
11.9	0.39	1.29	0.04	1.49
10.6	0.36	0.65	0.08	0.64
8.8	0.56	0.55	0.53	0.59

a The pH value in phase A was 7.3. Initial concentration of methyl orange in phases A and C was 1.0×10^{-5} mol dm^{-3}.

phase C). Transport of MO^- from phase A to phase C was measured as a function of time after 10 min of UV irradiation. As shown in Table 2.2, the concentration of MO^- in phase C increased with time. The mechanism of this photo-induced active anion transport was proposed to involve the initial photo-conversion of neutral **3c** to cationic **4c** in organic phase B. Complexation of the latter species with MO^- at the A/B interface leads to extraction of one molecule of MO^- from phase A into phase B and transfer of one hydroxide ion from region B to region A (i.e., anion exchange at the A/B interface, Figure 2.7). The newly formed **4c**+MO^- complex can diffuse to the B/C interface, where a second anion exchange can occur. Here, the cationic component (**4c**) is trapped by hydroxide from phase C to regenerate **3c** in phase B. During this process, MO^- is transferred from phase B to phase C. This active transport process proceeds until the concentration of ions in regions A and C becomes equivalent. Given that the process is mediated by hydroxide ion one would expect a pH dependence. Indeed, decreasing the pH in phase C lead to a buildup of MO^- in phase B presumably through inhibition of **4c** + OH^- combination (i.e., regeneration of **3c**) at the B/C interface (Figure 2.7, right).

A few years after Osa's report, Willner and coworkers described the photo-induced transport of carboxylate and phenolate anions across a liquid membrane mediated by a copolymer of styrene and **3a** (compound **5**, Figure 2.8) [37]. In this system, anion transport was very slow in the dark; ca. 3-fold enhancement was observed upon 330–370 nm irradiation.

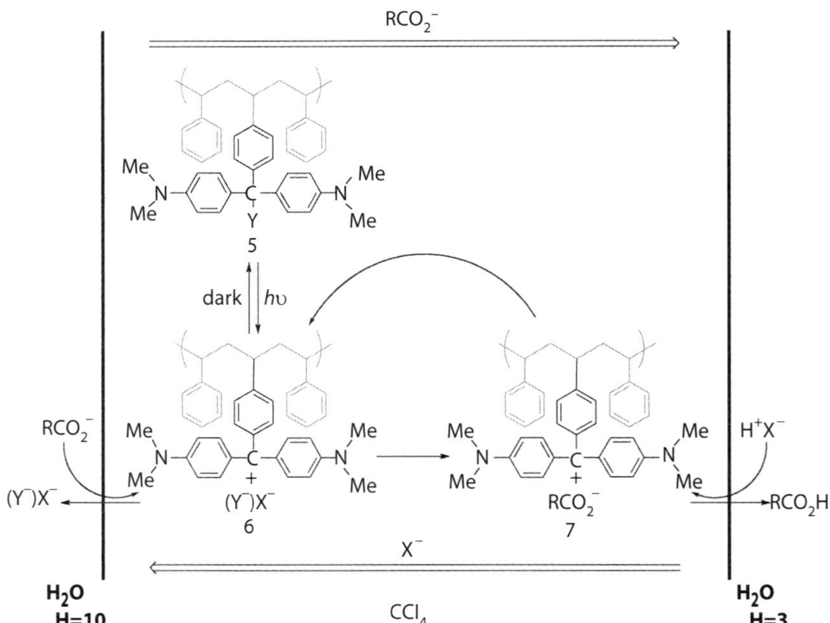

Figure 2.8 Scheme for photo-induced transport of carboxylates using copolymer **5** as carrier. The molar ratio of **3a** to styrene in the **5** is 1 : 200. Illumination of the organic phase generates the dark-green color of the triphenylmethyl ionic compound **6**. Adapted with permission from Ref. [37]. Copyright (1992) The Royal Society of Chemistry.

A representative example of carboxylate transport is shown in Figure 2.8. As in the case above, the transport process begins with photo-ionization of **5** to generate cationic polymer **6** and hydroxide ions. The hydroxide ions are exchanged with carboxylate anions in the source phase (pH = 10) to form complex **7**. The latter polymer diffuses to the sink phase (pH = 3) whereby the carboxylate is protonated and transferred to the sink as the carboxylic acid. This process occurs with a concurrent antiport of X^- ions from the sink to the source. From an application point of view, it is important to mention that this system displayed a degree of substrate selectivity imparted by difference in the binding properties of various anions to the polymer **6** (Table 2.3). Furthermore, this system is truly "photo-controlled": anion transport stops when the light source is removed, and begins again upon irradiation. This control arises from the reversibility of the initial photoionization step. When the light source is switch off, cationic **6** will be rapidly quenched with hydroxide from the high pH source phase to regenerate polymer **5**. Re-exposure to light can regenerate **6** and induce further anion transport.

The above studies separately established triphenylmethane leuco derivatives novel photoactive components of hydrogels and as promising ion transporters in liquid membranes. The intersection of these applications, i.e., hydrogel membranes, had not been reported until Rethwisch and coworkers introduced a new class of TPMLH-containing PAAm hydrogel membranes (**8**, Figure 2.9) [38–39]. In these studies, the membranes were prepared by free-radical copolymerization of suitable monomers and crosslinkers between two glass plates with a 0.65 mm spacer that served as a mold. The membrane thickness after demolding and swelling in deionized water for 48 h was 0.055 ± 0.005 cm. As expected for polymer gels, the swelling was reduced for solutions with higher ionic strength due to osmotic pressure effects. However, in this case UV irradiation had a negligible effect on the swelling, which was in sharp contrast to previous observations made by Irie and coworkers [32]. This apparent contradiction could originate from the higher TPMLH loading (ca. 7-fold higher) in Irie's work, and the fact the samples were studied in pure water (pH 6.6) instead of highly acidic or basic medium. In addition, in Rethwisch's case it was estimated that ca. 90% of the UV irradiation is absorbed in the first 20 μm of the membrane, which would limit bulk swelling.

A diffusion cell with quartz windows was used for the permeation experiments (Figure 2.10). The set up included a UV light source (500 W Hg lamp) fitted with a water filter for removing IR radiation and an inline cut-off filter to eliminate $\lambda < 270$ nm. The hydrogel membrane (diffusion area = 4.052 cm^2) was placed between the cell compartments (each cell had

Table 2.3 Data for photo-induced transport of organic anions across a liquid-liquid membrane using copolymer **5** as the ion carrier. Adapted with permission from Ref. [37]. Copyright (1992) The Royal Society of Chemistry. *a*

Substrate chemical name	Substrate chemical structure	*v*(dark)/ mol dm^{-3} min^{-1}	*v*(illumination)/ mol dm^{-3} min^{-1}	Switching efficiency *b*
Cinnamic acid	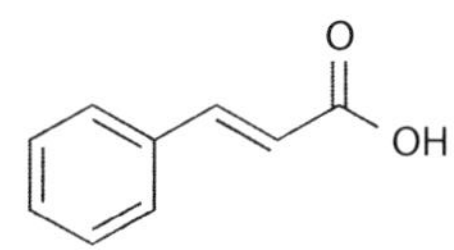	2.60×10^{-7}	7.00×10^{-7}	2.70
p-Hydroxycinnamic acid	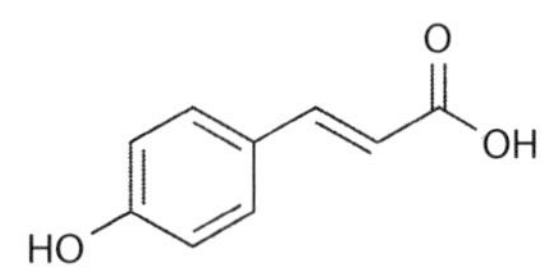	4.93×10^{-7}	1.61×10^{-6}	3.26
p-Nitrophenol	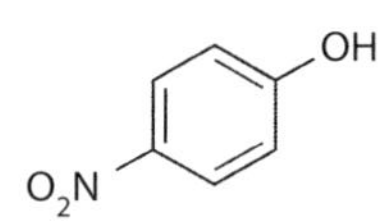	1.89×10^{-8}	7.43×10^{-8}	3.93
p-Hydroxybenzaldehyde	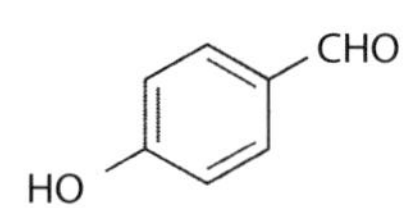	8.78×10^{-9}	7.53×10^{-8}	8.60

a The liquid-liquid membrane system is composed of an aqueous solution, 2.5 mL, pH 10, that includes the anion, 0.1 mol dm^{-3}, separated from a second aqueous phase, 2.5 mL, pH 3, by a carbon tetrachloride solution, 5 mL. In the organic phase the photo-chromic copolymer **5** (MW = 25000–35000 Da), 30 mg, is solubilized as carrier. *b* Switching efficiency is defined as *v*(illumination)/*v*(dark).

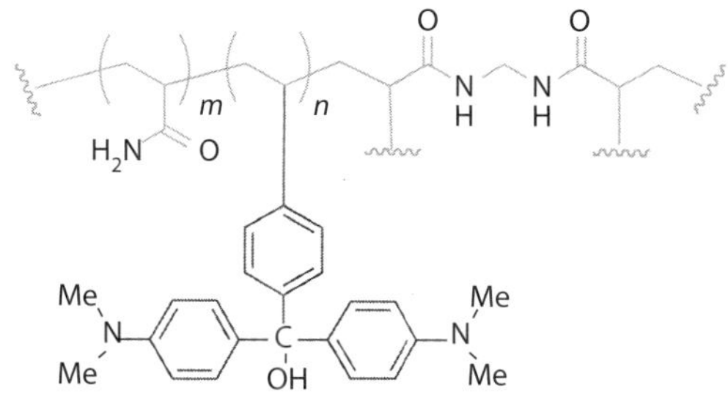

8

Figure 2.9 Representative structure of PAAm/TPMLH cross-linked polymer with *m* and *n* representing the relative amounts of acrylamide and TPMLH groups in the structure. In a typical synthesis, acrylamide (10 equiv), **3a** (1 equiv), MBBAm (0.12 equiv), and AIBN (0.43 equiv) were allowed to react in DMSO at 60 °C for 1.5 h under nitrogen. Adapted with permission from Ref. [38]. Copyright (1999) Elsevier.

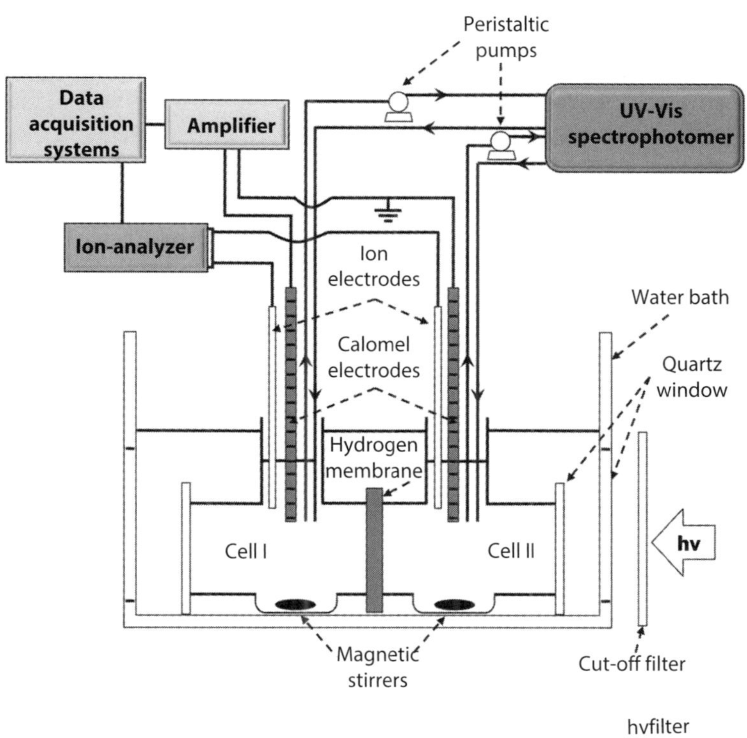

Figure 2.10 Schematic representation of the apparatus used for determining membrane potentials and measuring permeabilities. Adapted with permission from Ref. [38]. Copyright (1999) Elsevier.

Table 2.4 Summary of observed and predicted permeabilities for methyl orange and DMAP in a membrane prepared from 0.5 mol% PPAm/TPMLH **8**. Adapted with permission from Ref. [38]. Copyright (1999) Elsevier.

		Permeability × 10^5 (cm/s)			
Permeant	pH	Non-irradiated		Irradiated	
		Observed	Predicted	Observed	Predicted
Methyl orange	10	1.06 ± 0.05	0.75	1.73 ± 0.05	1.16
Methyl orange	4	6.46 ± 0.02	5.57	–	–
DMAP	10	0.28 ± 0.02	0.44	0.27 ± 0.02	0.44

a The swelling ratio for 0.5 mol% TPMLH/polyacrylamide in pure water was 6.9 ± 0.1 and 5.7 ± 0.2 in the presence of methyl orange before irradiation.

a volume of 120 mL), where solutions of the permeants at the desired pH were stirred at 25 °C.

These studies revealed that the transport of an anionic permeants (i.e., methyl orange, MO^-) was ca. 2-fold enhanced by the photo-induced generation of fixed cationic charges in the membrane (Table 2.4). In contrast, the flux of uncharged permeants (i.e., 4-dimethylamino pyridine (DMAP)) remained essentially constant after irradiation. This observation indicated that the changes in permeability were mainly due to changes in the fixed charge concentration and charge interactions within the hydrogel matrix, and not simply due to the modification of its porosity. It is worth to mention that theoretical modeling of the fluxes, assuming that diffusion occurred in the aqueous phase of an immobile polymer scaffold with fixed charges, gave a very good correlation with the experimental results.

2.2.3 Photo-Responsive Moiety: Azobenzene

Molecularly imprinted membranes prepared from stimuli-responsive functional monomers are a special class of polymeric materials that exhibit specific and reversible substrate binding affinity in response to environmental stimuli. The key difference between imprinted materials and traditional membranes is that imprinted materials *release and encapsulate* cargo in a switchable fashion. In contrast, traditional membranes that release cargo in response to a stimuli-induced phase transition of the bulk material do not typically re-encapsulate their cargo even if the phase transition is reversed. For further information on molecular imprinted hydrogels we refer the reader to the excellent review from Byrne and Salian [40]. Here we focus on photo-regulation of imprinted hydrogels, which is possible via incorporation of suitable photo-responsive moieties. Azobenzene derivatives are perhaps the most widely used functional groups in this regard [41–43].

Lam and coworkers reported the use of photo-responsive molecularly imprinted hydrogels for regulatation of chemical transport in biocompatible aqueous media [44]. In their study, the authors synthesized a water-soluble azobenzene-containing monomer 4-[(4-methacryloyloxy)phenylazo] benzenesulfonic acid (MAPASA) by base-catalyzed coupling of 4-[(4-hydroxy) phenylazo]benzenesulfonic acid and methacrylic chloride (~ 64% isolated yield). Cross-linking of this functional monomer with different bisacrylamide and bismethacrylamide cross-linkers in the presence of drug molecules yielded drug imprinted, transparent PAAm-*co*-MAPASA hydrogels. As a proof of concept, *N*-(4-hydroxyphenyl)acetamide (paracetamol) was used as the molecular template for the imprinting process; hydrogels were prepared from 1 equiv of monomer, 0.2 equiv of template, and 5 equiv of cross-linker. The resulting bulk hydrogel was crushed, milled, wet sieved in methanol, collected by centrifugation, and subjected to Soxhlet extraction to remove the template and yield the desired imprinted material. Finally, the binding properties of imprinted and control hydrogel materials were studied using rebinding assays in aqueous HEPES buffers at ca. pH 7 in the dark (Figure 2.11). The affinity of the hydrogel for paracetamol could be photo-regulated: 83.6% of receptor bound paracetamol was released from

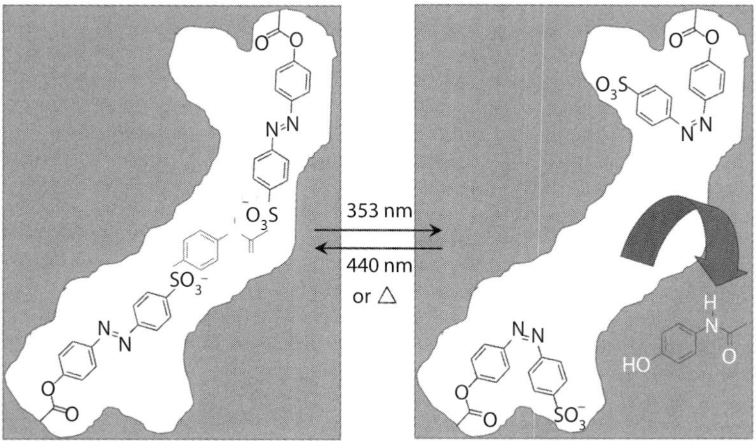

Figure 2.11 Schematic representation of reversible photo-regulated release and uptake of paracetamol with paracetamol imprinted MAPASA-containing polyacrylamide hydrogel. The concentration of substrate in solution was monitored by HPLC. Adapted with permission from Ref. [44]. Copyright (2008) American Chemical Society.

the imprinted gel upon irradiation at $\lambda = 353$ nm, whereas subsequent irradiation at $\lambda = 440$ nm caused 94.1% of the released drug to be rebound to the hydrogel matrix.

The photo-isomerization of sulfonated azobenzene units within these hydrogel matrices requires sufficient space. No photo-isomerization of the chromophores was observed when relatively rigid azobenzene-containing hydrogels were prepared using *N,N′*-methylenebisacrylamide (**1**-C). The use of cross-linkers with larger spacers from ethylene (**2**-C) to octylene (**8**-C) enhanced the flexibility of the hydrogel matrix and provided a gradual increase of the photo-isomerization rate of the chromophores (Table 2.5). No hydrogel formation was observed for *N,N′*-dodecylenebismethacrylamide (**12**-C) due to poor solubility of the cross-linker. The substrate-binding strength of the imprinted receptors decreased with decreasing network rigidity. *N,N′*-hexylenebismethacrylamide (**6**-C) provided the optimal balance of rigidity and response rate. Hydrogels prepared from 6-C showed the greatest difference between specific and non-specific substrate binding strengths (i.e., 1.96×10^5 and 747.0 M^{-1}, respectively).

Two sets of control experiments confirmed the substrate-specificity of the imprinted hydrogel receptor sites: (*1*) A control hydrogel that was prepared and processed in exactly the same way as the paracetamol-imprinted hydrogel except in the absence of paracetamol did not show any specific photo-regulated release/uptake behavior. (*2*) A minor degree of photo-regulated release/uptake was observed when using structural analogs of paracetamol (i.e., phenacetin, antifebrin) under similar experimental conditions. Though these materials are very promising, the photo-regulated uptake of substrate progressively reduced over multiple release and uptake cycles (only ca. 71% of released paracetamol was reabsorbed on third cycle). This phenomenon was ascribed to the gradual deformation of the imprinted receptors upon repetitive photo-switching. Work continues in this exciting field, and new advances are almost certainly forthcoming. Although bulk supramolecular hydrogels bearing azobenzene moieties have been also identified as promising photo-responsive soft materials for controlled release of drug molecules [45], more research is needed on these systems to achieve fully reversible and selective release/uptake of substrates.

2.2.4 Photo-Responsive Moiety: Spirobenzopyran

Spirobenzopyran moieties are widely studied photo-switchable functional groups. Sumaru and coworkers prepared the first dual photo- and thermo-responsive gate membrane (PTGM) for liquid transport by

Table 2.5 Photo-responsive and substrate binding properties of the polyacrylamide hydrogels fabricated from various bisacrylamide/ bismethacrylate cross-linkers. Adapted with permission from Ref. [44]. Copyright (2008) American Chemical Society.

Cross-linker for hydrogel fabrication		Photo-isomerization rate (s^{-1}) *a*	Substrate binding strength log K_B *b*	Binding site density (μmol g^{-1} of hydrogel) *c*
1-C	**1-C** 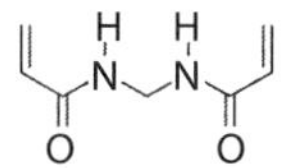	no photo-isomerization observed		N.D. *d*
2-C	**2-C** 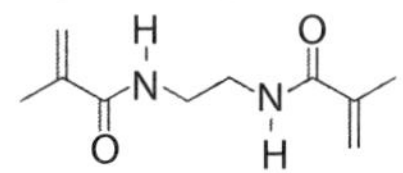	3.63×10^{-4}	5.82	0.20
3-C	**3-C** 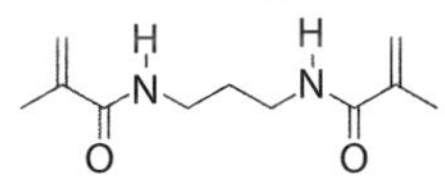	4.67×10^{-4}	5.49	0.31
4-C	**4-C** 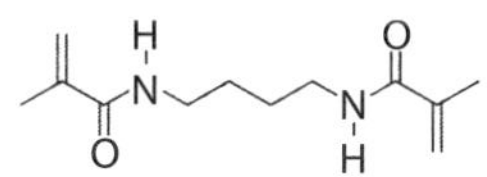	5.20×10^{-4}	5.47	0.24
6-C	**6-C** 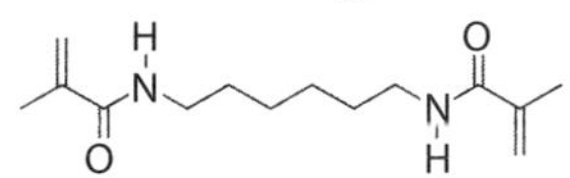	6.41×10^{-4}	5.29	0.47
8-C	**8-C** 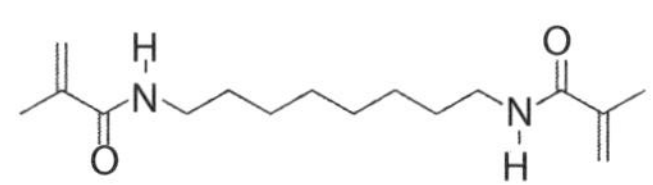	6.15×10^{-4}	5.12	1.73
12-C	**12-C** 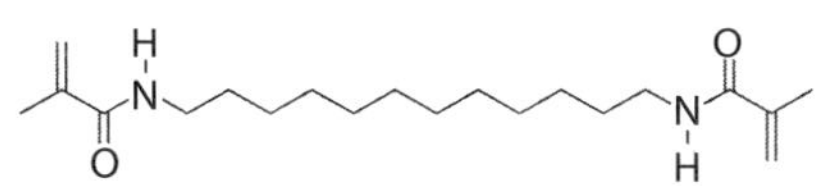		no hydrogel formed	

a Trans → *cis* photo-isomerization rate *k* of azo-benzene chromophores in the resultant hydrogel. *b* Substrate binding strength of the imprinted binding sites in hydrogel. *c* Imprinted binding site density in the hydrogel. *d* Not determined.

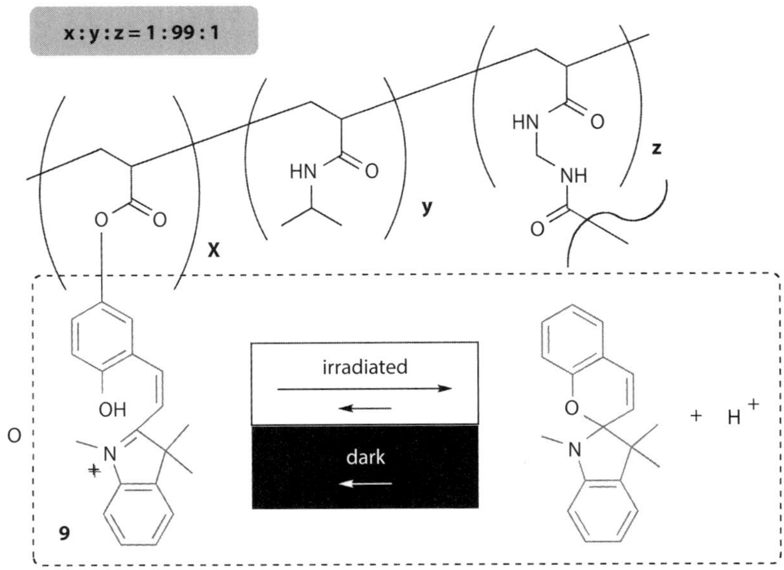

Figure 2.12 Chemical structure of cross-linked pSPNIPAAm polymer gelator and photo-isomerization process. Adapted with permission from Ref. [46]. Copyright (2006) American Chemical Society.

introducing a photo-responsive *N*-isopropylacrylamide (NIPAAm) hydrogel with pendant spirobenzopyrans to the surface of a porous hydrophilized poly(tetrafluoroethylene) (PTFE) membrane with a pore size of 1 mm [46]. In this case, the preparation of the hydrogel (**9**, Figure 2.12) was carried out by free-radical copolymerization of NIPAAm monomer (100 equiv), a spirobenzopyran acrylate derivative (1 equiv), and *N,N*-methylene-bis(acrylamide) (MBAAm) (1 equiv) in the presence of AIBN (6 equiv) in DMSO.

Transmission microscopy imaging and conductivity measurements demonstrated that **9** exhibited rapid volume shrinkage and proton dissociation upon irradiation with blue light (λ = 400–440 nm) under acidic conditions (1 mM HCl) (Figure 2.13). Specifically, the conductance increased ca. 20% upon irradiation, and gradually returned to the initial value (0.65 mS cm^{-1}) after the light was switched off. The dual responsive PTGM material was prepared by performing the *in situ* synthesis of **9** in the presence of a PTFE membrane that was submerged into the reaction medium. Microscopy data clearly showed a deformation process of the cross-liked pSPNIPAAm membrane under blue light irradiation in acidic conditions (Figure 2.14). The original yellow-tinted hydrogel, whose color arises from the presence of the protonated open-ring chromophore, shrunk and became colorless within seconds after irradiation, which was a direct consequence of the photo-isomerization to the closed-ring form and proton dissociation.

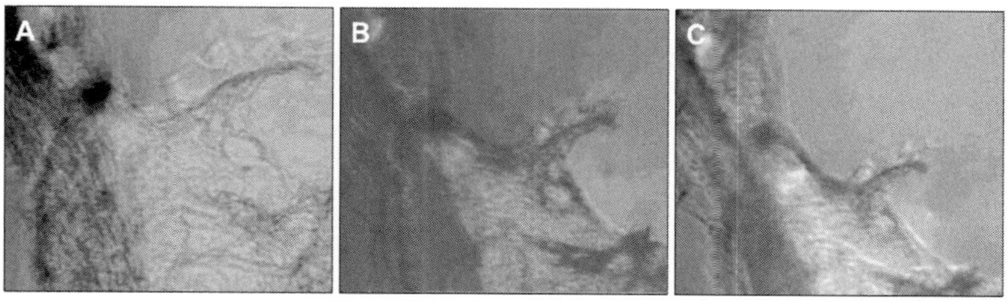

Figure 2.13 Photo-responsive deformation of a cross-linked pSPNIPAAm hydrogel in 1 mM HCl at 28 °C: (A) Before irradiation, (B) after irradiation for 9 s, and (C) after irradiation for 24 s. The volume of the gel decreased to ca. 1/5 of that of the initial state. Adapted with permission from Ref. [46]. Copyright (2006) American Chemical Society.

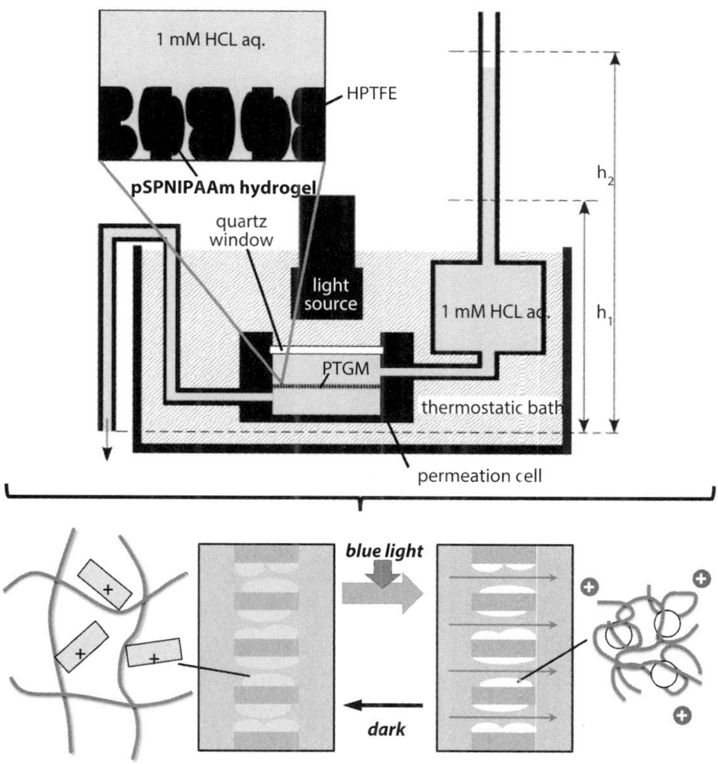

Figure 2.14 *Top*: Schematic illustrations showing the experimental set up for the permeability measurement of pSPNIPAAm-based hydrogel **9**. *Bottom*: Reversible volume shrinkage and proton dissociation of the gate membrane. In an acidic aqueous medium in the dark, most of the spirobenzopyran units are in the stable protonated open-ring form. Upon UV irradiation, the chromophores are isomerized to the free closed-ring form, dissociating protons and releasing cations. Adapted with permission from Ref. [46]. Copyright (2006) American Chemical Society.

Figure 2.14 illustrates the experimental set up used for permeability measurements of this novel PTGM. The PTGM was placed into a permeation cell equipped with a quartz window that allowed for intermittent irradiation. Both the permeation cell and the reservoir were placed in a thermostatic bath to ensure constant temperature. The permeability *P* of the PTGM was given by the equation $P = [r \ln(h_1/h_2)]/[rg(t\text{-}t_0)]$, where *r* is the ratio of the capillary cross section to the effective area of the PTGM in the permeation cell, *t* and *h* are related to the time and movement, respectively, of the meniscus of the HCl solution from a certain height h_1 to a different height h_2, *r* is the density of the HCl solution, *g* is the acceleration of gravity, and t_0 is the blank value of *t* obtained without the PTGM. The results showed that the *P* value of the PTGM in 1 mM HCl aqueous solution increased 2-fold upon blue light irradiation as a direct consequence of the photo-induced shrinking of the hydrogel. The *P* value decreased gradually to the former state after the light was turned off, completing a cycle that was confirmed to be repeatable. Interestingly, the permeability of the hydrogel membrane also increased drastically when the temperature was increased ubove the lower critical solution temperature of PNiPAAm-based hydrogel **9** (ca. 30 °C). Although this increase occurred regardless of the irradiation stage, the permeability was always larger under irradiation than in the dark; the difference maximized at around 30 °C.

2.2.5 A Comparative Example of Different Chromophores

Willner and coworkers carried out a significant study wherein they compared the membrane and catalytic properties of photo-responsive polyacrylamide-based copolymer hydrogels that possessed three of the above-described functional moieties (sections 2.2–2.4) and encapsulated enzymes [47]. Figure 2.15A shows the chemical structure of each of the photo-responsive components used in this study.

The authors used these materials to immobilize the enzyme a-chymotrypsin, which catalyzes the hydrolysis of *N*-(3-carboxypropionyl)-L-phenylalanine *p*-nitroanilide (**12**) (Figure 2.15B). Immobilization was achieved via gelation in the presence of the enzyme (in DMSO solvent). Gelation took place within a few minutes; the material was left undisturbed for 1 h before it was washed with water and isolated. Roughly 2 mm thick membranes with encapsulated enzyme were used for catalysis assays in buffered media that contained substrate **12**. The consumption of **12** (or formation of nitroaniline (**14**)) could be followed spectroscopically as illustrated in Figure 2.16 [47].

These materials were found to have photo-switchable "on-off" catalytic activities [47]. The enzyme activity was low (position "off") for all

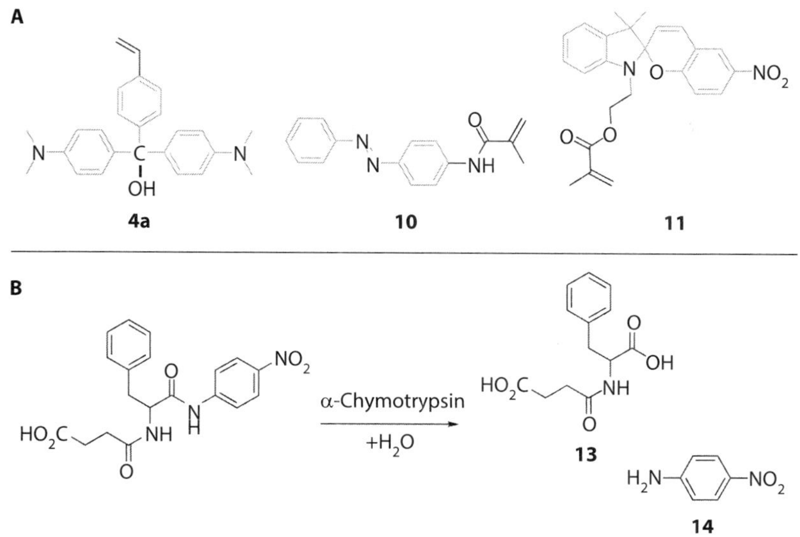

Figure 2.15 (A) Photo-isomerizable monomers bis-[4-(dimethylamino)phenyl](4-vinylphenyl)methyl leucohydroxide (**4a**), 4-(methacryloylamino)azobenzene (**10**), and 1-[(β-(methacryloxy)ethyl]-3,3-dimethyl-6'-nitropiro[indoline-2,2'-[2*H*-1]benzopyran] (**11**) used to prepare photo-responsive polyacrylamide copolymer membranes. Irradiation wavelengths for the various photo-responses of the difference chromophores could be adjusted as following: λ = 330–370 nm and λ > 400 nm for *trans* → *cis* and *cis* → *trans isomerization* of azobenzene moieties, respectively; λ = 300–400 nm and λ > 475 nm for *ring opening* and *ring closure* of spirobenzopyran moieties, respectively; λ = 330–370 nm ionic for *heterolytic bond cleavage* of triphenylmethane leuco derivatives, which undergo ionic recombination in dark. (B) α-Chymotrypsin-catalyzed hydrolysis of **12**. Adapted with permission from Ref. [47]. Copyright (1993) American Chemical Society.

Figure 2.16 Flow-dialysis cell composed of two chambers. The upper chamber contained the substrate **12** (1×10^{-2} M) in a buffer solution (triethanolamine 0.2 M, pH 3.8). The same buffer without **12** flowed through the lower chamber (30 mL h^{-1}). The photo-responsive copolymer membrane was placed between the two chambers. The fractions of the eluted solution were collected and analyzed spectroscopically (For **12**: λ = 314 nm, ε = 10000 M^{-1} cm^{-1}). Adapted with permission from Ref. [47]. Copyright (1993) American Chemical Society.

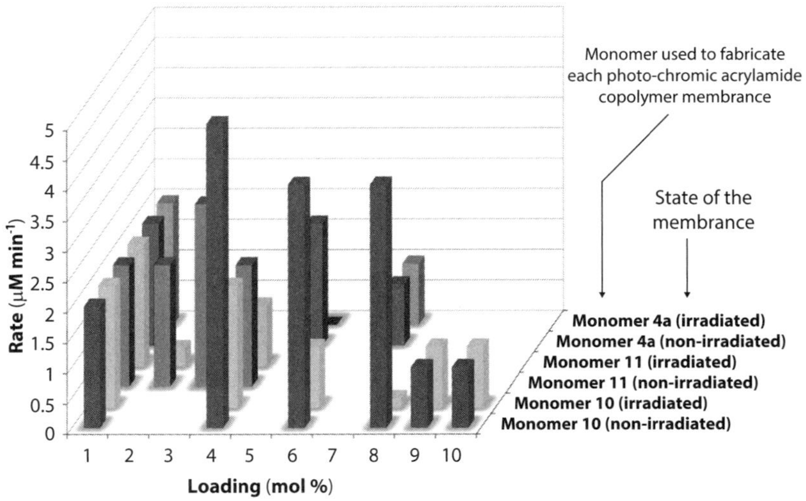

Figure 2.17 Representative normalized rates of hydrolysis of **12** by α-chymotrypsin immobilized in different photo-chromic acrylamide copolymers at different degrees of loading. Rate of 1 μM min^{-1} corresponds to the formation of 0.5 mM min^{-1} of **14**. Loading is expressed in terms of mol % of the photo-isomerizable component relative to the acrylamide monomer. Plot generated from published data with permission from Ref. [47]. Copyright (1993) American Chemical Society.

enzyme-copolymer hybrids before irradiation (i.e., **4a**, **10** and **11** moieties remained stable, and the corresponding copolymers exhibited poor permeability toward substrate **12**); the activity increased (turned "on") upon irradiation when presumably the photo-generated isomeric and/or ionic forms of the chromophores increased the membrane permeability of substrate **12** (Figure 2.17). The increased polarity of the matrices upon irradiation, together with possible electrostatic interactions, favors the transport of polar and charged substrates through the membranes.

This study represented a major advance in the development of photo-regulated protein systems. The authors compared the results to previous attempts based on the direct interaction of photo-isomerizable moieties with the corresponding proteins, where only partial deactivation of enzymes had been observed. In general, the relative permeabilities of these photo-responsive polymeric gels were in good agreement with previous studies [48].

2.3 Photo-Thermally Responsive Hydrogel Membranes

Optical absorbers that efficiently convert light into thermal energy can be incorporated into hydrogel membranes prepared from thermally responsive polymers to generate photo-thermally responsive hydrogel membranes. Much like the materials described above could be categorized by the identity of their chromophore, the photo-thermally responsive materials discussed here are categorized by the identity of their optical absorber.

2.3.1 Optical Absorber: Gold Nanoparticles

The development of nano- and microcapsules for the controlled encapsulation and triggered release of bioactive molecules in targeted areas *in vivo* represents a key goal in the field of biological delivery. Microfluidic emulsion polymerization techniques enable the synthesis of particles with precisely defined sizes and properties. As a very recent example, Kim and coworkers described a microfluidic approach for the *in situ* preparation of thermo- and photo-responsive PAAm hydrogel microcapsules with embedded gold nanorods that were capable of loading and release of hydrophilic molecules [49]. These materials were prepared via polymerization of a mixture of NiPAAm monomer and gold nanorods in the interior of double-emulsion drops of uniform size that were generated within a capillary microfluidic apparatus. Figure 2.18 shows the microfluidic device

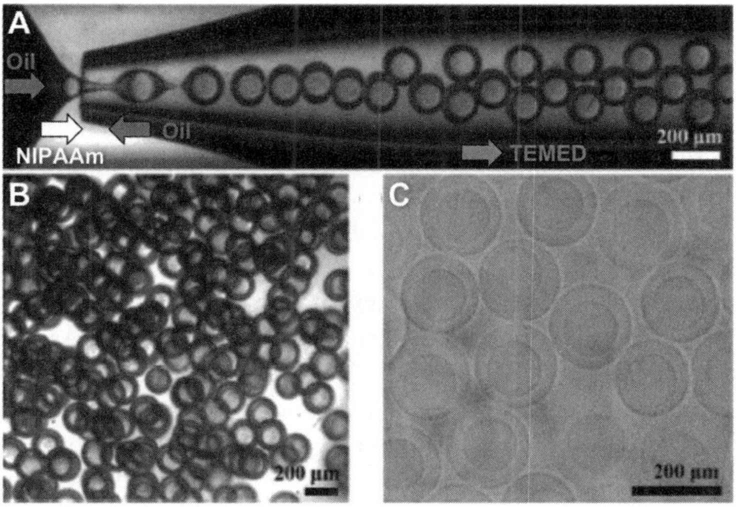

Figure 2.18 (A) Optical microscopy image of the microfluidic capillary device used for generation of O/W/O (oil-water-oil) double-emulsion drops. Oil refers to polydimetylsiloxane (PDMS) in this case. The water phase was composed of NIPAAm, *N,N′*-methylenebisacrylamide cross-linker, ammonium persulfate initiator, gold nanorods, and surfactants. Reaction accelerator, TEMED = *N,N,N',N'*-tetrametylethylenediamine, was added through the interstices between capillaries. (C-D) Optical microscope images of O/W/O double-emulsion drops and microcapsules composed of a water core and a pNIPPAm shell. Adapted with permission from Ref. [49]. Copyright (2013) Royal Society of Chemistry.

used for capsule synthesis. The set up consists of two junctions, one for the generation of double emulsion drops and the other for injection of a polymerization accelerator. Both the core and the continuous phase of the so-obtained hydrogel capsules could be replaced with water.

The particles exhibited reversible temperature-dependent deswelling and membrane permeability. The gold nanorods (35 nm length × 9 nm width) absorb infrared light very efficiently. The inclusion of gold nanorods enabled photo-thermal heating of the membrane upon IR laser irradiation (λ = 810 nm), which provided control over the shell permeability and facilitated the photo-triggered release encapsulants (FITC-tabbed dextran).

2.3.2 Optical Absorber: Graphene Oxide

Miyako and coworkers reported that agarose or pNIPAAm hydrogels that possess nanostructured carbons (e.g., single-walled carbon nanotubes and nanohorns) undergo marked phased transitions upon near-infrared (NIR) irradiation [50]. The photo-induced gel response was further used to demonstrate the controlled release of DNA-nanocarbon conjugates entrapped within the gel matrix. Compared to carbon nanotubes, carbon nanohorns, and gold nanorods, graphene oxide (GO) is more abundant, cheaper material that features high NIR light absorption. Chen, Yu and coworkers reported the preparation of a photo-thermally sensitive hydrogel by *in situ* γ irradiation-assisted polymerization of an aqueous solution of NIPAAm monomer in the presence of GO. Due to the high optical absorbance of the GO, the nanocomposite pNIPAAm/GO hydrogels showed remarkable photo-thermal properties; the phase-transitions could be controlled remotely and in a reversible manner via NIR irradiation [51]. Notably, the irradiation-induced temperature of the hydrogel increased more rapidly with increased GO loading. The temperature could also be fine-tuned by the irradiation time. These materials were used to construct a remotely actuated liquid microvalve. As shown in Figure 2.19, the hydrogel membrane (black object) originally blocked the flow of the pink solution from the left to the right side of the valve. Upon NIR light irradiation (λ = 808 nm), the hydrogel shrank and the pink liquid passed through the valve. In contrast, the color remained unchanged before and after laser irradiation when a pNIPAAm hydrogel that lacked GO was used.

2.4 Summary

The incorporation of pendant photo-active chromophores and optical absorbing moieties within hydrogel matrices has allowed the synthesis of

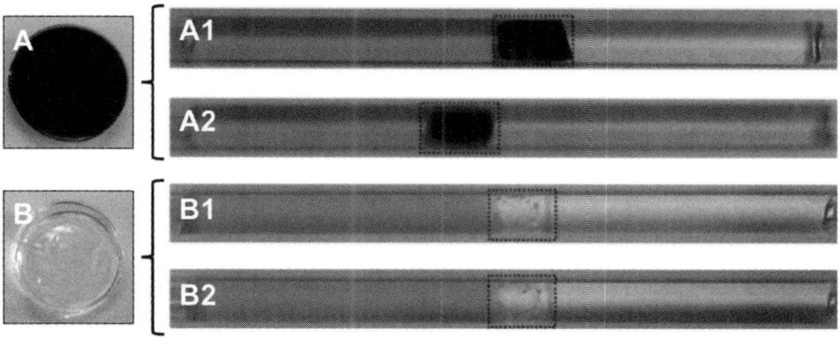

Figure 2.19 Liquid microvalves actuated with pNIPAAm/GO (A, [GO] = 1.0 g L^{-1}) and a pNIPAAm hydrogels (B). The photographs show the microvalves before (A1, B1), and after (A2, B2) NIR (808 nm, 2 W) irradiation for 2 min. The positions of the hydrogels are marked by the rectangular boxes. The solutions encapsulated in the left and right sides are aqueous rhodamine B solution and $NiSO_4$, respectively. Adapted with permission from Ref. [51]. Copyright (2012) John Wiley & Sons, Inc.

responsive membranes for the controlled and selective uptake/release of water-soluble substrates upon irradiation. In the case of photo-active chromophores, light drives actuation through initiation of reversible photo-isomerization or photo-ionization reactions of cromophoric groups bound to the materials, which leads to changes in the material's macroscopic properties such as permeability, density and viscosity, among others. For photo-thermally responsive matrials, optical absorbers convert light energy to heat, which induces a bulk thermal response.

Cinnamylidene (*a*), triphenylmethane leuco (*b*), azobenzene (*c*), and spiropyran (*d*) derivatives are common chromophores for the fabrication of tunable photo-responsive hydrogel membranes, in which [2+2] dimerization (for *a*), heterolytic bond cleavage (for *b*), and isomerization (for *c* and *d*) are the primary photochemical processes that ultimately define the photo-response of the membrane. Permeability measurements of membranes bearing these chromophores have established inherent relationships between mesh size and molecular composition of the hydrogel membrane, with irradiation time, irradiation wavelength, osmotic pressure (in the case of ionic gradients), charge interactions within the hydrogel matrix, swellability, size and charge of permeating solutes, and their diffusion coefficients.

The examples provided in this chapter demonstrate the feasibility of using photo-responsive hydrogels as switchable membranes for selective mass transfer. Light-induced swelling–shrinking of a hydrogel composite could be also transformed into an optical signal. All these systems possess inherent potential for advanced applications that include, among others, photo-regulated microfluidic and nanofluidic valves, miniaturized (bio) sensors, autonomous pharmaceutical-delivery systems, dynamic cell-culture substrates, 'lab-on-a-chip' systems, antimicrobial barriers, (bio) catalytic reactors, electrochemical regulators in (bio)fuel cells, and soft robotics. The development of practical and reliable synthetic methodologies for the preparation of new hydrogelators, the fabrication of supramolecular photo-responsive hydrogels with superior mechanical properties, and flow theoretical models will assist the future design of new photo-responsive gel membranes with a broader range of pore sizes that mimic in greater detail the hierarchical complexity and multifunctional responses of biological systems.

2.5 Acknowledgements

Universität Regensburg (Anschubfinanzierung von Wissenschaftlichen Projekten-2011) and the MIT Department of Chemistry are acknowledged for financial support. This work was supported in part by the MRSEC Program of the National Science Foundation under award number DMR – 0819762. D.D.D. thanks DFG for the Heisenberg Professorship Award.

Abbreviations

AIBN	2,2'-azobisisobutyronitrile
b-PEG	branched polyethylene glycol
CA	cinnamylidene acetyl
Đ	dispersity index (M_w / M_n)
DMAP	4-dimethylamino pyridine
DS	degree of swelling
FITC	fluorescein-isothiocyanate
Hb	hemoglobin
LDH	lactate dehydrogenase-L
MAPASA	4-[(4-methacryloyloxy)phenylazo] benzenesulfonic acid
Mb	myoglobin
MBAAm	*N,N*-methylene-bis(acrylamide)
MW	nominal molecular weight
NIPAAm	*N*-isopropylacrylamide
NIR	near-infrared
pNIPAAm	poly(*N*-isopropylacrylamide)

PAAm	polyacrylamide
PDMS	polydimethylsiloxane
PEG	poly(ethylene glycol)
b-PEG	four-armed poly(ethylene glycol) star polymer with hydroxyl end groups
PMMA	poly(methyl methacrylate)
pSPNIPAAm	pNIPAAm partly modified with spirobenzopyran
PTFE	poly(tetrafluoroethylene)
PTGM	photo- and thermo-responsive gate membrane
STDV	standard deviation
TEMED	*N,N,N',N'*-tetrametylethylenediamine
TPMLC	triphenylmethane leucocyanide
TPMLH	triphenylmethane leucohydroxide

References

1. M. Ulbricht, *Polymer*, Vol. 47, p. 2217, 2006.
2. N.A. Peppas, K.M. Wood and J.B. Thomas, "Membranes in Controlled Release", in K.V. Peinemann and S. Pereira Nunes, eds, "*Membrane Technology, Vol 1., Membranes for the Life Sciences*", 175–190, Wiley-VCH, Weinheim, 2008, and references therein.
3. D. He, H. Susanto, and M. Ulbricht, *Progress in Polymer Science*, Vol. 34, p. 62, 2009.
4. I. Tokarev, and S. Minko, *Advanced Materials*, Vol. 21, p. 241, 2009.
5. D. Wandera, S.R. Wickramasinghe, and S.M. Husson, *Journal of Membrane Science*, Vol. 357, p. 6, 2010.
6. P. Stroeve, and N. Ileri, *Trends in Biotechnology*, Vol. 29, p. 259, 2011.
7. W.A. Laftah, S. Hashim, and A.N. Ibrahim, *Polymer-Plastics Technology and Engineering*, Vol. 50, p. 1475, 2011.
8. C. Zhao, S. Nie, M. Tang, and S. Sun, *Progress in Polymer Science*, Vol. 36, p. 1499, 2011.
9. F.P. Nicoletta, D. Cupelli, P. Formoso, G. De Filpo, V. Coletta, and A. Gugliuzza, *Membranes*, Vol. 2, p. 134, 2012.
10. D. Wandera, S.R. Wickramasinghe, and S.M. Hussona, *Journal of Membrane Science*, Vol. 357, p. 6, 2010.
11. S.K. Kumar, and J.D. Hong, *Langmuir*, Vol. 24, p. 4190, 2008.
12. S. Sortino, *Journal of Materials Chemistry*, Vol. 22, p. 301, 2012.
13. A. Döring, W. Birnbaum, and D. Kuckling, *Chemical Society Reviews*, Vol. 42, p. 7391-7420, 2013.
14. C.M. Kirschner, and K.S. Anseth, *Acta Materialia*, Vol. 61, p. 931, 2013, and references therein.

15. D. Kuckling, *Colloid and Polymer Science*, Vol. 287, p. 881, 2009, and references therein.
16. I. Gibas, and H. Janik, *Chemistry and Chemical Technology*, Vol. 4, p. 297, 2010.
17. N.A. Peppas, J.Z. Hilt, A. Khademhosseini, and R. Langer, *Advanced Materials*, Vol. 18, p. 1345, 2006.
18. P. Kim, L.D. Zarzar, X. He, A. Grinthal, and J. Aizenberg, *Current Opinion in Solid State and Materials Science*, Vol. 15, p. 236, 2011.
19. E. Wang, M.S. Desai, and S.-W. Lee, *Nano Letters*, Vol. 13, p. 2826, 2013, and references therein.
20. N.A. Peppas, "Intelligent Hydrogels and their Biotechnological and Separation Applications," in G. Güven, ed., "*Radiation Synthesis of Intelligent Hydrogels and Membranes for Separation Purposes*," 1–14, IAEA, Vienna, 2000, and references therein.
21. Q. Yang, N. Adrus, F. Tomicki, and M. Ulbricht, *Journal of Materials Chemistry*, Vol. 21, p. 2783, 2011.
22. B.V. Slaughter, S.S. Khurshid, O.Z. Fisher, A. Khademhosseini, and N.A. Peppas, *Advanced Materials*, Vol. 21, p. 3307, 2009.
23. R. Barbey, L. Lavanant, D. Paripovic, N. Schuwer, C. Sugnaux, S. Tugulu, and H. A. Klok, *Chemical Reviews*, Vol 109, p. 5437, 2009.
24. D. T. Eddington, and D. J. Beebe, *Advanced Drug Delivery Reviews*, Vol. 56, p. 199, 2004.
25. C. Zhao, S. Nie, M. Tang, and S. Sun, *Progress in Polymer Scicne*, Vol. 36, p. 1499, 2011, and references therein.
26. S. Rahimi, E.H. Sarraf, G.K. Wong, and K. Takahata, *Biomedical Microdevices*, Vol. 13, p. 267, 2011.
27. A. Suzuki, T. Tanaka, *Nature*, Vol. 346, p. 345, 1990.
28. I. Tomatsu, K. Peng, and A. Kros, *Advanced Drug Delivery Reviews*, Vol. 63, p. 1257, 2011, and references therein.
29. Y. Ito, and Y.S. Park, *Polymers for Advanced Technologies*, Vol. 11, p. 136, 2000.
30. F.M. Andreopoulos, E. Beckman, and A.J. Russel, *Journal of the American Chemical Society*, Vol. 118, p. 6235, 1996.
31. F.M. Andreopoulos, E. Beckman, and A.J. Russel, *Biomaterials*, Vol. 19, p. 1343, 1998.
32. M. Irie, and D. Kunwatchakun, *Macromolecules*, Vol. 19, p. 2476, 1986.
33. A. Mamada, T. Tanaka, D. Kungwatchakun, and M. Irie, *Macromolecules*, Vol. 19, p. 1517, 1990.
34. M. Ishikawa, N. Kitamura, H. Masuhara, and M. Irie, *Makromolekulare Chemi, Rapid Communications*, Vol. 12, p. 887, 1991.
35. H. Sasaki, A. Ueno, and T. Osa, *Bulletin of the Chemical Society of Japan*, Vol. 61, p. 2321, 1988.
36. T. Tanaka, D. Fillmore, S.T. Sun, I. Nishio, G. Swislow, and A. Shah, *Physical Review Letters*, Vol. 45, p. 1636, 1980.
37. I. Willner, S. Sussan, and S. Rubin, *Journal of the Chemical Society, Chemical Communications*, Vol. 2, p. 100, 1992.

38. M.G. Kodzwa, and D.G. Rethwisch, *Polymeric Materials Science and Engineering*, Vol. 76, p. 279, 1997.
39. M.G. Kodzwa, M.E. Staben, and D.G. Rethwisch, *Journal of Membrane Science*, 1999, Vol. 158, p. 85, 1999.
40. M.E. Byrne, and V. Salian, *International Journal of Pharmaceutics*, Vol. 364, p. 188, 2008.
41. C. B. Gong, M.H.W. Lam, and H.X. Yu, *Advanced Functional Materials*, Vol. 16, p. 1759, 2006.
42. L. Fang, S. Chen, Y. Zhang, and H. Zhang, *Journal of Materials Chemistry*, Vol. 21, p. 2320, 2011, and references therein.
43. C.-B. Gong, Y.-Z. Yang, C. Gao, Q. Tang, C.-F. Chow, J.-D. Peng, and M. H.-W. Lam, *Journal of Sol-Gel Science and Technology*, Vol. 67, p. 442-450, 2013.
44. C.B. Gong, K.L. Wong, and M.H.-W. Lam, *Chemistry of Materials*, Vol. 20, p. 1353, 2008.
45. Y. Huang, Z. Qiu, Y. Xu, J. Shi, H. Lin, and Y. Zhang, *Organic and Biomolecular Chemistry*, Vol. 9, p. 2149, 2011, and references therein.
46. K. Sumaru, K. Ohi, T. Takagi, T. Kanamori, and T. Shinbo, *Langmuir*, Vol. 22, p. 4353, 2006.
47. I. Willner, S. Rubin, R. Shatzmiller, and T. Zor, *Journal of the American Chemical Society*, Vol. 115, p. 8690, 1993.
48. M. Irie, *Kagaku Kogyo*, Vol. 38, p. 613, 1987.
49. B. Kim, H.S. Lee, J. Kim, and S.-H. Kim, *Chemical Communications*, p. 1865, 2013, and references therein.
50. E. Miyako, H. Nagata , K. Hirano, and T. Hirotsu, *Small*, Vol. 4, p. 1711, 2008.
51. C.-H. Zhu, Y. Lu, J. Peng, J.-F. Chen, and S.-H. Yu, *Advanced Functional Materials*, Vol. 22, p. 4017, 2012.

3

Smart Vesicles: Synthesis, Characterization and Applications

Jung-Keun Kim[1,*], Chang-Soo Lee[2] and Eunji Lee[3,*]

[1]Corporate R&D Center, LG Chem Research Park, Daejeon, Republic of Korea
[2]Department of Chemical Engineering, Chungnam National University, Daejeon, Republic of Korea
[3]Graduate School of Analytical Science and Technology, Chungnam National University, Daejeon, Republic of Korea.

Abstract
Vesicles are representative mimic systems for natural lipid membranes. This chapter summarizes recent research in the area of synthesis of vesicular structures from self-assembly of lipids, polymers, and small molecules in solutions and synthesis of hard vesicles by using vesicles as templates, like hollow silica spheres. Then we provide an overview of the characterization tools of structures by microscopic and scattering method. The present study will focus on the discussion about the various types of self-assembled smart vesicles, which are revealed by their capability to respond to external stimuli such as temperature, pH, and others. In addition, the application of these smart vesicles toward molecular separation, chemical sensors, nanoreactors and microreactors, catalysts, and drug delivery vehicles are described.

Keywords: Self-assembly, vesicle, smart membrane, application of vesicle.

3.1 Introduction

The lipid membrane is a thin and semi-permeable membrane formed by two layers of lipid molecules [1,2]. Lipid bilayers in nature are usually composed of self-assembling phospholipids, which have hydrophobic tails and a hydrophilic head. In water, phospholipids form a two-layered sheet with

Corresponding author: 76jkkim@gmail.com, eunjilee@cnu.ac.kr

Annarosa Gugliuzza (ed.) Smart Membranes and Sensors, (53–104) 2014 © Scrivener Publishing LLC

their hydrophobic tails located at the center of the sheet interacting with one another and hydrophilic head and face turned toward the aqueous environment. Phospholipids bilayer is a structure element of the cell membrane common in almost all living organisms and many viruses; the membranes surround the cell nucleus to protect other sub–cellular units. which keep ions, proteins, and other molecules. Lipid bilayers are referred to as semipermeable membranes because specific substances can pass through the membrane in a selective manner under certain conditions. [3,4]

Vesicular structures are representative systems of natural lipid membranes. [5,6] Artificial systems exhibiting hollow structures are formed from the self-assembly of a wide variety of synthetic platforms consisting of hydrophilic and hydrophobic segments. Synthetic molecules may self–assemble into aggregate structures when they are exposed to a selective solvent that is good for one segment and a poor for another segment. The creation of vesicular structures through the self–assembly of synthetic molecules has been one of the major research topics in the field of materials science, nanochemistry, and biomimic chemistry. [7,8]

Furthermore, the precise control of vesicles of a well–defined size and design in self–organizing materials is of paramount importance to acquire smart vesicles with the desired functions and properties. [9] The important issue about smart membranes regarding the development of the hollow structures for specific applications is their ability to control the containment and release of the encapsulated species under the desired conditions.

This chapter includes a summary about recent research on the synthesis of soft and hard vesicular structures. Then we will provide an overview of the characterization tools of vesicles using microscopy and scattering, focusing particularly on the discussion about the various types of self–assembled smart vesicles that respond to external stimuli. Finally, we will describe the application of these smart vesicles in various fields, such as separation, sensors, reactors, catalysts, and drug delivery vehicles.

3.2 Synthesis of Soft Vesicles

During the past decade, various synthetic approaches for producing vesicles in solutions have been studied. With regard to the compartment of forming, several main classes of vesicles by synthetic approach can be distinguished by liposome, polymersome, and vesicles based on small molecules. (Fig. 3.1). [10] First, we will discuss a major factor for forming vesicles and describe how liposome, polymersome, and vesicles are formed by synthetic small molecules.

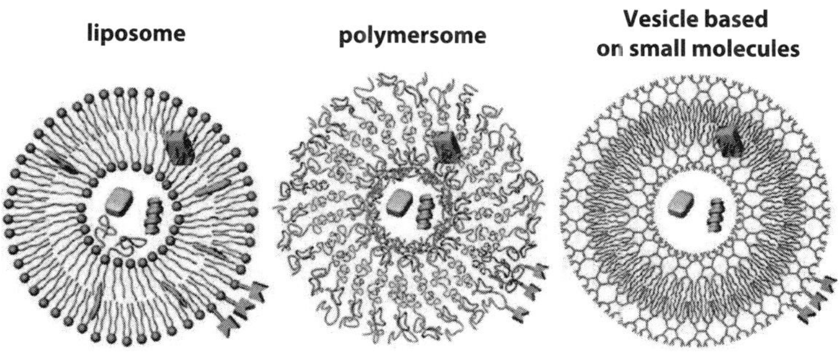

Figure 3.1 Summary of stable vesicles by forming units. Liposomes co-assembled from phospholipids (left). Polymersomes assembled from block–copolymers (middle). Vesicles based on small molecules (right).

3.2.1 Self-assembly into Vesicles

Supramolecular morphology of amphiphiles depends highly on the relative volume fraction of hydrophilic species to that of hydrophobic species. It has been proposed that self-assembled aggregates of amphiphiles consisting of hydrophobic and hydrophilic segments can be predicted using the critical packing parameter, $P = v/a_0 l$, where v is the volume as the hydrophobic part, a_0 is the optimal interfacial area per molecule and l is the hydrophobic length normal to the interface (Fig. 3.2). When the packing parameter is less than 1/3, the amphiphile has a strong tendency to form spherical micelles; when $v/a_0 l_c$ is between 1/3 and 1/2, cylindrical micelles will be favored; and when the parameter is between 1/2 and 1, flexible bilayers are observed. If $v/a_0 l_c$ is approximately equal to 1, planar bilayers are manipulated via the self–assembly process. These flexible and planar bilayers could convert into spherical bilayers or vesicles to minimize free energy by eliminating energetically unfavorable edges. But, there are some exceptions to this rule because the amphiphiles have a different packing parameter as a function of the condition. Packing parameters are not definitive design factors of morphology because the aggregation shape of molecules should be balanced between all the free energy contributions to the self–assembly and also kinetic factors to the final structure.

3.2.2 Liposomes

Liposomes are spherical bilayer structures that enclose an inner water pool formed by lipid molecules. Preparation of liposomes in the laboratory was

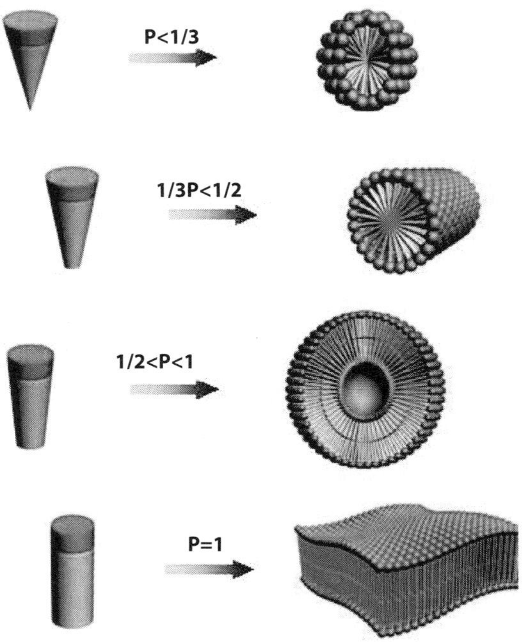

Figure 3.2 Various self–assembled nanostructures depending on the relative volume fraction of hydrophobic and hydrophilic blocks.

first reported by Bangham et al. in the 1960s [11,12]. Since then, it has been studied extensively as a model system for natural lipid membranes due to their structure and semi permeability ability. [13–19] But liposomes have some limitations that hinder their ability to successful mimic structures of nature. A major limitation is lack of stability over a long period of time, which prevents the wider application of the liposomes. Normally, liposomes are formed by applying simple external physical forces like sonication or extrusion to phospholipid bilayers. Therefore, liposomes are kinetically trapped metastable structures and do not exist in a thermodynamically stable state.

To overcome these limitations, lipid–like synthetic vesicles can be formed from surfactants, as first reported by Kaler in 1989. [20] Spontaneous equilibrium vesicles can be prepared from aqueous mixtures of single–tailed cationic and anionic surfactants. Vesicle size, surface charge, or permeability could be readily controlled by varying the mixing ratio of anionic to cationic surfactant. Vesicle formation apparently results from the production of anion–cation surfactant pairs that then act as double–tailed zwitterionic surfactants. These vesicles are quite stable in comparison to conventional vesicles prepared by mechanical disruption of insoluble liquid crystalline dispersions.

Liposome is the oldest and the most common type of lipid-based structure used for the manufacture of pharmaceuticals, cosmetics, and biochemicals. [21] Extensive research has been carried out during past few decades on the ability of liposomes to carry drugs, peptides, proteins, DNA, antisense oligonucleotides, or ribozymes that are highly biocompatible. [22] Cationic liposomes can electrostatically bind anionic nucleic acids to their surface to carry nucleic acids efficiently, especially to the skin, because they have the advantage of being able to encapsulate hydrophilic drugs inside the lipid bilayer, or hydrophobic materials within their own shell.

3.2.3 Polymersomes

Polymer vesicles, also called polymersomes, are spherical shaped structures formed by amphiphilic block copolymers [23–25]. In water, polymersomes consist of hydrophilic blocks exposed to the outer and inner water and hydrophobic blocks that form in the shell of vesicles. Polymersomes can encapsulate an aqueous part and water soluble compartment, enclosed by a bilayer membrane. Compared to lipid-based vesicular structures, polymer vesicles are more stable and robust, and they can easily be chemically or physiologically functionalized to be utilized in a miscellaneous field. [26–30]

Pioneer work on polymersomes was conducted by Discher et al. (1999). They reveal that vesicles are made from amphiphilic diblock copolymers and characterized by micromanipulation. [31] The average molecular weight of polyethyleneoxide–polyethylethylene is several times greater than that of typical phospholipids in natural membranes. Consequently, polymersomes from block copolymers proved have higher mechanical properties than normal liposomes, and polymersome membranes are also at least 10 times less permeable to water than common phospholipid bilayers because more spacing exists between polymers than lipid bilayers. These results suggest a new class of synthetic thin–shelled vesicle based on block copolymer chemistry.

Biocompatible polymersomes can be obtained by connecting biomolecules directly to the shell of vesicles. Heise et al. (2012) report a versatile and facile route to bioactive polymersomes fully based on amino acid and carbohydrate building blocks (Fig. 3.3). [32] Polypeptide block copolymers with different block length ratios were obtained by sequential ring-opening polymerization of benzyl–L–glutamate and propargylglycine (PG) N–carboxyanhydrides. Glycosylation of the poly(PG) block was obtained by "click" reaction using azide–functionalized galactose.

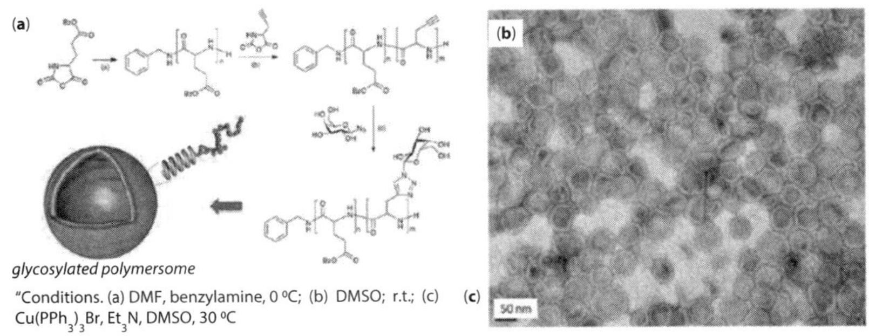

Figure 3.3 (a) Synthesis of PBLG-*b*-PGG Glycopeptide Block Copolymers. (b) High-magnification TEM image of polymersomes made of PBLG20-*b*-PGG25.

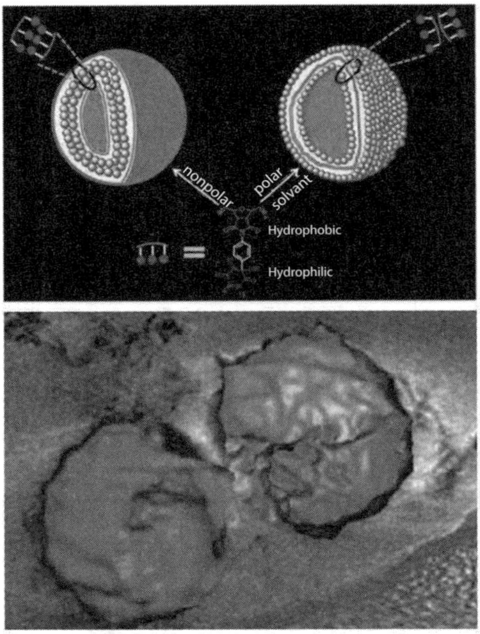

Figure 3.4 Cartoon representation of the newly designed amphiphilic homopolymer and its self-assembled vesicle structures in both a hydrophobic and a hydrophilic environment (top). Tomographic 3D view of vesicles (bottom).

Polymersomes are usually manipulated by block copolymers, but Shunmugam et al. (2012) report the synthesis of a new molecular architecture, a norbornene-derived thiobarbiturate amphiphilic homopolymer, by ring-opening metathesis polymerization using Grubb's catalyst and its characterization (Fig. 3.4). [33] The designed amphiophilic homopolymer shows a self–assembled vesicle formation in aqueous solution. Dynamic light scattering and critical aggregation concentration studies confirm the aggregate formation in solution with an average diameter of around 100 nm.

Recently, Lecommandoux et al. (2012) reported on an emulsion–centrifugation method that allows polymersomes in polymersomes or polymer vesosomes to be prepared. [34] This simple approach to generate biomimetic compartmentalized structures offers a way to encapsulate several different and even mutually incompatible active substances, thereby opening avenues in combinatory drug release and providing exquisite control of permeation properties.

3.2.4 Vesicles Based on Small Molecules

Small molecules consisting of antagonistic amphiphilic blocks in the same molecular architecture are excellent candidates for creating well defined supramolecular structures. In a selective solvent for one of the blocks, these amphiphilic molecules can form aggregates of vesicles as a result of the association of the soluble and insoluble blocks. [35–42] In addition, small molecules can form highly ordered and complex vesicular structures even at low molecular weight because their molecular shape and the conformation of the organization. [43–48]

One of the examples is amphiphilic dendrimers, which are novel macromolecules with well–defined architectures that use building blocks to form vesicles in solution. Wegner et al. (2005) revealed the synthesis of a codendrimer (g3–PBE–b–g3–PMDC), composed of a third-generation poly(benzyl ether) (PBE) monodendron and an aliphatic polyether (PMDC) monodendron (Fig. 3.5). [49] In THF/dilospropyl ether (1:1) the PMDC block functioned as a "hydrophilic" block, while the PBE acted as a "hydrophobic" block. The codendrimer could form interdigitated layers leading to vesicle formation. They proposed that the different solubility of the PBE and PMDC blocks in the mixed solvent was the driving force for the formation of the vesicles.

The supramolecular concept, like the host–guest interaction, can be introduced to vesicles to enhance their capacity and widen the application field of vesicles forming by small molecules. Kim et al. (2005) introduced the supramolecular character in vesicular structure (Fig. 3.6). [50] They demonstrate a direct approach for the synthesis of polymer nanocapsules without using any preorganized structure, emulsifier, or template, which appears to be applicable to any monomers with a flat core and multiple polymerizable groups at the periphery. Polymer nanocapsules is directly synthesized by the thiol–ene photopolymerization of $(allyloxy)_{12}$cucurbit[6]uril, a rigid disk–shaped

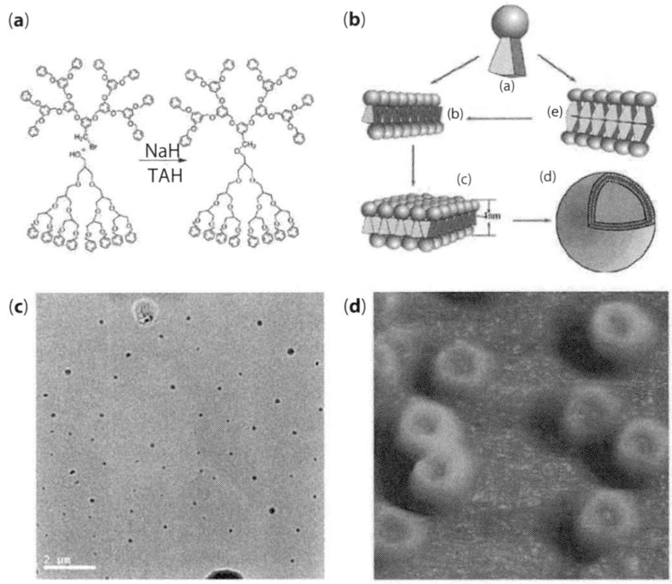

Figure 3.5 (a) Synthetic route to the g3–PBE–*b*–g3–PMDC block codendrimer. (b) Schematic illustration depicting the formation of a vesicle with an interdigitated layer of block codendrimer. The pink spheres are the PMDC blocks, while the blue fans represent the PBE blocks. (c) TEM and (d) AFM (670 nm × 670 nm) images of the vesicles.

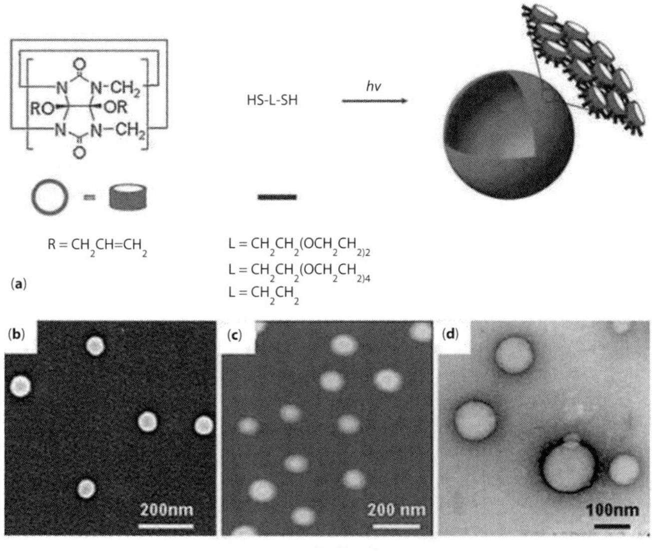

Figure 3.6 (a) Synthesis of nanocapsules by the thiol–ene photopolymerization of cucurbit[6]uril and dithiol. Microscopy images of vesicles. (b) SEM, (c) AFM, and (d) TEM images of vesicles prepared in methanol.

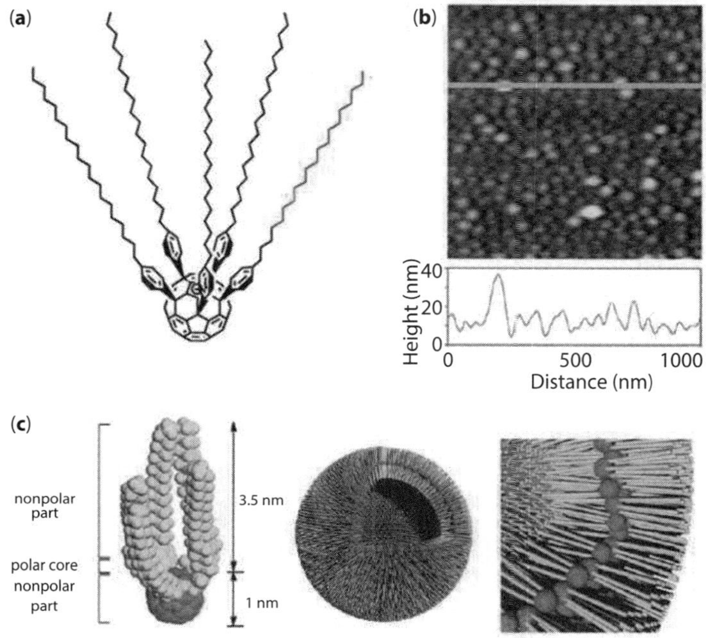

Figure 3.7 (a) Chemical structures of potassium complexes of fullerene anions. (b) AFM images of the vesicles after drop-casting of a 2 mM solution. The cross-section profiles along the light-blue line in each figure are shown below the AFM images. (c) Schematic images of vesicles are made of fullerene anions.

molecule with a cavity and 12 polymerizable allyl groups at the periphery, and dithiol. UV irradiation of a mixture of cucurbit and thiol in methanol for 20 h followed by dialysis produced polymer nanocapsule in good yield.

Vesicles from small molecules show relatively strong interaction from π–π interaction between the rigid block and the ability of microphase separation of each block. Nakamura et al. (2011) report the synthesis of 20 potassium complexes of penta- [(4-substituted) phenyl] fullerene anions and examine their ability to form bilayer vesicles in water (Fig. 3.7). [51] The overall structure of the amphiphiles can be described as a nonpolar/polar/nonpolar motif as opposed to the usual polar/ nonpolar motif of normal lipid amphiphiles. Despite the hydrophobicity of the fullerene moiety and alkyl/perfluoroalkyl chains, the amphiphiles are soluble in water because of the centrally located polar part of fullerene cyclopentadienide and they spontaneously form a vesicle with a narrow unimodel size distribution. The vesicles appear highly stability upon heating to a high temperature, almost boiling point of water, or standing over one year in air, as well as on a solid substrate in vacuum, keeping their spherical form.

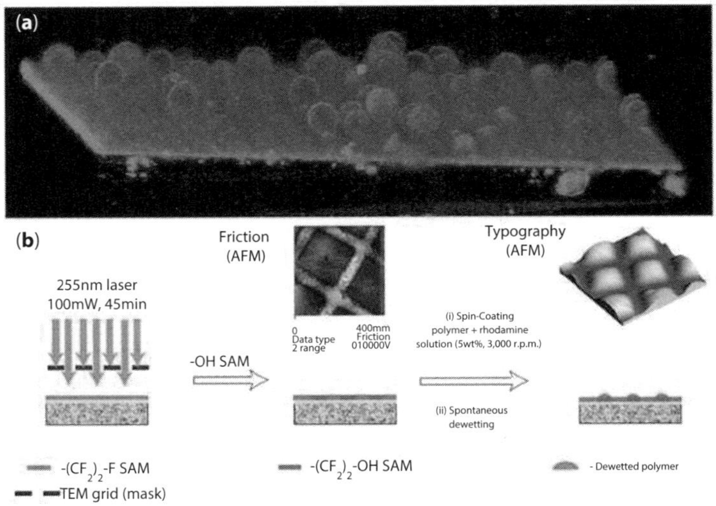

Figure 3.8 (a) 3D image (generated from a series of vertical slices) of the vesicle–forming surface showing the swollen exterior bilayer before detachment. (b) Schematic representations of the procedure of the polymer island formation.

3.2.5 Direct Synthesis

It is now routine to produce vesicles using lipid, surfactant, copolymer, and small molecules. They form lamellar phases in solution by their amphiphilic character. These lamellar phases can be converted into vesicles, through a variety of energy–intensive processes such as electroformation, extrusion, and sonication. [52–58] Unilamellar vesicles form when molecules self–assemble to a single bilayer structure and the resulting membrane overlaps and rearranges to a spherical shape. [59–63] But this mechanism of forming vesicles is weak of selectivity of size and distribution, only the width of the shell of vesicular aggregation maintains constant.

Many efforts have been made to manipulate the uniform size of distribution in preparation of vesicles. One modified method for uniform vesicles is reported by Howse et al. (2009) (Fig. 3.8). [64] They report a method for the production of size-controlled distributions of micrometre–sized vesicles combining the "top–down" control of micrometre–sized features by photolithography and dewetting with the "bottom–up" control of nanometre–sized features by molecular self–assembly. It enables the spontaneous creation of unilamellar vesicles with a narrow size distribution that has applications in drug and gene delivery, nanoreactors and microreactors, substrates for macromolecular crystallography, and model systems for studies of membrane function.

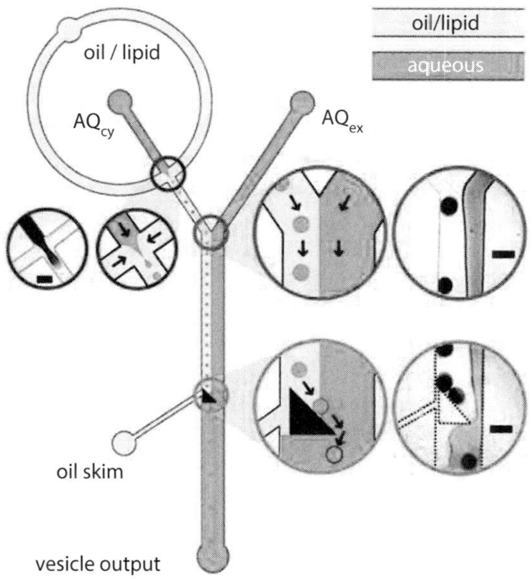

Figure 3.9 Circuit schematic and operation of the microfluidic assembly line. The oil/lipid input is introduced at the top left, focusing the cytoplasmic aqueous input (AQcy) to generate uniform, lipid–stabilized droplets. The droplet flow merges with an extracellular aqueous input (AQex) to form a lipid–stabilized oil/water interface adjacent to the droplet flow.

Many groups have made an effort to acquire the uniformed vesicular structures using the microfluidic system. [65–67] The advantage of this system is the production of monodisperse particles that are of similar size and distribution. Paegel (2011) reports a microfluidic assembly line that produces uniform cellular compartments from droplets, lipids, and oil/water interface starting materials (Fig. 3.9). [68] Droplets form in oil that contains lipid and flows to a junction where the confluence of oil and extracellular aqueous media establishes a flow–patterned interface that is both stable and reproducible. A triangular post-mediated phase transfer bilayer assembly by deflecting droplets from oil, through the interface, and into the extracellular aqueous phase to yield a continuous stream of unilamellar phospholipid vesicles of uniform and tunable sizes. The size of the droplet precursor dictates the vesicle size, the encapsulation of the small–molecule cargo is highly efficient, and the single bilayer promotes functional insertion of a bacterial transmembrane pore.

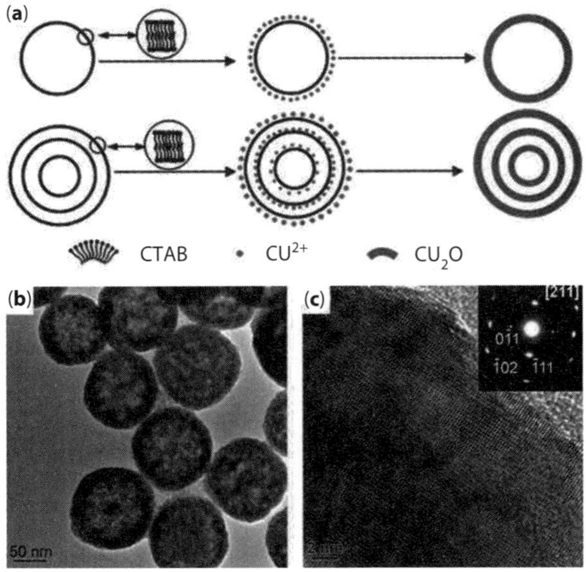

Figure 3.10 (a) The formation of different Cu_2O hollow structures in the presence of a CTAB template: single–lamellar vesicle, and multilamellar vesicle. (b) TEM and (c) HRTEM images of the Cu_2O hollow spheres prepared at a CTAB concentration of 0.1 M.

3.3 Synthesis of Hard Vesicles

The soft vesicle itself has been challenging and makes it difficult to construct hard materials directly. Otherwise, vesicles could prove useful as templates for the synthesis of hard materials because of their characteristic morphology. [69–71] The templating method with soft vesicles allows for simple fabrication processes under mild conditions with variable templates, and easy removal of the templates with less damage to hard vesicles. [72–74] First, we will discuss how to synthesize hard vesicles by templating soft vesicles and, more precisely, the hollow silica particles.

3.3.1 "Soft" Templates for the Synthesis of Hard Vesicles

The templating of soft vesicles provides a powerful tool to tune the size, shape, and configuration of the resulting hard vesicles in the solution–phase synthesis with various inorganic materials. [75–81] Wang et al. (2007) report on inorganic hard vesicle using a soft vesicle template. They have developed a facile method to synthesize inorganic nanostructures with the assistance of cetyltrimethylammonium bromide (CTAB) vesicles and multilamellar vesicles (Fig. 3.10). [82] The structure (single–, double–, triple–, and quadruple–shelled) of these Cu_2O hollow spheres could be

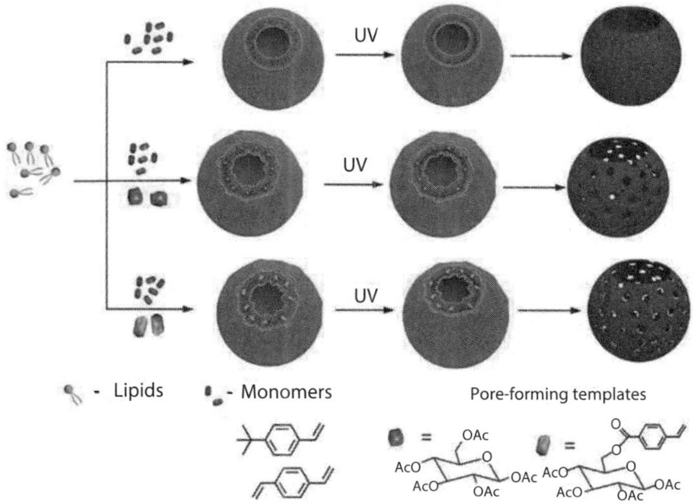

Figure 3.11 Scheme of vesicles formation; formation of liposomes with loaded monomers and irradiation with UV light, and formation of fortified liposomes. Lipid removal and formation of hollow nanocapsules with thin walls.

easily controlled by adjusting the concentration of the CTAB surfactant. The TEM images reveal that the shells of these hollow spheres are single–crystalline, which might improve their stability. Their report confirms the feasibility of vesicle and multilamellar vesicle directed synthesis of inorganic multishelled hollow spheres. More importantly, it might prove possible to create multifunctional materials by filling different target materials into different chambers of the vesicular structures.

Another synthetic pathway to the manipulation of hard vesicles is based on the polymerization within the hydrophobic part of soft vesicles. The hard vesicles were obtained easily through the controlled polymerization inside of the outer shell. Meier et al. (2000) describe the synthesis and the characterization of a poly(2-methyloxazoline)–*b*–poly(dimethylsiloxane)–blockpoly(2–methyloxazoline) triblock copolymer carrying polymerizable groups at both chain ends. [83] This copolymer forms vesicular structures in a diluted aqueous solution, the size of which could be controlled. The groups at the methacrylate end of the triblock copolymer are polymerized in the vesicular aggregates using an UV-induced free radical polymerization. The cross-linking polymerization does not lead to morphological changes in the underlying vesicles; moreover, due to their cross-linked structure, the nanocapsules are shape persistent, thus maintaining their integrity even after their isolation from the aqueous solution.

Another example is featured in the paper by Pinkhassik et al. (2010) (Fig. 3.11). [84] Hollow polymer nanocapsules are produced by the

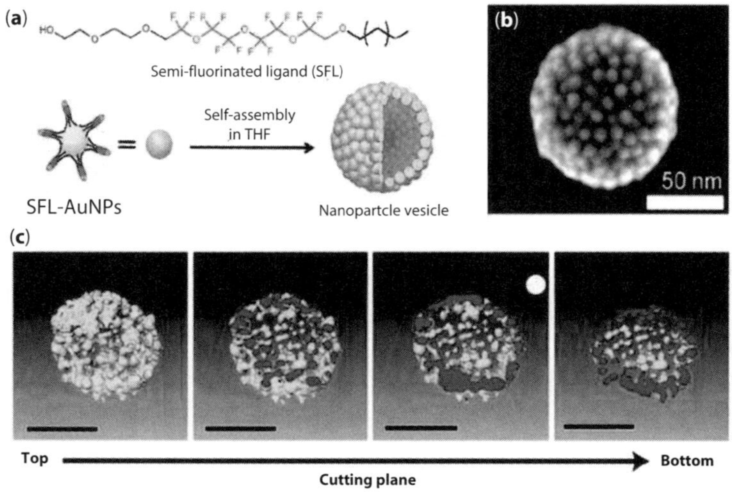

Figure 3.12 (a) Structure of the semifluorinated ligand (SFL) and self-assembly of SFL-AuNPs to generate vesicles. (b) SEM images of vesicles. (c) Reconstructed images (XY plane) of vesicles embedded in epoxy resin. Scale bars are 50 nm.

polymerization within hydrophobic interior of lipid bilayers that act as temporary self-assembled scaffolds. Pore-forming templates are codissolved with monomers in the bilayers to create pores in a size and chemically-controlled environment. High-resolution magic angle spinning NMR characterization provides detailed structural information about nanocapsules, and spherical shape is confirmed by electron microscopy.

Niikura et al. (2012) report the production of hard vesicles without templates (Fig. 3.12). [85] They state that gold nanoparticles (AuNPs) coated with semi fluorinated oligo (ethylene glycol) ligands efficiently provide sub-100 nm hollow vesicular assemblies in THF. Semi fluorinated-coated AuNPs are synthesized through ligand exchange from citric acid to semi fluorinated oligo(ethylene glycol) ligands in THF. Their strategy is based on the unique function of the fluorinated region, which induces the bundling of ligand molecules on nanoparticles.

3.3.2 Hollow Silica Spheres

Hollow silica spheres are utilized as materials with light weight, low dielectric constant, and thermal insulation from typical silica properties. Since hollow silica spheres using soft vesicles were reported by Pinnavaia (1998), [86] many papers have reported on their diverse developments. [87–92]

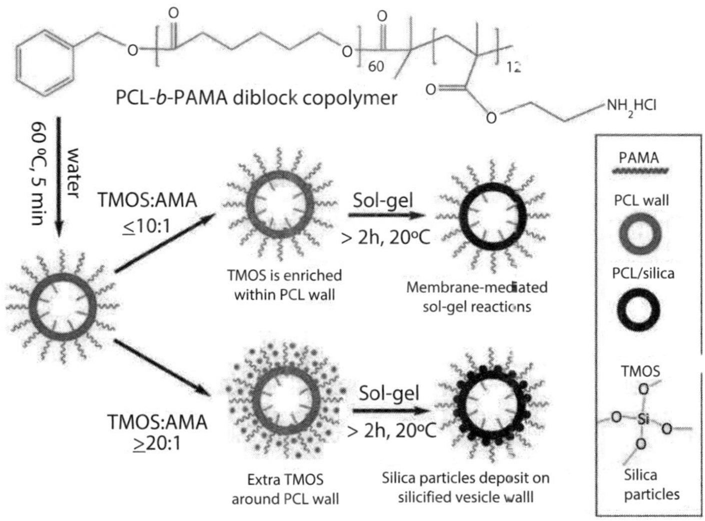

Figure 3.13 Formation of PCL–*b*–PAMA diblock copolymer vesicles by direct dissolution and their subsequent silicification using TMOS.

Preparation of hollow silica sphere is simply done by organosiloxane–based vesicles and complexation of tetramethylorthosilicate (TMOS) or tetraethylorthosilicate (TEOS) with soft vesicles. Organosiloxane–based vesicles can be silicified by sol–gel chemistry within the vesicle walls using TMOS or TEOS as a water soluble silica precursor to coat either the surfactant vesicles or the polymer vesicles. [93–98] One such example is this is described by Zheng et al. (2010), who showed the hollow silica sphere based on catanionic surfactant system, composed of a long-chain imidazolium ionic liquid and sodium dodecyl sulfate. [99] Silica hollow spheres, with diameters 30-60 nm and a wall thickness of 8-10 nm, are prepared by using the vesicles as the templates and TEOS as a silica precursor. The hollow silica spheres are characterized by TEM, scanning electron microscopy (SEM), and nitrogen adsorption-desorption.

Du et al. (2008) show a new class of copolymer vesicles using poly(ε–caprolactone)–*b*–poly(2–aminoethyl methacrylate), PCL–*b*–PAMA, obtained by the spontaneous self–assembly of a new primary amine-based block copolymer in a purely aqueous solution (Fig. 3.13). [100] Due to their primary amine functionalized exterior, these vesicles could be readily surface functionalized, as demonstrated by the gold nanoparticle decoration. Moreover, the hydrophobic PCL vesicle membrane could be readily silicified using TMOS to form colloidal stable organic/inorganic hybrid vesicles. Unusually, the hydrophobic PCL membrane acts as a scaffold by solubilizing the poorly

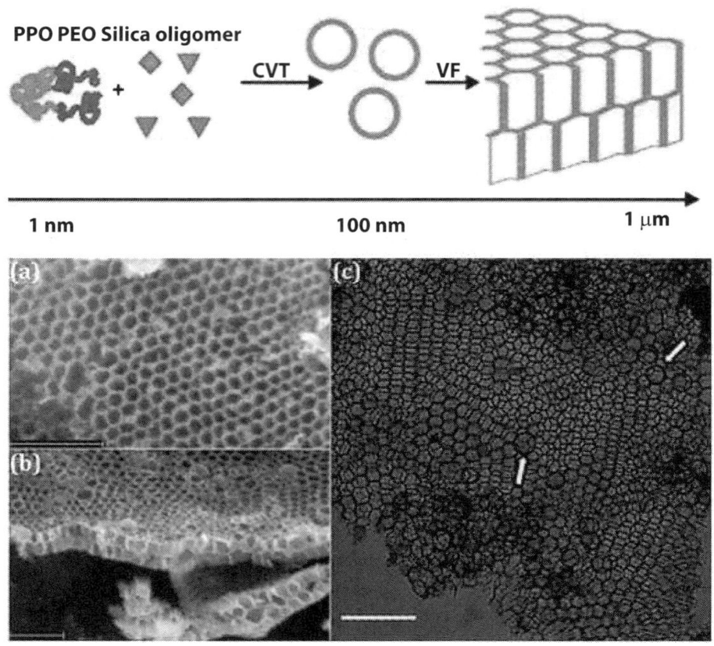

Figure 3.14. SEM (a, b) and TEM (c) images of MOSF. Scale bar is 500 nm in all images.

water soluble TMOS precursor, while the cationic PAMA chains act as an efficient catalyst for silicification at the vesicle/aqueous solution interface.

Yu et al. (2006) used silica hollow spheres as a building block of secondary structures. They report a new approach to produce macroporous (110 nm in diameter) ordered siliceous foams by utilizing vesicles of poly(ethylene oxide)–poly(propylene oxide)–poly(ethylene oxide) copolymers (Fig. 3.14). [101] When the concentration of the base is adjusted between a certain area, a vesicle to a tightly aggregated vesicle, and finally macroporous siliceous foams structural transition is observed. They demonstrate that molecular assembled compound vesicles with diameters larger than 100 nm can be further utilized to assemble elegant structures, similar to those of diatoms and honey bees in nature. Their strategy provides a simple and new pathway to synthesize ordered macroporous materials.

3.4 Characterization of Vesicular Structures

Vesicles have very thin and fragile walls that are difficult to visualize using a traditional microscope, so characterization of bilayers was very challenging before technical improvements were made in microscopy. Advanced skills like light or x-ray scattering and various types of electron microscopy had been required in experiments on vesicles. This section will consist of a short discourse on several techniques that are most commonly used in the characterization of vesicles.

3.4.1 Microscopy

The growth in size and number of vesicles is an interesting problem from the point of view of the physicochemical behavior, and it has a particular meaning since vesicles are generally considered to be suitable models for protocells. Growth, budding, fission, and other vesicle transformations have been studied theoretically, and experimentally direct investigations had been presented using mainly giant vesicles and light microscopy. [102–108] In the case of submicrometer–sized vesicles, the lack of experimental data may be mostly due to the lack of a suitable methodology to follow such processes, until a study about the formation and transformation of vesicles is possible

using a new methodology that involves the use of cryotransmission electron microscopy (cryo–TEM). Water et al. (1989) revealed the possibility of real visualization of vesicle in a solution using the cryo–TEM method. [109] Cryo–TEM is the only experimental tool that provides direct micro structural information of complex fluids over the 1–1,000 nm-size range. The insight this allows toward the structural elucidation of self–assemblies and dynamics in liquids is evident. The samples are prepared as a thin film in a controlled temperature and humid environment, rapidly quenched in liquid ethane at its freezing point, and examined in the vitrified hydrated state. [110–116] This technique, which avoids the artifacts of staining and drying procedures, permits observations of relatively undistorted samples. Meanwhile, the cryo–TEM method has become established despite the reporting of several other methods with vesicular characterization, like atomic force microscopy [117–120] and fluorescence microscopy, [121–126] but until now no powerful methods have been able to replace the cryo–TEM method in the microscopic field. In Figure 3.15, we demonstrate the vesicular structures using fluorescence microscopy, TEM, and cryo–TEM. [127] As described in this section, vesicles can be strongly visualized by cryo–TEM with no artifacts during evaporation of the solvent, and the shell of the vesicle is clearly observed in a high resolution.

3.4.2 Scattering

Static and dynamic light scattering experiments with the use of sophisticated instrumentation has emerged as the main tool for studying aggregation in

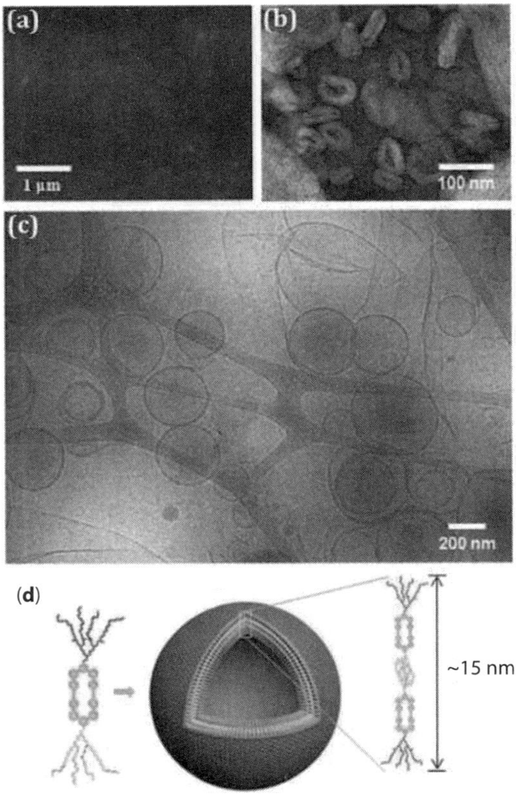

Figure 3.15 (a) Fluorescence microscopy, (b) TEM, and (c) cryo–TEM images of an aqueous solution of molecules (0.01 wt.–%) showing the vesicular objects. (d) Schematic representation of bilayered vesicular objects formed from molecules.

solutions. [128–133] Laser light scattering can probe aggregates in the size range of 1 nm to 1 μm. After the beam of laser light passes through a polymer solution in a probe cell, most of the light will pass through the sample, but a small portion will be scattered by aggregation in a solution and from scattering intensity.

In dynamic light scattering (DLS), fluctuations in the intensity of the scattered light in the microsecond timescale appear because of diffusive motions of particles in the solution. From static light scattering (SLS), structural properties are available, such as weight averaged molecular weight of aggregator, and aggregative shape and size and the second viral coefficient can reveal the information of particle–particle interactions as well as particle–solvent interactions of particles in a diluted solution. Wagenknecht et al. (2010) [134] state that the CONTIN analysis of the

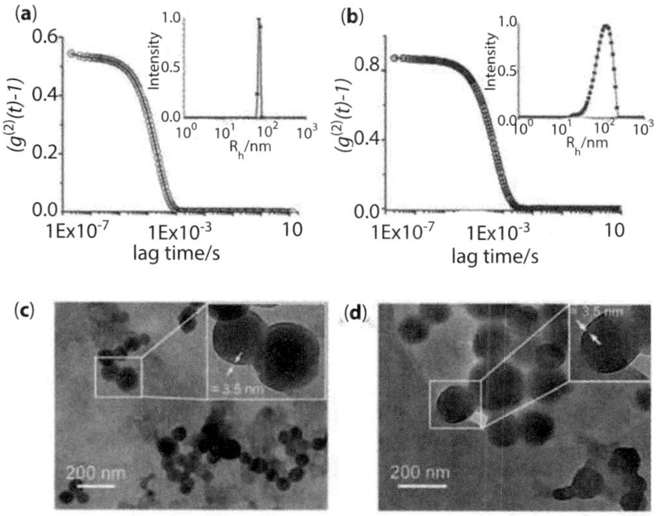

Figure 3.16 DLS autocorrelation function of vesicles (a) in toluene and (b) in water at a scattering angle of 90°, and (c) and (d) are the corresponding TEM images. The insets in (a) and (b) show the corresponding distribution functions.

autocorrelation function obtained from the DLS analysis of vesicles from a non–polar solvent shows a narrow and unimodel distribution of spherical aggregates with a hydrodynamic radius (R_H) of ~70 nm (Fig. 3.16a). Interestingly, the corresponding CONTIN analysis of vesicular structures from water also shows unimodel distribution of spherical aggregates (RH= ~110 nm) (Fig. 3.16b). These observations clearly point to the formation of equilibrated spherical aggregates of molecules in solution.

High throughput scattering methods, such as combinatorial small–angle X–ray scattering (SAXS)/wide–angle X–ray scattering (WAXS), which provide information about structural features of colloidal size, have been used to study phase behavior over a concentration gradient of aggregator in solution. [135–142] From SAXS or WAXS data, using the Fourier transform to obtain the form factor of the electron density of the aggregator in a solution state, the shape of aggregate can be estimated from the form factor. Otherwise, the small–angle neutron scattering (SANS) technique is another scattering method in the study of solution state structures. [143–150] SANS data reveal molecular conformations and interaction parameters, so it allows investigations of the morphology and thermodynamics of molecules during the formation of vesicles. But there are limitations in the exchange of deuterium and it often causes the morphological instability.

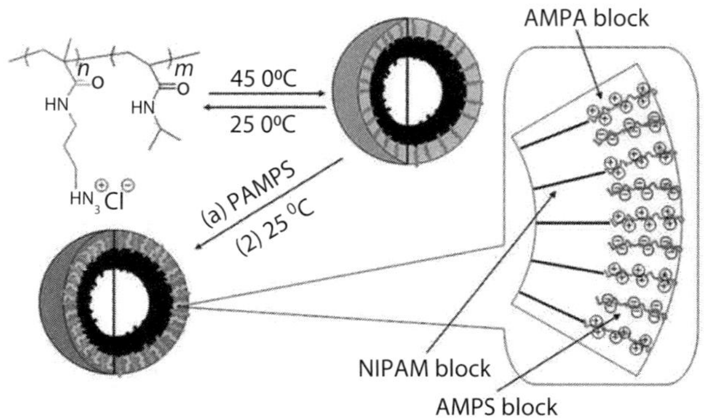

Figure 3.17 Schematic illustration of the formation of vesicles from PAMPA–PNIPAM diblock copolymers and their subsequent ionic crosslinking.

3.5 Stimuli–Responsive Behaviors of Vesicular Structures

An important issue in self–assembled structures is their capability to respond to external stimulus. To obtain external stimulus responsive hollow spheres, more elaborate corresponding blocks are required in designing of synthesis of vesicles. Along this line, this section will introduce smart vesicles that respond to certain stimuli, such as temperature, pH, light, gas, and redox.

3.5.1 Thermo–Responsive Vesicles

Temperature has been used as a successful tool to gauge the responsiveness of self–assembled structures for several years. [151–157] The molecule and solvent interactions that are tuned by temperature change are important to induce the self–assembly and morphological change of secondary structures. For example, the aggregate structures consist of a hydrophilic poly ethylene oxide (PEO) or poly(*N*–isopropylacrylamide) (PNIPAM) exterior that is well–known to exhibit a lower critical solution temperature (LCST) behavior in aqueous media. [158–160] Above the LCST, thermally sensitive PEO, or PNIPAM are dehydrated to collapse into a molecular globule. Consequently, a significant structural change can be expected to happen by changing temperature.

McCormick et al. (2006) report the temperature–responsive vesicle containing PNIPAM block using the diblock copolymer poly(N–(3–aminopropyl) methacrylamide hydrochloride)–*b*–(N–isopropylacrylamide) (PAMPA–*b*–PNIPAM) is synthesized using the reversible addition–fragmentation chain–transfer (RAFT) polymerization technique (Fig. 3.17).

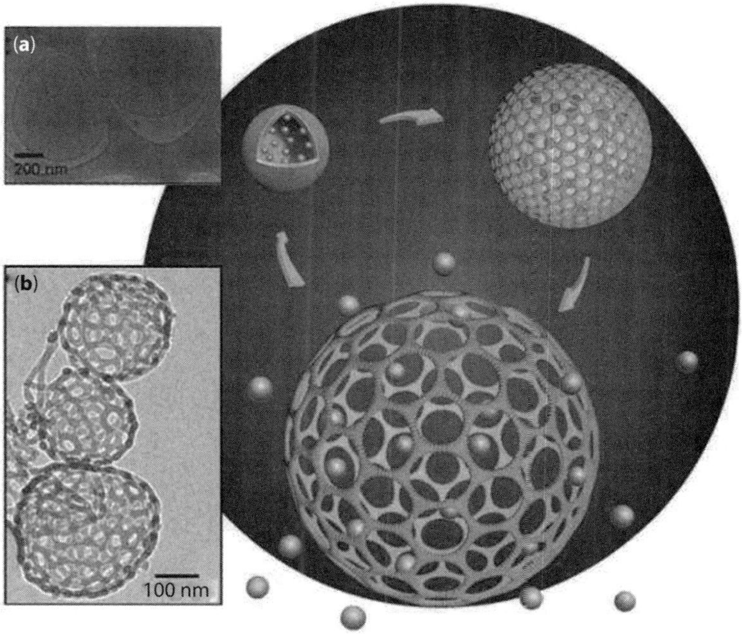

Figure 3.18 Schematic representation of a reversible open/closed gating motion in the lateral nanopores of capsule and cryo-TEM image (a) of closed capsule structures and (b) nanoporous capsules.

[161] The diblock copolymers were dissolved in an aqueous solution at room temperature. Increasing the solution temperature caused the formation of uniform aggregates with diameters of approximately 280 nm and wall width of 7 nm. Since the contour length of the diblock copolymers is approximately 35 nm, vesicular structures are found. At room temperature, these block copolymers exist as unimers in aqueous solution and are self-assembled into vesicles when the solution temperature is increased.

Another moiety for temperature responsive vesicles is PEO. The present paper reports several systems concerning morphological control of aggregation by temperature change in homologous series. [162–164] When synthesizing with a conjugated rod segment that is grafted by hydrophilic polyether dendrons at one end and hydrophobic branches at the other end, it leads the spontaneous formation of a capsule structure with lateral nanopores, which is based on the two-dimensional self-assembly of dumbbell-shaped amphiphiles (Fig. 3.18). [165] In particular, the porous capsules undergo a structural transformation in which the nanopores in the shell are reversibly closed upon heating. This gating behavior of the pores can be explained by the fact that the oligo (ethylene oxide) dendritic exterior exhibits a lower

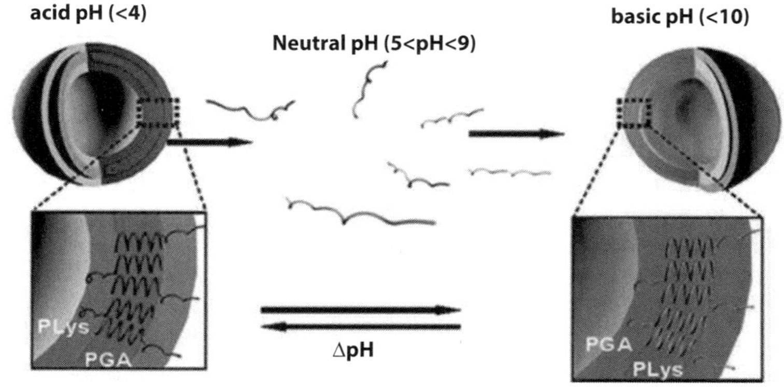

Figure 3.19 Schematic representation of inside-out micellization of pH-responsive and water soluble vesicle from polypeptide diblock copolymer PGA-*b*-PLys.

critical solution temperature (LCST) behavior in aqueous media. Above the LCST, the ethylene oxide segments are dehydrated to collapse into molecular globules, which leads to a decrease in the effective hydrophilic volume as the hydrodynamic volume of the polyether dendrons decreases.

3.5.2 pH-Responsive Vesicles

Changing the pH is particularly attractive for applications involving biological systems because of the large number of pH gradients that can be found in normal or pathophysiological states. [166,167] pH is an important environmental factor in the biomedical field, because pH changes occur in many specific or pathological compartments in nature. Therefore, one of the most significant recent highlights in the field of vesicles is the development of pH-responsive nanostructures. [168–177]

Along this line, self-assembly of pH-responsive polypeptide diblock copolymer with biocompatible character has been reported by Lecommandoux et al. (2005) (Fig. 3.19). [178] Polypeptide consisting of a diblock copolymer poly(L-glutamic acid)-*b*-poly(L-lysine) is synthesized by sequential ring-opening polymerization. Upon dissolution of the polypeptide diblock copolymer in aqueous basic and acid solutions, self-assembly occurs spontaneously. The zwitterionic polypeptide is obtained with positive charges of the protonated lysine and negative charges of the deprotonated glutamic acid at neutral pH. At acidic pH, the copolymer poly(L-glutamic acid) block is neutralized, thereby α-helical structure

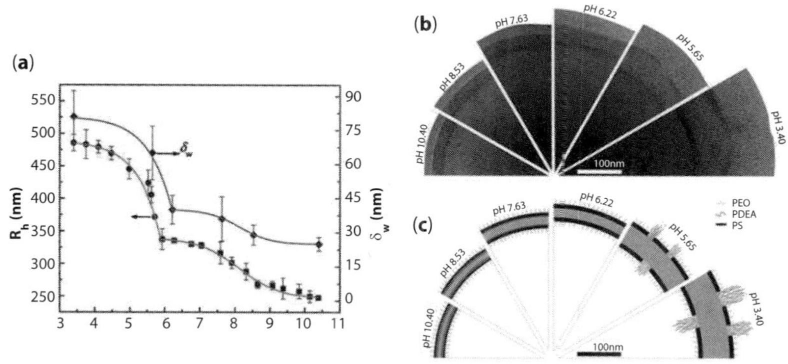

Figure 3.20 (a) Dependence of Rh and wall thickness δw on the solution pH. Δw was measured from cryo–TEM images with the wall thickness in each vesicle measured many times in different places. (b) Cryo–TEM images of vesicle wall structures at corresponding pH values. (c) Schematic illustrations of the vesicle structures

forming the core of the vector while the poly(L–lysine) block forms the shell. Under basic conditions, by contrast, the protonated poly(L–lysine) block is transformed into a neutral amine group, forming the core of the aggregates.

Eisenberg et al. (2009) describe a vesicle system that possesses a pH–induced "breathing" feature and consists of a three–layered wall structure (Fig. 3.20). [179] The "breathing" feature consists of a highly reversible vesicle volume change by a factor of pH, accompanied by diffusion of species into and out of the vesicles with a slow relaxation time. The triblock copolymer poly(ethylene oxide)$_{45}$–block–polystyrene$_{130}$–block–poly(2–diethylaminoethyl methacrylate)$_{120}$ (PEO_{45}–b–PS_{130}–b–$PDEA_{120}$) is synthesized via ATRP. Self–assembly into vesicles is carried out at a pH of ca.10.4. As the pH decreases, both the vesicle size and the thickness of all three layers increase. These changes between pH 10.4 and 3.4 are highly reversible after a certain relaxation time, performed repeatedly. The change in the wall structure not only dramatically increases the wall permeability to water, but also greatly expands the rate of proton diffusion.

Zhou et al. (2012) prepare a novel class of polymeric vesicles that exhibit similar pH–induced "breathing" behavior accompanied by jellyfish–like fluorescence that can be switched on and off through the aqueous self–assembly of a diblock copolymer of PEG–b–PDMA–Azo (Fig. 3.21). [180] At an initial pH value of 7, the average vesicle diameter is about 120 nm, and the vesicle–wall thickness is about 15 nm. When the pH value decreases to 4, a diameter change is seen of approximately 240 nm. Thus, a remarkable expansion of the vesicle volume occurs by a factor of 7, the

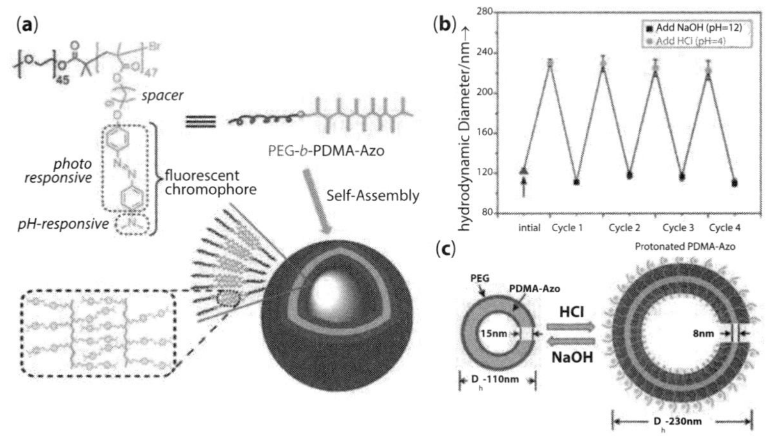

Figure 3.21 (a) Schematic representation of the amphiphilic diblock copolymer PEG–*b*–PDMA–Azo and the vesicle structure. (b) Reversible size change of the vesicles as detected by DLS upon the alternate addition of HCl (pH 4) and NaOH (pH 12). (c) Schematic representation of the pH–induced reversible "breathing" process of the vesicles upon the protonation and deprotonation of PDMA–Azo blocks.

vesicle–wall thickness is halved by approximately same. The vesicles possess a compact or thick bilayer wall structure at pH 7. When the pH value is decreased to 4, the PDMA–Azo segments are protonated and change from hydrophobic to hydrophilic. As a result, the interchain electrostatic repulsive force compels the vesicles to expand to some extent to lower their interaction free energy.

3.5.3 Others

The success of temperature and pH–responsive vesicles has inspired the development of vesicular systems responsive to dual or multiple stimuli. [181–189] Combining responsiveness to different stimuli provides a more efficient system to apply to various fields of application.

Other tools in this direction of the engineering of vesicles possibly possess remote stimulus such as light, gas, and redox. First, irradiation with a specific wavelength light, which involves an easily controllable photochemical process, can tune vesicular structures. [190–196] Vamvakaki et al. (2012) report a light–driven supramolecular engineering of water–dispersible nanocapsules with random copolymer brushes, containing photosensitive molecules nonpolar spiropyrans, which upon UV irradiation isomerize to polar merocyanines (Fig. 3.22). [197] The fabrication of

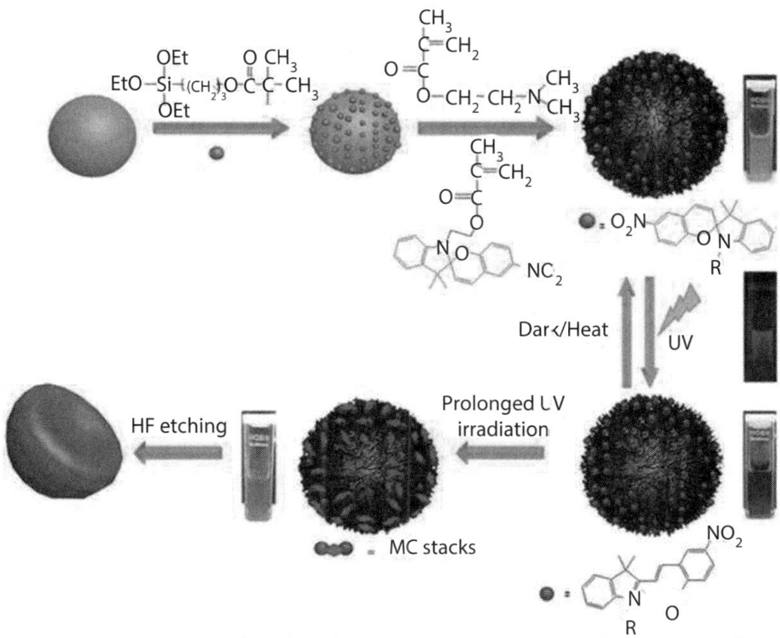

Figure 3.22 Synthetic procedure for fabricating the water–dispersible nanocapsules consisting of 2–(dimethylamino)ethyl methacrylate (DMAEMA) and light–sensitive spiropyran (SP) moieties.

the nanocarriers is based on the formation of H–type π–π interactions between merocyanine isomers within the sterically crowded environment of the polymer brushes upon UV irradiation, which enable the spiropyrans–to–merocyanine isomerization of the photosensitive species.

Another stimulus responsive system reacts to gas due to a new type of vesicle that self–assembles from amidine–containing diblock copolymer. Yuan et al. (2011) discusses amidine–containing diblock copolymers that can spontaneously form vesicles in aqueous media on the basis of their amphiphilicity (Fig. 3.23). [198] Specific functionalities in the polymer endow these vesicles with unique gas–responsibility. CO_2 can tune the size of these vesicles over a wide range by controlling the degree of protonation of the amidine species. Alternating treatment with CO_2 and Ar realizes a smart expansion and contraction cycle of these vesicles.

The redox–responsive vesicles are caused by either a change in polarity or cleavage of the bonds between hydrophobic and hydrophilic blocks, induced during the redox processes. [199–204] Park et al. (2011) discuss an aqueous vesicular system that can be switched by electric potential without the addition of any chemical redox agents into the solution. This

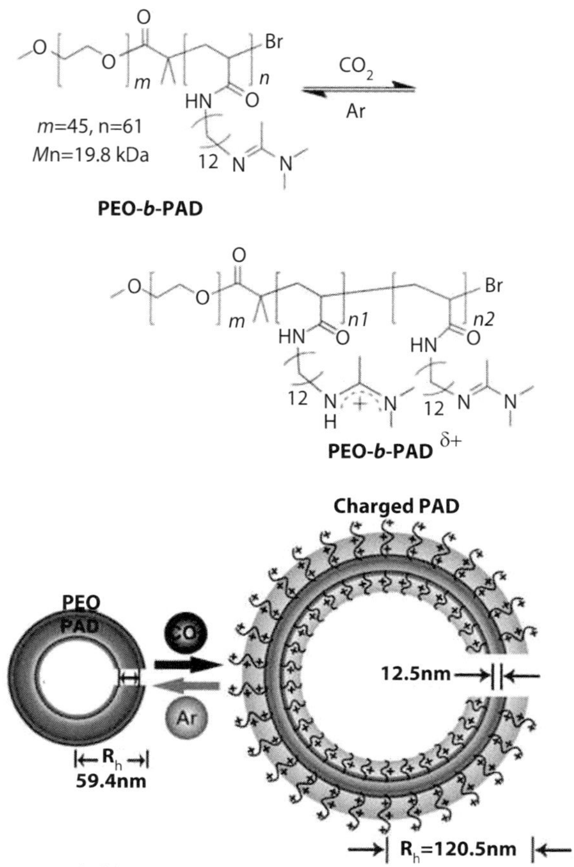

Figure 3.23 Gas-switchable structural change of amidine-containing diblock copolymer PEO-b-PAD (top) and schematic representation of self-assembly into vesicles and their reversible gas-responsive breathing in aqueous media (bottom).

is demonstrated using redox-responsive self-assembly of an amphiphilic molecule consisting of a tetraaniline and a poly(ethylene glycol) block (Fig. 3.24). [205] The vesicle membrane is tuned by an oxidizing voltage into smaller puck-like micelles that can reversibly assemble to form vesicles upon exposure to a reducing voltage. The switching mechanism is explained by the packing behavior of the tetraaniline units constituting the membrane core, which depend on their oxidation states.

3.6 Application of Vesicles

The application of smart vesicles to various fields is basically revealed by two major features of vesicle. Vesicular morphology has an interior pool

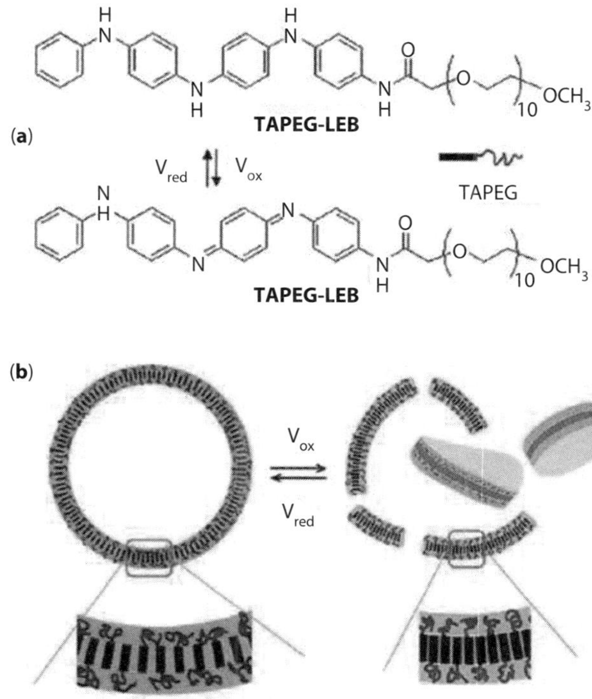

Figure 3.24 Redox-responsive TAPEG rodcoils in aqueous solution. (a) Chemical structures of TAPEG in the LEB and EB oxidation states. (b) Schematic representation of redox switching between vesicles and puck-like micelles of rodcoils that can be induced by the change in packing density in the membrane core.

that can encapsulate the cargo in the center, and an outer wall that can entrap the species within the shell. [206–213] Furthermore, as described in the previous section, when external stimulus is activated, the encapsulated material in the interior pool or the outer shell can be released or they can react with external stimulus. From these characteristic features like morphology and stimulus-responsibility, it is clear vesicles have applications in molecular separation, chemical sensors, reactors, catalysts, delivery vehicles, and others.

3.6.1 Molecular Separation by Vesicles

Self-assembly of amphiphilic molecules into a vesicle structure has attracted a great deal of interest due to mimicking biological membranes,

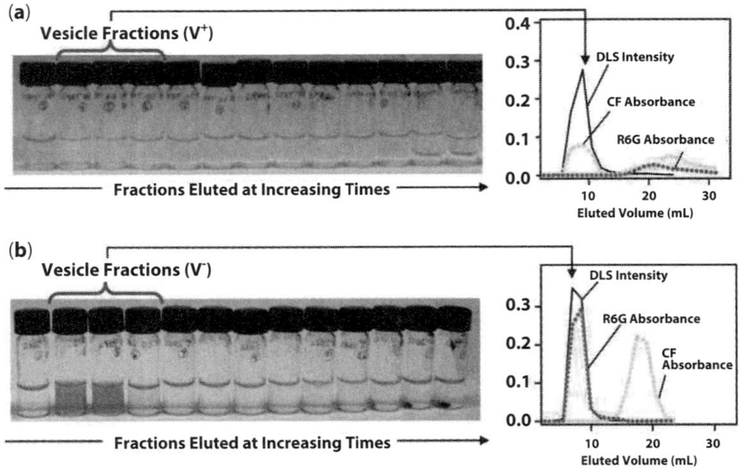

Figure 3.25 Vesicle–mediated separation of mixtures of R6G and CF dyes using (a) V+ vesicles and (b) V– vesicles. The plots on the right show the DLS intensity, which is proportional to vesicle concentration, and the UV absorbances, which are proportional to the concentrations of R6G and CF, as a function of eluted volume.

which have characteristic features like biphasic separation with trans-membrane permeability. [214–217] The permeation of hydrophilic solutes can be either size selective or substrate specific, depending mainly on the transport mechanism of the inserted proteins. English et al. (2007) revealed the results on electrostatic adsorption of organic solutes and DNA to the exterior surfaces of catanionic, unilamellar vesicles (Fig. 3.25). [218] They show that organic ions and polyelectrolytes bind to the exterior surface of oppositely charged catanionic vesicles through interactions with unpaired ionic surfactants present in the vesicle bilayer. The electrostatic sequestration of organic ions with catanionic vesicles is extremely efficient with excellent long–term stability and can be used to separate mixtures of charged organic solutes. They employ fluorescence correlation spectroscopy to make sensitive measurements of bilayer adsorption and compare the adsorption of a small molecular probe with that of a single–stranded, dye–labeled DNA molecule. The results show that DNA binds much more strongly to the exterior surface of positively charged catanionic vesicles, and could even stabilize vesicles at very low surfactant concentrations near the critical aggregation concentration.

Chiu et al. (2012) report that polymeric vesicles attained from the self–assembly of distearin (a diacylglycerol lipid)–conjugated poly(acrylic acid) with various distearin contents in the aqueous phase are capable of controlling

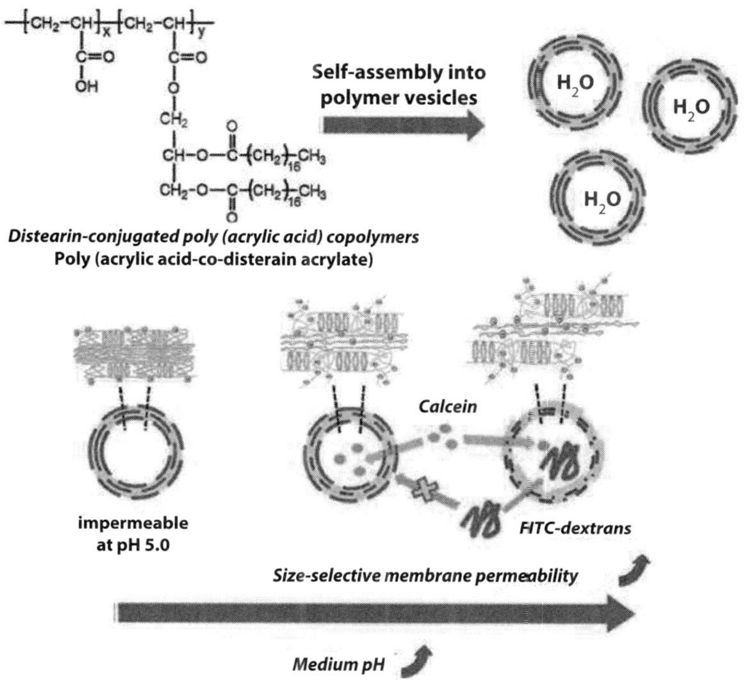

Figure 3.26 Schematic illustration of polymer-vesicle assemblies equipped with pH-evolved size-selective permeability from self-association of hydrophobically modified PAAc (poly(Aacco-DSA)).

the vesicular-wall permeability of the hydrophilic solutes of varying sizes by a simple manipulation of the external pH (Fig. 3.26). [219] The pH-evolved size-selective permeability of the vesicular membranes is virtually governed by the lipid content of copolymer and the addition of salt cations. With the addition of salt in aqueous vesicle suspensions, the pH-evolved assembled structure and the membrane permeability could be immobilized with promoted resistance to further pH alteration, along with an additional counterion screening effected that reduces the pH required for the onset of polar solutes of certain sizes to pass through the membranes. SAXS measurements of the vesicle structures at the aqueous phase indicate that the ion-induced permeability is governed by the extent of hydration and swelling of the vesicle membranes.

3.6.2 Chemical Sensors

When composed of a functional bilayer membrane that reacts to specific stimuli responses and releases guest molecules, vesicles can provide an application with smart sensors. [220–230] Furthermore, as a function

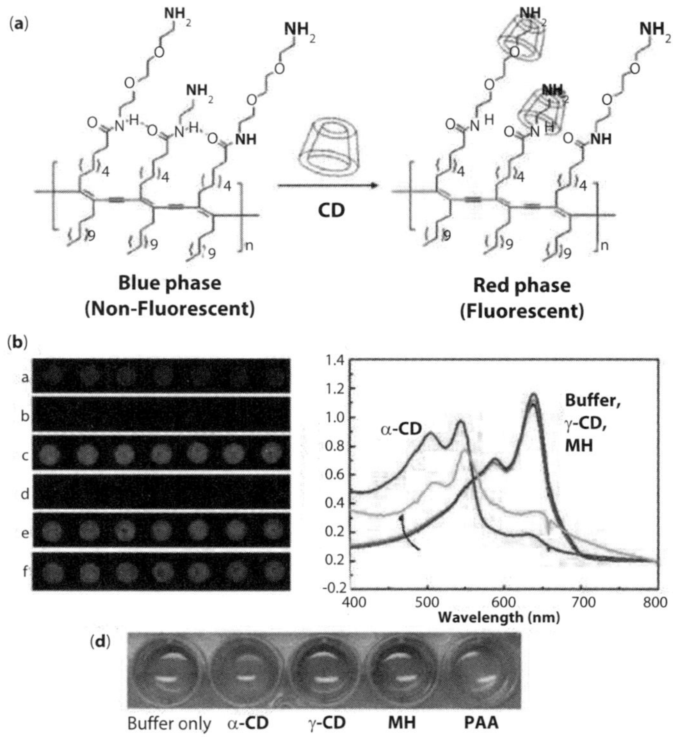

Figure 3.27 (a) A schematic representation of the interaction between polydiacetylene and cyclodextrin. (b) Microarray-based fluorescence profiles of PDA-immobilized glass substrates after treatment. (c) Visible spectra of PDA vesicle solutions in the presence of carbohydrates and PAA. (d) Photographs of vesicle solutions as in B.

of external stimulus, vesicular structures can tune their morphology or character to act like sensors. Jelinek et al. (2006) report "Naked eye" color detection of proteins by embedding calixarene receptors within vesicles comprising phospholipids and the chromatic polymer polydiacetylene (Fig. 3.27). [231] Dramatic visible absorbance changes were induced through electrostatic interactions between the protein surface and the vesicle incorporated hosts. The colorimetric responses could be induced by micromolar protein concentrations; furthermore, specific protein fingerprints could be obtained by incorporating different receptors within the vesicles. Fluorescence and circular dichroism experiments confirm the relationship between the colorimetric phenomena and protein docking on the surface of the chromatic vesicles. The colorimetric assay constitutes a generic platform for high-sensitivity detection of soluble proteins and call for an evaluation of protein surface charge distribution.

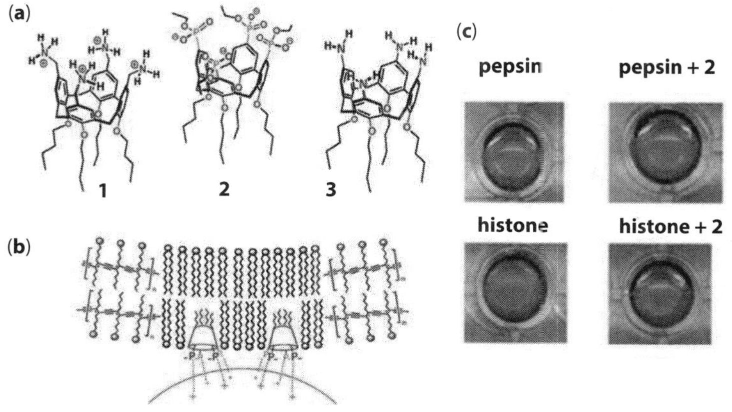

Figure 3.28 (a) Structures of the three receptors. (b) Schematic structure of the receptor/phospholipid/PDA assemblies (depicting 2 as an example). (c) Pictures of DMPC/PDA vesicle solutions following addition of pepsin and histone.

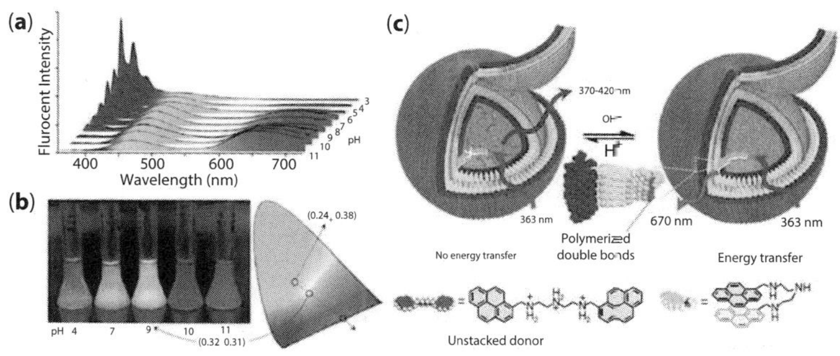

Figure 3.29 (a) Fluorescence spectra of donor-loaded polymerized vesicles in aqueous solution. (b) Photograph of vesicles at different pH, and CIE 1931 chromaticity diagram. (c) Schematic illustration of the vesicles with pH-tunable energy transfer.

Kim et al. (2005) developed a new approach for the construction of PDA-based fluorescent chemosensor systems that is compatible with conventional microarray technologies (Fig. 3.28). [232] Patterned fluorescence profiles are generated by taking advantage of specific ligand–receptor interaction occurring between cyclodextrins and PDA vesicles. Combining this methodology with modern array-based sensing technologies, the stress-induced self-fluorescent nature of the PDAs should be widely applicable to the development of new and interesting PDA-based chemosensor systems.

Würthner et al. (2009) describe that vesicles functionalize with water-soluble perylene bisimide loaded with bispyrene-based energy donors in their aqueous interior (Fig. 3.29). [233] Owing to their outstanding optical and redox properties, perylene bisimide chromophores make invaluable

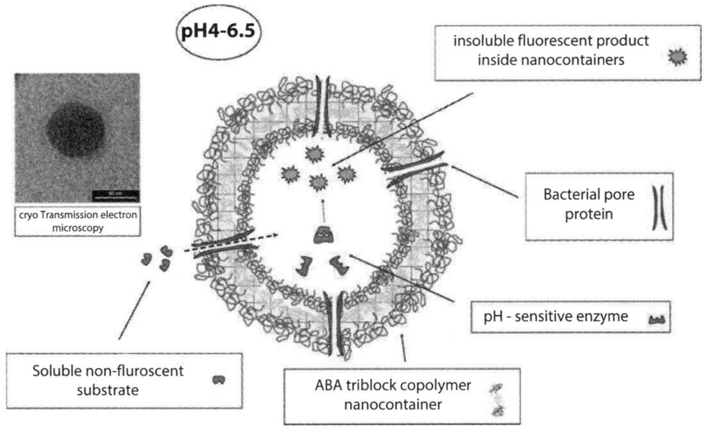

Figure 3.30 Two-dimensional outline and visualization of the nanoreactor system.

building blocks for functional bilayer membranes in a water environment. These loaded vesicles are stabilized by *in situ* photopolymerization of the shell of vesicles to make nanocapsules that are stable over the entire aqueous pH range. On the basis of pH–tunable spectral overlap of donors and acceptors, the donor–loaded polymerized vesicles display pH–dependent fluorescence resonance energy transfer from the encapsulated donors to the bilayer dye membrane, providing ultrasensitive pH information on their aqueous environment with fluorescent color changes covering the whole visible light range.

3.6.3 Nanoreactors and Microreactors

Chemical reactions in biological cells occur on a tiny scale according to the dimensions of cells and their compartments. Self–assembled vesicles have evolved to enclose small volumes as living cells, which can lead to biomimetic nanoreactors and microreactors with encapsulated species. [234–241] Van Hest et al. (2007) developed the controlled nanoscale nanoreactor by combining synthetic triblock copolymer membranes with complex biological components with a pH–switchablity (Fig. 3.30). [242] The nanovesicles are equipped with bacterial transmembrane pore proteins and the pH–sensitive enzyme acid phosphatase, resulting in a switchable substrate processing at pH 4–6.5. The nanoreactor is able to change its state of activity as demonstrated by producing a water–insoluble fluorescent dye inside the polymeric vesicle. The bacterial pores that were integrated into

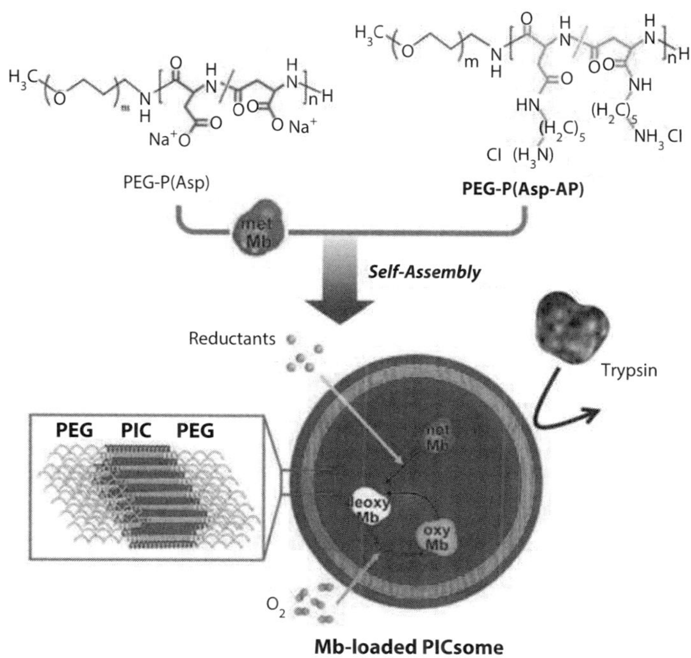

Figure 3.31 Reversible Mb oxygenation inside the PICsome self-assembled from a pair of oppositely charged block ionomers.

the polymer membrane remain functional and allow the passive diffusion of both protons for activity control of the encapsulated acid phosphatase.

Kataoka et al. (2007) report the preparation and functionality of myoglobin (Mb), which forms stable oxygen adducts in the muscle, in which loaded vesicles are demonstrated as the first successful pathway to fabricate functional containers for a variety of proteins through the self-assembled vesicular formation of a pair of oppositely charged block ionomers (Fig. 3.31). [243] The myoglobin-loaded vesicle is indeed readily prepared in an aqueous medium by simple mixing of the block ionomer solutions containing myoglobin. Loaded myoglobin is smoothly reduced to deoxymyoglobin by $S_2O_4^{2-}$ that permeates through vesicular membranes, and reversible oxygenation/deoxygenation of the Mb in the container is revealed even in the presence of trypsin in the outer medium. The biocompatible nature of this Mb-loaded vesicle, composed of poly(amino acid)s and a bioinert PEG shell, may also be feasible for further development of a new oxygen carrier for use *in vivo*.

Caruso et al. (2011) report the synthesis and use of a novel block copolymer, poly(N-vinyl pyrrolidone)-block-(cholesteryl acrylate), to anchor

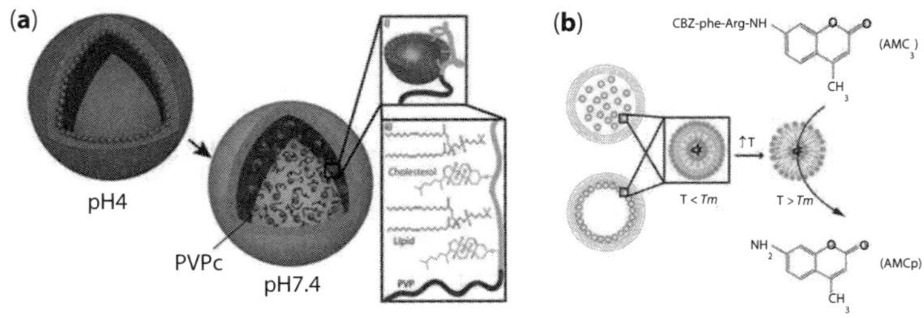

Figure 3.32 (a) Schematic illustration of a capsosome containing "free floating" liposomal sub-compartments. (b) Schematic illustration of the enzymatic conversion using capsosomes loaded with the protease upon increasing the temperature above T_m.

liposomes to a polymer film allowed control over the spatial positioning of liposomes entrapped within a poly(methacrylic acid) (PMA) hydrogel carrier capsule (Fig. 3.32). [244] They demonstrate that, with the appropriate use of cholesterol–modified polymers and sacrificial layers, capsosomes with "free–floating" and polymer membrane associated sub-compartments can be prepared. The triggered enzymatic reaction by converting the coumarine peptidase substrate N-CBZ–L–phenylalanyl–L–arginine–7–amido–4–methylcoumarin into its hydrolyzed product 7–amino–4–methylcoumarin demonstrates the preservation of cargo functionality when the enzyme subtilisin is encapsulated into the "free–floating" or membrane associated liposomal compartments of capsosomes.

3.6.4 Catalysts

Many studies show vesicles as catalysts with encapsulated reactive counterions can accelerate chemical reactions because of high local concentrations of reactant. [245–253] Another possible way of catalytic application of vesicles is by making hard metallic vesicles based on soft vesicles. Li et al. (2010) illustrate the promise of approaches in which a catalyst such as a metal ion is used as the counterion (Fig. 3.33). [254] The synthesis and characterization of hollow Pd–Co bimetallic nanospheres is reported. The bimetallic hollow chamber structure is prepared through a vesicle–assisted chemical reduction method. During Sonogashira–type coupling reactions between aryl halides and terminal alkynes in aqueous medium, these hollow materials exhibit much higher activity than the solid counterpart nanoparticles. The enhanced activity is attributed to both the hollow chamber structure and the promotional effect of Co–dopants, which provide more Pd active sites for the reactants. This catalytic structure can be easily separated from the reaction solution via centrifugation and can be used repetitively.

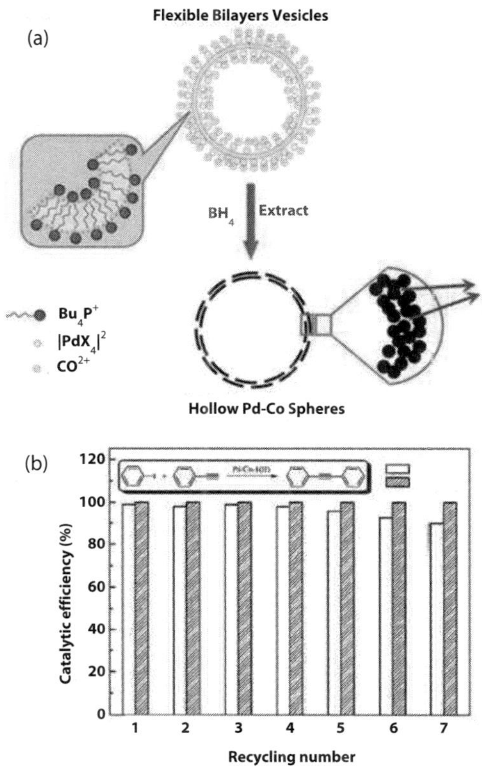

Figure 3.33 (a) The possible formation process of hollow Pd–Co nanospheres. (b) Recycling test of Pd–Co bimetallic nanospheres.

Uozumi et al. (2011) developed an architecture-based system of transition-metal catalysis using an amphiphilic pincer palladium complex bearing hydrophilic and hydrophobic chains (Fig. 3.34). [255] This system involves the self-assembly of bilayer vesicles, the self-concentration of organic substrates within the hydrophobic region of the bilayer membrane, and the catalytic transformation of the substrate with the palladium species located within close diffusion proximity, all of which occur sequentially in water.

Vesicles themselves can also be used as biological catalysts by entrapping bioactive species inside vesicle pools. Van Hest et al. (2012) report the successful construction of a polymersome-stabilized Pickering emulsion for application in biphasic enzymatic catalysis (Fig. 3.35). [256] This type of Pickering emulsion is stabilized by fully packed crosslinked polymersomes at the water/oil interface. *Candida Antarctica* lipase B (CalB),

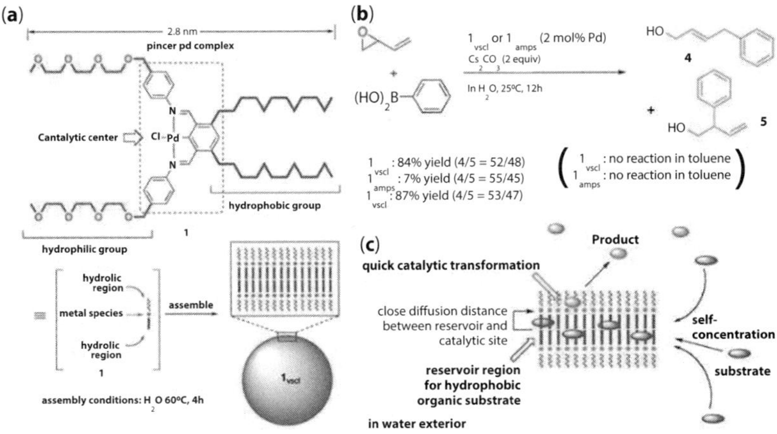

Figure 3.34 (a) Formation of vesicle 1_{vscl} by self–assembly of the pincer palladium complex 1. (b) Palladium–catalyzed oxirane ring opening with $PhB(OH)_2$. 1_{vscl} was generated *in situ* without being isolated in water. (c) Schematic image of the concept of catalysis within the bilayer membrane of the 1_{vscl}.

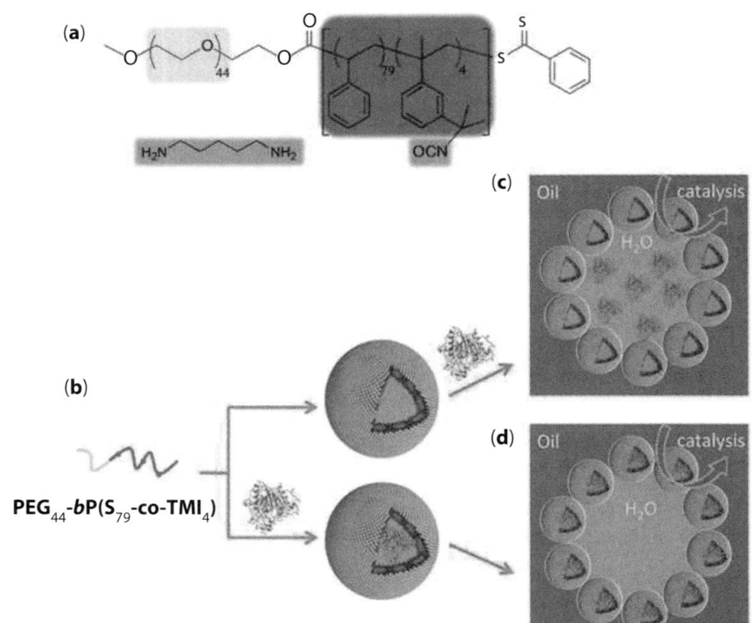

Figure 3.35 (a) The chemical structure of the block copolymer of which the polymersome is constructed. (b) Representation of the crosslinking process to prepare stabilized polymersomes. (c,d) Schematic representation of a Pickering emulsion with the enzyme in the water phase.

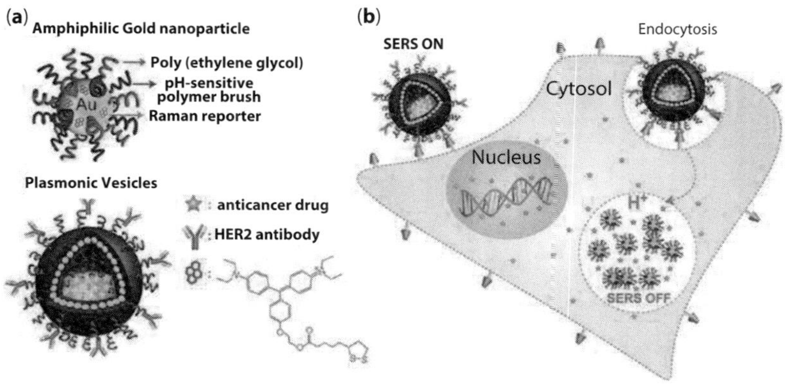

Figure 3.36 (a) Schematic illustration of the amphiphilic gold nanoparticle coated with Raman reporter BGLA and mixed polymer brushes of hydrophilic PEG and pH-sensitive hydrophobic PMMAVP grafts. (b) The cellular binding, uptake, and intraorganelle disruption of the SERS-encoded pH-sensitive plasmonic vesicles.

as a model enzyme, is loaded either in the water phase or in the lumen of the polymersomes of the Pickering emulsion, which highly enhances its catalytic performance. Furthermore, the recyclability of CalB in the polymersome Pickering emulsion system can be effectively realized. Since the special structure of the polymersome Pickering emulsion naturally creates a system with different compartments, different enzymes or other catalysts can be loaded in separate spaces.

3.6.5 Drug Delivery Vehicles

The vesicle-based drug carriers can significantly enhance the solubility of rarely dissolved drugs, reduce cytotoxicity toward normal tissues, and improve efficiency for disease. [257–264] Furthermore, stimuli-responsiveness provides vesicles capabilities of controlled release of cargo species in the cavity. Zumbuehl et al. (2012) show that vesicles made from an artificial 1,3-diaminophospholipid are stable under static conditions but release their encapsulated species with increasing shear stress. [265] Since these vesicles in the shape of a lens, it is the characteristic morphology that could potentially lead to react to the external shear modulus.

The stimuli-responsiveness can be introduced by incorporating stimuli-sensitive blocks or groups into the vesicle-forming molecules. Duan et al. (2012) report the development of bioconjugated plasmonic vesicles assembled from surface-enhanced Raman scattering (SERS)-encoded amphiphilic gold nanoparticles for cancer-targeted drug delivery (Fig. 3.36).

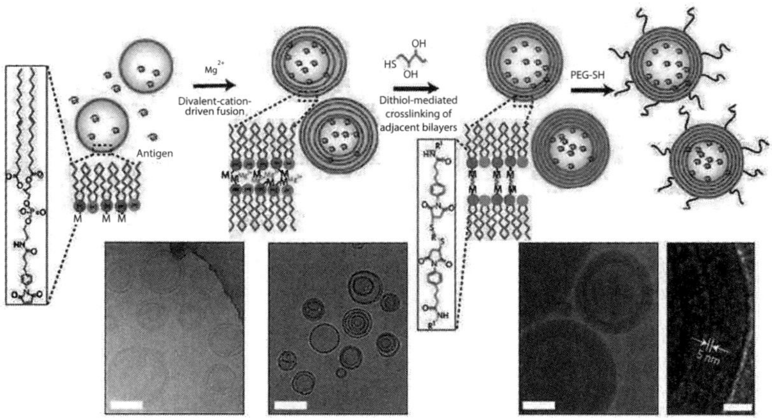

Figure 3.37 Schematic illustration of ICMV synthesis and cryo–electron–microscope images.

[266] This new type of plasmonic assembly with a hollow cavity could play multifunctional roles as delivery carriers for anticancer drugs and SERS–active plasmonic imaging probes to specifically label targeted cancer cells. They show that the pH–responsive disassembly of the vesicle, stimulated by the hydrophobic–to–hydrophilic transition of the hydrophobic brushes in acidic intracellular compartments, allow for triggered intracellular drug release. Because self–assembled plasmonic vesicles exhibit significantly different plasmonic properties and greatly enhanced SERS intensity in comparison with single gold nanoparticles due to strong interparticle plasmonic coupling, disassembly of the vesicles in endocytic compartments leads to dramatic changes in scattering properties and SERS signals.

Vesicles can act like artificial vaccines with the capability of encapsulation of hydrophilic and hydrophilic cargos spontaneously with increasing stability. Irvine et al. (2012) describe interbilayer–crosslinked vesicles formed by crosslinking headgroups of adjacent lipid bilayers within multilamellar vesicles (Fig. 3.37). [267] Interbilayer–crosslinked vesicles stably entrap protein antigens in the vesicle core and lipid–based immunostimulatory molecules in the vesicle walls under extracellular conditions, but exhibit rapid release.

When trapped with imaging maker, vesicles can apply to direct visualize of the delivery process. [268,269] Gong et al. (2012) reported a multifunctional stable and pH–responsive polymer vesicle nanocarrier system. They combine tumor–targeted delivery and superparamagnetic iron oxide nanoparticles, encapsulated into the aqueous core of the stable vesicles,

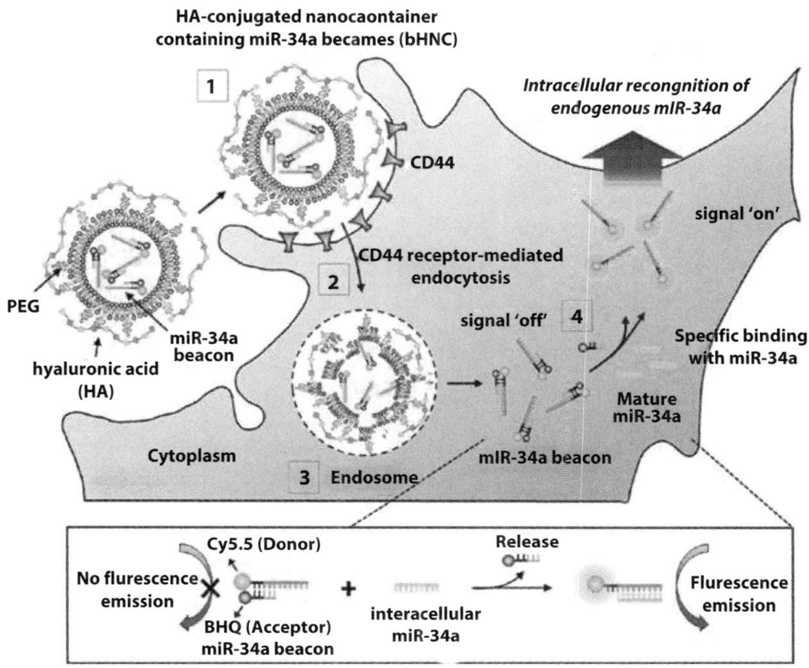

Figure 3.38 Schematic illustration of miR–34a beacon delivery system for targeted intracellular recognition of miR–34a based on HA–coated nanocontainers that encapsulate the miR–34a beacons (bHNCs).

allowing for ultrasensitive magnetic resonance imaging (MRI) detection. [270] Sandre et al. (2011) report MRI detective vesicles using hydrophobically modified maghemite nanoparticles, embedded within the membrane of poly(trimethylene carbonate)-*b*-poly(L-glutamic acid) block copolymer vesicles. [271] Huh et al. (2012) report smart nanoprobe, hyaluronic acid (HA)–based nanocontainers containing miR–34a beacons (bHNCs), for the intracellular recognition of miR–34a levels in metastatic breast cancer cells, which is distinct from the imaging of biomarkers such of cell membrane receptors (Fig. 3.38). [272] They demonstrate that a nanoscale vesicle that couples a targeting endocytic route, and a molecular imaging probe, enable the efficient detection of specific miRNAs.

3.7 Conclusions

This chapter will include a discussion about synthesis and characterization of smart vesicles that react to external stimulus toward various applications.

First, we will summarize recent research in the area of synthesis of vesicular structures from self-assembly of soft vesicles. With regard to the compartment of forming, several classes of vesicles by synthetic approach can be distinguished by liposome, polymersome, and vesicles based on small molecules. Then we will discuss the synthesis of hard vesicles by using vesicles for template, like hollow silica spheres. The templating method with soft vesicles shows a wide variety of available templates, simple fabrication processes under mild conditions, and easy removal of the templates with less damage to the hard vesicles. This paper will provide an overview of characterization tools of structures by microscopic and scattering method, like cryo-TEM, and DLS. We will discuss the various types of self-assembled smart vesicles. The smart vesicles are revealed by their capability to respond to external stimuli, such as temperature, pH and others. In addition, the application of these smart vesicles toward molecular separation, chemical sensors, nanoreactors and microreactors, catalysts, and drug delivery vehicles are described.

Acknowledgment

This work was supported by research fund of Chungnam National University

References

1. G. Gabrielli, *Adv. Colloid Interface Sci.*, Vol. 34, p. 31, 1991.
2. P.R. Cullis, D.B. Fenske, and M.J. Hope, *Biochemistry of lipids, lipoproteins and membranes*, Elsevier Science, Amsterdam, 1996.
3. Z.I. Cabanchik, and A. Darmon, *In Structure and Properties of Cell Membranes*, CRC Press, Boca Raton, 1985.
4. O.G. Mouritsen, *Life–as a Matter of Fat: The Emerging Science of Lipidomics*, Springer, Berlin, 2005.
5. R. Lipowsky, and E. Sackmann, *Structure and Dynamics of Membranes–from Cell to Vesicles*, Elsevier Science, Amsterdam, 1995.
6. S. Šegota, and Đ. Težak, *Adv. Colloid Interface Sci.*, Vol. 121, p. 51, 2006.
7. P. Walde, K. Cosentino, H. Engel, and P. Stano, *ChemBioChem*, Vol. 11, p. 848, 2010.
8. K. Kita–Tokarczyk, J. Grumelard, T. Haefele, and W. Meier, *Polymer*, Vol. 46, p. 3540, 2005.
9. J. Du, and R.K. O'Reilly, *Soft Matter.*, Vol. 5, p. 3544, 2009.
10. M. Peterca, V. Percec, P. Leowanawat, and A. Bertin, *J. Am. Chem. Soc.*, Vol. 133, p. 20507, 2011.

11. A.D. Bangham, and R.W. Horne, *J. Mol. Biol.*, Vol. 8, p. 660, 1964.
12. A.D. Bangham, M.M. Standish, and J.C. Watkins, *J. Mol. Biol.*, Vol. 13, p. 238, 1965.
13. P. Harvie, F.M. Wong, and M.B. Bally, *Biophys J.*, Vol. 75, p. 1040, 1998.
14. M.C. Woodle, and D.D. Lasic, *Biochim. Biophys. Act*, Vol. 1113, p. 171, 1992.
15. V.P. Torchilin, *Nat. Rev. Drug. Discov.*, Vol. 4, p. 145, 2005.
16. H. Zhang, J.R. Joubert, and S.S. Saavedra, *Adv. Polym. Sci.*, Vol. 224, p. 1, 2010.
17. V.U. Miyamoto, and W. Stoeckenius, *J. Membrane Biol.*, Vol. 4, p. 252, 1971.
18. P.L Luisi, P. Walde, and T. Oberholzer, *Curr. Opin. Colloid Interface Sci.*, Vol. 4, p. 33, 1999.
19. R. Lipowsky, *Nature*, Vol. 349, p. 475, 1991.
20. E.W. Kaler, A.K. Murthy, B.E. Rodriguez, and J.A. Zasadzinski, Science, Vol. 245, p. 1371, 1989.
21. Y. Namiki, T. Fuchigami, N. Tada, R. Kawamura, S. Matsunuma, Y. Kitamoto, and M. Nakagawa, Acc. Chem. Res., Vol. 44, p. 1080, 2011.
22. A.S. Ulrich, Biosci. Rep., Vol. 22, p. 129, 2002.
23. D.E. Discher, and A. Eisenberg, Science, Vol. 297, p. 967, 2002.
24. H. Kukula, H. Schlaad, M. Antonietti, and S. Fōrster, J. Am. Chem. Soc., Vol. 124, p. 1658, 2002.
25. L. Luo, and A. Eisenberg, J. Am. Chem. Soc., Vol. 123, p. 1012, 2001.
26. S. Chen, B. Yang, C. Guo, J.-H. Ma, L.-R. Yang, X. Liang, C. Hua, and H.-Z. Liu, *J. Phys. Chem. B*, Vol. 112, p. 15659, 2008.
27. X. Lian, D. Wu, X. Song, and H. Zhao, *Macromolecules*, Vol. 43, p. 7434, 2010.
28. B.M. Discher, H. Bermudez, D.A. Hammer, D.E. Discher, Y.-Y. Won, and F.S. Bates, J. Phys. Chem. B, Vol. 106, p. 2848, 2002.
29. A. Blanazs, S.P. Armes and A.J. Ryan, *Macromol. Rapid Commun.*, Vol. 30, p. 267, 2009.
30. W. Zhao, D. Chen, Y. Hu, G.M. Grason, and T.P. Russell, *ACS Nano*, Vol. 5, p. 486, 2011.
31. B.M. Discher, Y.-Y. Won, D.S. Ege, J.C.M. Lee, F.S. Bates, D.E. Discher, and D.A. Hammer, Science, Vol. 284, p. 1143, 1999.
32. J. Huang, C. Bonduelle, J. Thévenot, S. Lecommandoux, and A. Heise, *J. Am. Chem. Soc.*, Vol. 134, p. 119, 2012.
33. R.M. Shivshankar, R.N. Vijayakameswara, and S. Raja, *ACS Macro Lett.*, Vol. 1, p. 482, 2012.
34. M. Marguet, L. Edembe, and S. Lecommandoux, *Angew. Chem. Int. Ed.*, Vol. 51, p. 1173, 2012.
35. C. Huang, Y. Li, Y. Song, Y. Li, H. Liu, and D. Zhu, *Adv. Mater.*, Vol. 22, p. 3532, 2010.
36. D. R. Vutukuri, S. Basu, and S. Thayumanavan, *J. Am. Chem. Soc.*, Vol. 126, p. 15636, 2004.
37. T. Homma, K. Harano, H. Isobe, and E. Nakamura, *J. Am. Chem. Soc.*, Vol. 133, p. 6364, 2011.

38. A. Ajayaghosh, R. Varghese, S. Mahesh, and V.K. Praveen, *Angew. Chem. Int. Ed.*, Vol. 45, p. 7729, 2006.
39. D.-S. Guo, K. Wang, Y.-X. Wang, and Y. Liu, *J. Am. Chem. Soc.*, Vol. 134, p. 10244, 2012.
40. D. Jiao, J. Geng, X.J. Loh, D. Das, T.-C. Lee, and O.A. Scherman, *Angew. Chem. Int. Ed.*, Vol. 51, p. 9633, 2012.
41. M. Wang, A.R. Mohebbi, Y. Sun, and F. Wudl, *Angew. Chem. Int. Ed.*, Vol. 51, p. 6920, 2012.
42. A.J. van Hell, C.I.C.A. Costa, F.M. Flesch, M. Sutter, W. Jiskoot, D.J.A. Crommelin, W.E. Hennink, and E. Mastrobattista, *Biomacromolecules*, Vol. 8, p. 2753, 2007.
43. M.R. Molla, and S. Ghosh, *Chem. Eur. J.*, Vol. 18, p. 9860, 2012.
44. P. Bauduin, S. Prevost, P. Farràs, F. Teixidor, O. Diat, and T. Zemb, *Angew. Chem. Int. Ed.*, Vol. 50, p. 5298, 2011.
45. E. Lee, J.-K. Kim, and M. Lee, *Macromol. Rapid Commun.*, Vol. 31, p. 975, 2010.
46. S.H. Seo, J.Y. Chang, and G.N. Tew, *Angew. Chem. Int. Ed.*, Vol. 45, p. 7526, 2006.
47. H. Wang, L. Wang, X. Wang, J. Xu, Q. Luo, and J. Liu, *New J. Chem.*, Vol. 35, p. 2632, 2011.
48. Y. Li, X. Li, Y. Li, H. Liu, S. Wang, H. Gan, J. Li, N. Wang, X. He, and D. Zhu, *Angew. Chem. Int. Ed.*, Vol. 45, p. 3639, 2006.
49. M. Yang, W. Wang, F. Yuan, X. Zhang, J. Li, F. Liang, B. He, B. Minch, and G. Wegner, *J. Am. Chem. Soc.*, Vol. 127, p. 15107, 2005.
50. D. Kim, E. Kim, J. Kim, K.M. Park, K. Baek, M. Jung, Y.H. Ko, W. Sung, H.S. Kim, J.H. Suh, C.G. Park, O.S. Na, D.K. Lee, K.E. Lee, S.S. Han, and K. Kim, *Angew. Chem. Int. Ed.*, Vol. 46, p. 3471, 2007.
51. J. Liu, M. Morikawa, and N. Kimizuka, *J. Am. Chem. Soc.*, Vol. 133, p. 17370, 2011.
52. Y. Barenholz, D. Gibbes, B.J. Litman, J. Goll, T.E. Thompson, and F.D. Carlson, Biochemistry, Vol. 16, p. 2806, 1977.
53. J.P. Reeves, and R.M. Dowben, J. Cell. Physio., Vol. 73, p. 49, 1969.
54. P. Mueller, T.F. Chien, and B. Rudy, Biophys. J., Vol. 44, p. 375, 1983.
55. F. Szoka, and D. Papahadjopoulos, Proc. Natl. Acad. Sci. USA, Vol. 75, p. 4194, 1978.
56. E. Evans, and D. Needham, J. Phys. Chem., Vol. 91, p. 4219, 1987.
57. J.M.H. Kremer, M.W. Van der Esker, C. Pathmamanoharan, and P.H. Wiersema, Biochemistry, Vol. 16, p. 3932, 1977.
58. A. Moscho, O. Orwar, D.T. Chiu, B.P. Modi, and R.N. Zare, Proc. Natl. Acad. Sci. USA, Vol. 93, p. 11443, 1996.
59. Z. Cheng, and P.L. Luisi, J. Phys. Chem. B, Vol. 107, p. 10940, 2003.
60. T. Thorsen, R.W. Roberts, F.H. Arnold, and S.R. Quake, Phys. Rev. Lett., Vol. 86, p. 4163, 2001.

61. A. Jahn, W.N. Vreeland, M. Gaitan, and L.E. Locascio, J. Am. Chem. Soc., Vol. 126, p. 2674, 2004.
62. Jahn, S.M. Stavis, J.S. Hong, W.N. Vreeland, D.L. DeVoe, and M. Gaitan, ACS Nano, Vol. 4, p. 2077, 2010.
63. S. Ota, S. Yoshizawa, and S. Takeuchi, Angew. Chem. Int. Ed., Vol. 48, p. 6533, 2009.
64. J.R. Howse, R.A.L. Jones, G. Battaglia, R.E. Ducker, G.J. Leggett, and A.J. Ryan, Nat. Mater., Vol. 8, p. 507, 2009.
65. G. Kim, S. Song, J. Lee, and J.–M. Kim, Langmuir, Vol. 26, p. 17840, 2010.
66. J.C. Stachowiak, D.L. Richmond, T.H. Li, A.P. Liu, S.H. Parekh, and D.A. Fletcher, Proc. Natl. Acad. Sci. USA, Vol. 105, p. 4697, 2008.
67. Y.–C. Tan, K. Hettiarachchi, M. Siu, Y.–R. Pan, and A.P. Lee, J. Am. Chem. Soc., Vol. 128, p. 5656, 2006.
68. S. Matosevic, and B.M. Paegel, J. Am. Chem. Soc., Vol. 133, p. 2798, 2011.
69. R. Dong, W. Liu, and J. Hao, *Acc. Chem. Res.*, Vol. 45, p. 504, 2012.
70. C.A. McKelvey, E.W. Kaler, J.A. Zasadzinski, B. Coldren, and H.–T. Jung, *Langmuir*, Vol. 16, p. 8285, 2000.
71. D.H.W. Hubert, M. Jung, and A.L. German, *Adv. Mater.*, Vol. 12, p. 1291, 2000.
72. A. Song, X. Jia, M. Teng, and J. Hao, *Chem.Eur. J.*, Vol. 13, p. 496, 2007.
73. J. Hotz, and W. Meier, *Langmuir*, Vol. 14, p. 1031, 1998.
74. M. Teng, A. Song, L. Liu, and J. Hao, *J. Phys. Chem. B*, Vol. 112, p. 1671, 2008.
75. L. Zhang, P. Li, X. Liu, L. Du, and E. Wang, *Adv. Mater.*, Vol. 19, p. 4279, 2007.
76. X. Zhang, and D. Li, *Angew. Chem. Int. Ed.*, Vol. 45, p. 5971, 2006.
77. X.–J. Wu and D. Xu, *J. Am. Chem. Soc.*, Vol. 131, p. 2774, 2009.
78. X. Zhang, and D. Li, *Angew. Chem. Int. Ed.*, Vol. 45, p. 5971, 2006.
79. L. Wang, H. Liu, and J. Hao, *Chem. Commun.*, p. 1353, 2009.
80. M.P. Pileni, *Nat. Mater.*, Vol. 2, p. 145, 2003.
81. Y. Ma, L. Qi, J. Ma, H. Cheng, and W. Shen, *Langmuir*, Vol. 19, p. 9079, 2003.
82. H. Xu and W. Wang, *Angew. Chem. Int. Ed.*, Vol. 46, p. 1489, 2007.
83. C. Nardin, T. Hirt, J. Leukel, and W. Meier, *Langmuir*, Vol. 16, p. 1035, 2000.
84. S.A. Dergunov, K. Kesterson, W. Li, Z. Wang, and E. Pinkhassik, *Macromolecules*, Vol. 43, p. 7785, 2010.
85. K. Niikura, N. Iyo, T. Higuchi, T. Nishio, H. Jinnai, N. Fujitani, and K. Ijiro, *J. Am. Chem. Soc.*, Vol. 134, p. 7632, 2012.
86. S.S. Kim, W.Z. Zhang, and T.J. Pinnavaia, *Science*, Vol. 282, p. 1302, 1998.
87. J. Yuan, X. Bai, M. Zhao, and L. Zheng, *Langmuir*, Vol. 26, p. 11726, 2010.
88. H.–P. Hentze, S.R. Raghavan, C.A. McKelvey, and E.W. Kaler, *Langmuir*, Vol. 19, p. 1069, 2003.
89. L. Han, C. Gao, X. Wu, Q. Chen, P. Shu, Z. Ding, and S. Che, *Solid State Sci.*, Vol. 13, p. 721, 2011.
90. J. Liu, S.B. Hartono, Y.G. Jin, Z. Li, G.Q. Lu and S.Z. Qiao, *J. Mater. Chem.*, Vol. 20, p. 4595, 2010.

91. Y.-S. Lin, S.-H. Wu, C.-T. Tseng, Y. Hung, C. Chang, and C.-Y. Mou, *Chem. Commun.*, p. 3542, 2009.
92. V.S. Murthy, J.N. Cha, G.D. Stucky, and M.S. Wong, *J. Am. Chem. Soc.*, Vol. 126, p. 5292, 2004.
93. X. Gu, C. Li, X. Liu, J. Ren, Y. Wang, Y. Guo, Y. Guo, and G. Lu, *J. Phys. Chem. C*, Vol. 113, p. 6472, 2009.
94. Q. Sun, P.C.M.M. Magusin, B. Mezari, P. Panine, R.A. van Santen and N.A.J.M. Sommerdijk, *J. Mater. Chem.*, Vol. 15, p. 256, 2005.
95. Y.-Q. Yeh, B.-C. Chen, H.-P. Lin, and C.-Y. Tang, *Langmuir*, Vol. 22, p. 6, 2006.
96. P.T. Tanev, and T.J. Pinnavaia, *Science*, Vol. 271, p. 1267, 1996.
97. J. Li, J. Liu, D. Wang, R. Guo, X. Li, and W. Qi, *Langmuir*, Vol. 26, p. 12267, 2010.
98. M. Kepczynski, F. Ganachaud, and P. Hemery, *Adv. Mater.*, Vol. 16, p. 1861, 2004.
99. J. Yuan, X. Bai, M. Zhao, and L. Zheng, *Langmuir*, Vol. 26, p. 11726, 2010.
100. J. Du, and S.P. Armes, *Langmuir*, Vol. 24, p. 13710, 2008.
101. H. Wang, X. Zhou, M. Yu, Y. Wang, L. Han, J. Zhang, P. Yuan, G. Auchterlonie, J. Zou, and C. Yu, *J. Am. Chem. Soc.*, Vol. 128, p. 15992, 2006.
102. F.Z. Zernike, *techn. Physik.*, Vol. 16, p. 454, 1935.
103. L. Chen, H. Shen, and A. Eisenberg, *J. Phys. Chem. B*, Vol. 103, p. 9488, 1999.
104. H.G. Kapitza, *Microscopy from the very beginning*, GmbH, Carl Zeiss, 1997.
105. A. Saitoh, K. Takiguchi, Y. Tanaka, and H. Hotani, *Proc. Natl. Acad. Sci. USA*, Vol. 95, p. 1026, 1998.
106. F.M. Menger, and K.D. Gabrielson, *Angew. Chem. Int. Ed.*, Vol. 34, p. 2091, 1995.
107. G.H. Lei, and R.C. MacDonald, *Biophys. J.*, Vol. 85, p. 1585, 2003.
108. R. Wick, M.I. Angelova, P. Walde, and P.L. Luisi, *Chem. Biol.*, Vol. 3, p. 105, 1996.
109. P.K. Vinson, Y. Talmon, and A. Walter, *Biophys. J.*, Vol. 56, p. 669, 1989.
110. H. Jeon, and G.G. Shipley, *J. Biol. Chem.*, Vol. 275, p. 30458, 2000.
111. B. Böttcher, U. Lücken, and P. Gräber, *Biochem. Soc. Trans.*, Vol. 23, p. 780, 1995.
112. M. Silvander, G. Karlsson, and K. Edwards, *J. Colloid Interface Sci.*, Vol. 179, p. 104, 1996.
113. Y. Zheng, Z. Lin, J.L. Zakin, Y. Talmon, H.T. Davis, and L.E. Scriven, *J. Phys. Chem. B*, Vol. 104, p. 5263, 2000.
114. N. Berclaz, M. Müller, P. Walde, and P.L. Luisi, *J. Phys. Chem. B*, Vol. 105, p. 1056, 2001.
115. Y.-Y. Won, A.K. Brannan, H.T. Davis, and F.S. Bates, *J. Phys. Chem. B*, Vol. 106, p. 3354, 2002.
116. B. Coldren, R. van Zanten, M.J. Mackel, J.A. Zasadzinski, and H.-T. Jung, *Langmuir*, Vol. 19, p. 5632, 2003.

117. S. Kumar, and J.H. Hoh, *Langmuir*, Vol. 16, p. 9936, 2000.
118. G. Binnig, C.F. Quate, and C. Gerber, *Phys. Rev. Lett.*, Vol. 56, p. 930, 1986.
119. K. Schillén, K. Bryskhe, and Y.S. Mel'Nikova, *Macromolecules*, Vol. 32, p. 6885, 1999.
120. F. Ahmed, A. Hategan, D.E. Discher, and B.M. Discher, *Langmuir*, Vol. 19, p. 6505, 2003.
121. J.J. Park, and Y.P. Loh, *Methods Mol. Biol.*, Vol. 789, p. 137, 2011.
122. E. Galperin, and A. Sorkin, *J. Cell. Sci.*, Vol. 116, p. 4799, 2003.
123. S. Nappini, T.A. Kayal, D. Berti, B. Nordén, and P. Baglioni, *J. Phys. Chem. Lett.*, Vol. 2, p. 713, 2011.
124. H. Li, X. Xin, T. Kalwarczyk, E. Kalwarczyk, P. Niton, R. Hołyst, and J. Hao, *Langmuir*, Vol. 26, p. 15210, 2010.
125. E. Elizondo, J. Larsen, N.S. Hatzakis, I. Cabrera, T. Bjørnholm, J. Veciana, D. Stamou, and N. Ventosa, *J. Am. Chem.Soc.*, Vol. 134, p. 1918, 2012.
126. T. Portet, and R. Dimova, *Biophys. J.*, Vol. 99, p. 3264, 2010.
127. J.-K. Kim, E. Lee, and M. Lee, *Macromol. Rapid Commun.*, Vol. 31, p. 980, 2010.
128. J. Pencer, and F.R. Hallett, *Langmuir*, Vol. 19, p. 7488, 2003.
129. M. Warner, *Colloid. Polym. Sci.*, Vol. 261, p. 508, 1983.
130. J.C. Selser, Y. Yeh, and R.J. Baskin, *Biophys J.*, Vol. 16, p. 337, 1976.
131. F.R. Hallett, J. Watton, and P. Krygsman, *Biophys J.*, Vol. 59, p. 357, 1991.
132. J.H. van Zanten, and H.G. Monbouquette, *J. Colloid Interface Sci.*, Vol. 165, p. 512, 1994.
133. S. Hupfeld, A.M. Holsæter, M. Skar, C.B. Frantzen, and M. Brandl, *J. Nanosci. Nanotechnol.*, Vol. 6, p. 1, 2006.
134. R. Varghese, and H.-A. Wagenknecht, *Chem. Eur. J.*, Vol. 16, p. 9040, 2010.
135. M. Foldvari, I. Badea, S. Wettig, D. Baboolal, P. Kumar, A.L. Creagh, and C.A. Haynes, *Mol. Pharm.*, Vol. 7, p. 751, 2010.
136. M.R. Brzustowiczb, and A.T. Brunger, *J. Appl. Cryst.*, Vol. 38, p. 126, 2005.
137. G. Pabst, M. Rappolt, H. Amenitsch, and P. Laggner, *Phys. Rev. E*, Vol. 62, p. 4000, 2000.
138. J. Valério, M.H. Lameiro, S.S. Funari, M.J. Moreno, and E. Melo, *J. Phys. Chem. B*, Vol. 116, p. 168, 2012.
139. R.M. Fernandez, K.A. Riske, L.Q. Amaral, R. Itri, and M.T. Lamy, *Biochim. Biophys. Acta*, Vol. 1778, p. 907, 2008.
140. J.A. Bouwstra, G.S. Gooris, W. Bras, and H. Talsma, *Chem. Phys. Lipids*, Vol. 64, p. 83, 1993.
141. S. Schmölzer, D. Gräbner, M. Gradzielski, and T. Narayanan, *Phys. Rev. Lett.*, Vol. 88, p. 258301, 2002.
142. Z. Varga, S. Berényi, B. Szokol, L. Őrfi, G. Kéri, I. Peták, A. Hoell, and A. Bóta, *J. Phys. Chem. B*, Vol. 114, p. 6850, 2010.
143. M. Bergström, J.S. Pedersen, P. Schurtenberger, and S.U. Egelhaaf, *J. Phys. Chem. B*, Vol. 103, p. 9888, 1999.

144. F.P. Hubbard Jr., and N.L. Abbott, *Soft Matter,* Vol. 4, p. 2225, 2008.
145. J.S. Pedersen, S.U. Egelhaaf , and P. Schurtenberger, *J. Phys. Chem.*, Vol. 99, p. 1299, 1995.
146. N. Kučerka, M.A. Kiselev, and P. Balgavý, *Eur. Biophys. J.*, Vol. 33, p. 328, 2004.
147. H. Schmiedel, P. Jörchel, M. Kiselev, and G. Klose, *J. Phys. Chem. B*, Vol. 105, p. 111, 2001.
148. T.S. Davies, A.M. Ketner, and S.R. Raghavan, *J. Am. Chem. Soc.*, Vol. 128, p. 6669, 2006.
149. R.D. Harvey, R.K. Heenan, D.J. Barlow, and M.J. Lawrence, *Chem. Phys. Lipids*, Vol. 133, p. 27, 2005.
150. K. Bressel, M. Muthig, S. Prévost, I. Grillo, and M. Gradzielski, *Colloid. Polym. Sci.*, Vol. 288, p. 827, 2010.
151. S. Kessel, N.P. Truong, Z. Jia, and M.J. Monteiro, *J. Polym. Sci. Part A: Polym. Chem.*, Vol. 50, p. 4879, 2012.
152. L. Martín, E. Castro, A. Ribeiro, M. Alonso, and J.C. Rodríguez-Cabello, *Biomacromolecules*, Vol. 13, p. 293, 2012.
153. S. Qin, Y. Geng, D.E. Discher, and S. Yang, *Adv. Mater.*, Vol. 18, p. 2905, 2006.
154. M. Hales, C. Barner-Kowollik, T.P. Davis, and M.H. Stenzel, *Langmuir*, Vol. 20, p. 10809, 2004.
155. H. Yin, Y. Lin, J. Huang, and J. Ye, *Langmuir,* Vol. 23, p. 4225, 2007.
156. Y. Zhou, D. Yan, W. Dong, and Y. Tian, *J. Phys. Chem. B*, Vol. 111, p. 1262, 2007.
157. R. Liu, P. De Leonardis, F. Cellesi, N. Tirelli, and B.R. Saunders, *Langmuir*, Vol. 24, p. 7099, 2008.
158. E. E. Dormidontova, *Macromolecules*, Vol. 35, p. 987, 2002.
159. M. Heskins, and J.E. Guillet, *J. Macromol. Sci. Chem.*, Vol. 2, p. 1441, 1968.
160. H.G. Schild, *Prog. Polym. Sci.*, Vol. 17, p. 163, 1992.
161. Y. Li, B. S. Lokitz, and C. L. McCormick, *Angew. Chem. Int. Ed.*, Vol. 45, p. 5792, 2006.
162. J.-K. Kim, E. Lee, Z. Huang, and M. Lee, *J. Am. Chem. Soc.*, Vol. 128, p. 14022, 2006.
163. J.-K. Kim, E. Lee, Y.-H. Jeong, J.-K. Lee, W.-C. Zin, and M. Lee, *J. Am. Chem. Soc.*, Vol. 129, p. 6082, 2007.
164. E. Lee, Y.-H. Jeong, J.-K. Kim, and M. Lee, *Macromolecules*, Vol. 40, p. 8355, 2007.
165. J.-K. Kim, E. Lee, Y.-b. Lim, and M. Lee, *Angew. Chem. Int. Ed.*, Vol. 47, p. 4662, 2008.
166. K. Eigin, D.B. Leeper, J.R. Cater, A.J. Thistlethwaite, L. Tupchong, and J.D. McFarlane, *Int. J. Hyperthermia*, Vol. 11, p. 211, 1995.
167. I. Mellman, R. Fuchs, and A. Helenius, *Annu. Rev. Biochem.*, Vol. 55, p. 663, 1986.
168. A. Blanazs, M. Massignani, G. Battaglia, S. P. Armes, and A.J. Ryan, *Adv. Funct. Mater.*, Vol. 19, p. 2906, 2009.

169. E.G. Bellomo, M.D. Wyrsta, L. Pakstis, D.J. Pochan, and T.J. Deming, *Nat. Mater.*, Vol. 3, p. 244, 2004.
170. E.R. Gillies, T.B. Jonsson, and J.M.J. Fréchet, *J. Am. Chem. Soc.*, Vol. 126, p. 11936, 2004.
171. I. Lagzi, D. Wang, B. Kowalczyk, and B.A. Grzybowski, *Langmuir*, Vol. 26, p. 13770, 2010.
172. D.L. Roberts, Y. Ma, S.E. Bowles, C.M. Janczak, J. Pyun, S.S. Saavedra, and C.A. Aspinwall, *Langmuir*, Vol. 25, p. 1908, 2009.
173. H.-C. Chiu, Y.-W. Lin, Y.-F. Huang, C.-K. Chuang, and C.-S. Chern, *Angew. Chem. Int. Ed.*, Vol. 47, p. 1875, 2008.
174. R. Sigel, M. Łosik, and H. Schlaad, *Langmuir*, Vol. 23, p. 7196, 2007.
175. A. Kishimura, S. Liamsuwan, H. Matsuda, W. Dong, K. Osada, Y. Yamasaki, and K. Kataoka, *Soft Matter*, Vol. 5, p. 529, 2009.
176. J. Du, L. Fan, and Q. Liu, *Macromolecules*, Vol. 45, p. 8275, 2012.
177. M.S. Kim, and D.S. Lee, *Chem. Commun.*, Vol. 46, p. 4481, 2010.
178. J. Rodriguez-Hernandez, and S. Lecommandoux, *J. Am. Chem. Soc.*, Vol. 127, p. 2026, 2005.
179. S. Yu, T. Azzam, I. Rouiller, and A. Eisenberg, *J. Am. Chem. Soc.*, Vol. 131, p. 10557, 2009.
180. R. Dong, B. Zhu, Y. Zhou, D. Yan, and X. Zhu, *Angew. Chem. Int. Ed.*, Vol. 51, p. 11633, 2012.
181. C.B. Minkenberg, F. Li, P. van Rijn, L. Florusse, J. Boekhoven, M.C. Stuart, G.J. Koper, R. Eelkema, and J.H. van Esch, *Angew. Chem. Int. Ed.*, Vol. 50, p. 3421, 2011.
182. A. Klaikherd, C. Nagamani, and S. Thayumanavan, *J. Am. Chem. Soc.*, Vol. 131, p. 4830, 2009.
183. F. Wang, A. Klaikherd, and S. Thayumanavan, *J. Am. Chem. Soc.*, Vol. 133, p. 13496, 2011.
184. Q. Yan, R. Zhou, C. Fu, H. Zhang, Y. Yin, and J. Yuan, *Angew. Chem. Int. Ed.*, Vol. 50, p. 4923, 2011.
185. K. Wang , D.-S. Guo, X. Wang, and Y. Liu, *ACS Nano*, Vol. 5, p. 2880, 2011.
186. E. Amstad, S.-H. Kim, and D.A. Weitz, *Angew. Chem. Int. Ed.*, Vol. 51, p. 12499, 2012.
187. J. Zhang, L. Wu, F. Meng, Z. Wang, C. Deng, H. Liu, and Z. Zhong, *Langmuir*, Vol. 28, p. 2056, 2012.
188. W.-H. Chiang, V.T. Ho, W.-C. Huang, Y.-F. Huang, C.-S. Chern, and H.-C. Chiu, *Langmuir*, Vol. 28, p. 15056, 2012.
189. S.S. Naik, J.G. Ray, and D.A. Savin, *Langmuir*, Vol. 27, p. 7231, 2011.
190. L. Paasonen, T. Laaksonen, C. Johans, M. Yliperttula, K. Kontturi, and A.J. Urtti, *J. Control. Release*, Vol. 122, p. 86, 2007.
191. T. Ikeda, J. Mamiya, and Y. Yu, *Angew. Chem. Int. Ed.*, Vol. 46, p. 506, 2007.
192. G.H. Wu, A. Milkhailovsky, H.A. Khant, C. Fu, W. Chiu, and J.A. Zasadzinski, *J. Am. Chem. Soc.*, Vol. 130, p. 8175, 2008.
193. B. Yan, X. Tong, P. Ayotte, and Y. Zhao, *Soft Matter.*, Vol. 7, p. 10001, 2011.

194. H. Jin, Y. Zheng, Y. Liu, H. Cheng, Y. Zhou, and D. Yan, *Angew. Chem. Int. Ed.*, Vol. 50, p. 10352, 2011.
195. L. Lin, Z. Yan, J. Gu, Y. Zhang, Z. Feng, and Y. Yu, *Macromol. Rapid. Commun.*, Vol. 30, p. 1089, 2009.
196. G.P. Robbins, M. Jimbo, J. Swift, M.J. Therien, D.A. Hammer, and I.J. Dmochowski, *J. Am. Chem. Soc.*, Vol. 131, p. 3872, 2009.
197. D.S. Achilleos, T.A. Hatton, and M. Vamvakaki, *J. Am. Chem. Soc.*, Vol. 134, p. 5726, 2012.
198. Q. Yan, R. Zhou, C. Fu, H. Zhang, Y. Yin, and J. Yuan, *Angew. Chem. Int. Ed.*, Vol. 50, p. 4923, 2011.
199. A. Napoli, M. Valentini, N. Tirelli, M. Muller, and J.A. Hubbell, *Nat. Mater.*, Vol. 3, p. 183, 2004.
200. N. Ma, Y. Li, H.P. Xu, Z.Q. Wang, and X. Zhang, *J. Am. Chem. Soc.*, Vol. 132, p. 442, 2010.
201. K.N. Power-Billard, R.J. Spontak, and I. Manners, *Angew. Chem. Int. Ed.*, Vol. 43, p. 1260, 2004.
202. Y.J. Ma, W.F. Dong, M.A. Hempenius, H. Mohwald, and G.J. Vancso, *Nat. Mater.*, Vol. 5, p. 724, 2006.
203. W.F. Dong, A. Kishimura, Y. Anraku, S. Chuanoi, and K. Kataoka, *J. Am. Chem. Soc.*, Vol. 131, p. 3804, 2009.
204. Q.A. Yan, J.Y. Yuan, Z.N. Cai, Y. Xin, Y. Kang, and Y.W. Yin, *J. Am. Chem. Soc.*, Vol. 132, p. 9268, 2010.
205. H. Kim, S.-M. Jeong, and J.-W. Park, *J. Am. Chem. Soc.*, Vol. 133, p. 5206, 2011.
206. A.V. Kabanov, T.K. Bronich, V.A. Kabanov, K. Yu, and A. Eisenberg, *J. Am. Chem. Soc.*, Vol. 120, p. 9941, 1998.
207. S.A. Walker, M.T. Kennedy, and J.A. Zasadzinski, *Nature*, Vol. 387, p. 61, 1997.
208. P.S. Pramod, K. Takamura, S. Chaphekar, N. Balasubramanian, and M. Jayakannan, *Biomacromolecules*, Vol. 13, p. 3627, 2012.
209. X. Wang, E.J. Danoff, N.A. Sinkov, J.-H. Lee, S.R. Raghavan, and D.S. English, *Langmuir*, Vol. 22, p. 6461, 2006.
210. T. Maruyama, Y. Fukui, E. Tsuchiya, and A. Fujii, *RSC Adv.*, Vol. 2, p. 11672, 2012.
211. B. Lohse, P. Bolinger, and D. Stamou, *J. Am. Chem. Soc.*, Vol. 130, p. 14372, 2008.
212. R.J. Hickey, A.S. Haynes, J.M. Kikkawa, and S.-J. Park, *J. Am. Chem. Soc.*, Vol. 133, p. 1517, 2011.
213. B. Sun, and D.T. Chiu, *Anal. Chem.*, Vol. 77, p. 2770, 2005.
214. S.L. Veatch, and S.L. Keller, *Biophys J.*, Vol. 85, p. 3074, 2003.
215. M. Hong, B.S. Weekley, S.J. Grieb, and J.P. Foley, *Anal. Chem.*, Vol. 70, p. 1394, 1998.
216. H. Nabika, B. Takimoto, and K. Murakoshi, *Analyt. Bioanalyt. Chem.*, Vol. 391, p. 2497, 2008.

217. T. Motegi, H. Nabika, and K. Murakoshi, *Langmuir*, Vol. 28, p. 6656, 2012.
218. E.J. Danoff, X. Wang, S.-H. Tung, N.A. Sinkov, A.M. Kemme, S.R. Raghavan, and D.S. English, *Langmuir*, Vol. 23, p. 8965, 2007.
219. W.-C. Huang, W.-H. Chiang, S.-J. Lin, Y.-J. Lan, H.-L. Chen, C.-S. Chern, and H.-C. Chiu, *Adv. Funct. Mater.*, Vol. 22, p. 2267, 2012.
220. S. Kolusheva, L. Boyer, and R. Jelinek, *Nat. Biotechnol.*, Vol. 18, 225, 2000.
221. A. Grover, B.F. Schmidt, R.D. Salter, S.C. Watkins, A.S. Waggoner, and M.P. Bruchez, *Angew. Chem. Int. Ed.*, Vol. 51, p. 4838, 2012.
222. Z. Cheng, and C.A. Aspinwall, *Analyst.*, Vol. 131, p. 236, 2006.
223. G. Ma, and Q. Cheng, *Langmuir*, Vol. 21, p. 6123, 2005.
224. J.-M. Kim, E.-K. Ji, S.M. Woo, H. Lee, and D.J. Ahn, *Adv. Mater.*, Vol. 15, p. 1118, 2003.
225. H.-Y. Lee, K.R. Tiwari, and S.R. Raghavan, *Soft Matter*, Vol. 7, p. 3273, 2011.
226. D.L. Huber, R.P. Manginell, M.A. Samara, B.-I. Kim, and B.C. Bunker, *Science*, Vol. 301, p. 352, 2003.
227. M.A. Rahman, H.B. Noh, and Y.B. Shim, *Anal. Chem.*, Vol. 80, p. 8020, 2008.
228. J. Hu, and S. Liu, *Macromolecules*, Vol. 43, p. 8315, 2010.
229. M.D. Kim, S.A. Dergunov, E. Lindner, and E. Pinkhassik, *Anal. Chem.* Vol. 84, p. 2695, 2012.
230. A. Manna, D. Sahoo, and S. Chakravorti, *J. Phys. Chem. B*, Vol. 116, p. 2464, 2012.
231. S. Kolusheva, R. Zadmard, T. Schrader, and R. Jelinek, *J. Am. Chem. Soc.*, Vol. 128, p. 13592, 2006.
232. J.-M Kim, Y.B. Lee, D.H. Yang, J.-S. Lee, G.S. Lee, and D. Ahn, *J. Am. Chem. Soc.*, Vol. 127, p. 17580, 2005.
233. X. Zhang, S. Rehm, M.M. Safont-Sempere, and F. Würthner, *Nat. Chem.*, Vol. 1, p. 623, 2009.
234. V. Noireaux, and A. Libchaber, *Proc. Natl. Acad. Sci. USA*, Vol. 101, p. 17669, 2004.
235. K. Renggli, P. Baumann, K. Langowska, O. Onaca, N. Bruns, and W. Meier, *Adv. Funct. Mater.*, Vol. 21, p. 1241, 2011.
236. P.-Y. Bolinger, D. Stamou, and H. Vogel, *Angew. Chem. Int. Ed.*, Vol. 47, p. 5544, 2008.
237. J. Gaitzsch, D. Appelhans, L. Wang, G. Battaglia, and B. Voit, *Angew. Chem. Int. Ed.*, Vol. 51, p. 4448, 2012.
238. S.F.M. van Dongen, M. Nallani, J.L.L.M. Cornelissen, R.J.M. Nolte, and J.C.M. van Hest, *Chem. Eur. J.*, Vol. 15, p. 1107, 2009.
239. T.-J.M. Luo, R. Soong, E. Lan, B. Dunn, and C. Montemagno, *Nat. Mater.*, Vol. 4, p. 220, 2005.
240. P. Broz, S. Driamov, J. Ziegler, N. Ben-Haim, S. Marsch, W. Meier, and P. Hunziker, *Nano Lett.*, Vol. 6, p. 2349, 2006.
241. S. Hiyama, Y. Moritani, R. Gojo, S. Takeuchi, and K. Sutoh, *Lab Chip.*, Vol. 10, 2741, 2010.

242. D.M. Vriezema, P.M.L. Garcia, N.S. Oltra, N.S. Natzakis, S.M. Kuiper, R.J.M. Nolte, A.E. Rowan, and J.C.M. van Hest, *Angew. Chem. Int. Ed.*, Vol. 46, p. 7378, 2007.
243. A. Kishimura, A. Koide, K. Osada, Y. Yamasaki, and K. Kataoka, *Angew. Chem. Int. Ed.*, Vol. 46, p. 6085, 2007.
244. L. Hosta-Rigau, S.F. Chung, A. Postma, R. Chandrawati, B. Städler, and F. Caruso, *Adv. Mater.*, Vol. 23, p. 4082, 2011.
245. T. Rispens, and J.B.F.N. Engberts, *Org. Lett.*, Vol. 3, p. 941, 2001.
246. K. Manabe, X.-M. Sun, and S. Kobayashi, *J. Am. Chem. Soc.*, Vol. 123, p. 10101, 2001.
247. G. Feng, Y. Jia, L. Liu, W. Chang, and J. Li, *J. Polym. Sci. Part A: Polym. Chem.*, Vol. 48, p. 5992, 2010.
248. F. Wang, H. Liu, L. Cun, J. Zhu, J. Deng, and Y. Jiang, *J. Org. Chem.*, Vol. 70, p. 9424, 2005.
249. M.S. Goedheijt, B.E. Hanson, J.N.H. Reek, P.C.J. Kamer, and P.W.N.M. van Leeuwen, *J. Am. Chem. Soc.*, Vol. 122, p. 1650, 2000.
250. L. Yang, L. Luo, S. Zhang, X. Su, J. Lan, C.-T. Chen, and J. You, *Chem. Commun.*, Vol. 46, 3938, 2010.Affiliation Information *1. Department of Chemistry,National Chung Hsing University, Taichung 402, Taiwan*
251. Y. Murakami, J. Kikuchi, K. Akiyoshi, and T. Imori, *J. Chem. Soc., Perkin Trans. 2,* p. 1919, 1985.
252. J.T. Groves, and R. Neumann, *J. Am. Chem. Soc.*, Vol. 109, p. 5045, 1987.
253. H. Li, J. Liu, S. Xie, M. Qiao, W. Dai, Y. Lu, and H. Li, *Adv. Funct. Mat.*, Vol. 18, p. 3235, 2008.
254. H. Li, Z. Zhu, J. Liu, S. Xieb, and H. Li, *J. Mater. Chem.*, Vol. 20, p. 4366, 2010.
255. G. Hamasaka, T. Muto, and Y. Uozumi, *Angew. Chem. Int. Ed.*, Vol. 50, p. 4876, 2011.
256. Z. Wang, M.C.M. van Oers, F.P.J.T. Rutjes, and J.C.M. van Hest, Angew. Chem. Int. Ed., Vol. 51, p. 10746, 2012.
257. K.J. Pekarek, J.S. Jacob, and E. Mathiowitz, Nature, Vol. 367, p. 258, 1994.
258. H. Lomas, I. Canton, S. MacNeil, J. Du, S.P. Armes, A.J. Ryan, A.L. Lewis, and G. Battaglia, Adv. Mater., Vol. 19, p. 4238, 2007.
259. E. Soussan, S. Cassel, M. Blanzat, and I. Rico-Lattes, Angew. Chem. Int. Ed., Vol. 48, p. 274, 2009.
260. E.P. Holowka, V.Z. Sun, D.T. Kamei, and T.J. Deming, *Nat. Mater.*, Vol. 6, p. 52, 2007.
261. P.P. Ghoroghchian, P.R. Frail, K. Susumu, D. Blessington, A.K. Brannan, F.S. Bates, B. Chance, D.A. Hammer, and M.J. Therien, Proc. Natl. Acad. Sci. USA, Vol. 102, p. 2922, 2005.
262. A. Ranquin, W. Versées, W. Meier, J. Steyaert, and P.V. Gelder, Nano Lett., Vol. 5, p. 2220, 2005.
263. Q. Chen, H, Schönherr, and G.J. Vancso, Small, Vol. 5, p. 1436, 2009.
264. C.G. Palivan, O. Fischer-Onaca, M. Delcea, F. Itel, and W. Meier, Chem. Soc. Rev., Vol. 41, p. 2800, 2012.

265. M.N. Holme, I.A. Fedotenko, D. Abegg, J. Althaus, L. Babel, F. Favarger, R. Reiter, R. Tanasescu, P. Zaffalon, A. Ziegler, B. Müller, T. Saxer, and A. Zumbuehl, *Nat. Nanotechnol. Volume:*, Vol. 7,Pages: p. 536, 2012.
266. J. Song, J. Zhou, and H. Duan, *J. Am. Chem. Soc.*, Vol. 134, p. 13458, 2012.
267. J.J. Moon, H. Suh, A. Bershteyn, M.T. Stephan, H. Liu, B. Huang, M. Sohail, S. Luo, S.H. Um, H. Khant, J.T. Goodwin, J. Ramos, W. Chiu, and D.J. Irvine, Nat. Mater., Vol. 10, p. 243, 2011.
268. J.M. Criscione, B.L. Le, E. Stern, M. Brennan, C. Rahner, X. Papademetris, and T.M. Fahmy, *Biomaterials*, Vol. 30, p. 3946, 2009.
269. E.S. Levitan, F. Lanni, and D. Shakiryanova, *Nat. Protoc.*, Vol. 2, p. 1117, 2007.
270. X. Yang, J.J. Grailer, I.J. Rowland, A. Javadi, S.A. Hurley, V.Z. Matson, D.A. Steeber, and S. Gong, *ACS Nano*, Vol. 4, p. 6805, 2010.
271. C. Sanson, O. Diou, J. Thévenot, E. Ibarboure, A. Soum, A. Brûlet, S. Miraux, E. Thiaudière, S. Tan, A. Brisson, V. Dupuis, O. Sandre, and S. Lecommandoux, *ACS Nano*, Vol. 5, p. 1122, 2011.
272. E. Kim, J. Yang, J. Park, S. Kim, N.H. Kim, J.I. Yook, J.-S. Suh, S. Haam, and Y.-M. Huh, *ACS Nano*, Vol. 6, p. 8525, 2012.

Part 2

STIMULI-RESPONSIVE INTERFACES

4

Computational Modeling of Sensing Membranes and Supramolecular Interactions

Giacomo Saielli

CNR Institute on Membrane Technology, Unit of Padova, Padova, Italy

Abstract
In this chapter we present an overview of the most recent computational tools available for the description of weak interactions, ranging from high level ab-initio methods to coarse-grained molecular dynamics simulations. In the following Sections, we will first present a survey of the many types of weak interactions that dominate the world of self-assembled materials and supra-molecular structures and then we will describe the various computational protocols available and their pros and cons.

Keywords: Quantum chemistry, density functional theory, molecular dynamics simulation

4.1 Introduction

Supra-molecular interactions, which are at the heart of any sensing application, are largely responsible for the organization and structure of materials and represent a major challenge for computational chemistry. The main reason why supra-molecular interactions are difficult to be treated by means of calculations is that they are usually weak, at least compared with covalent interactions, but "many" both in type and in number, so that they eventually take control over the organization and structure of large assemblies. Thus, a computational approach to study supra-molecular interactions, self assembled materials, and, ultimately, smart sensors,

**Corresponding author*: giacomo.saielli@unipd.it

Annarosa Gugliuzza (ed.) Smart Membranes and Sensors, (107–144) 2014 © Scrivener Publishing LLC

should be capable to deal with many types of weak interactions extending their "jurisdiction" over large length and long time scales. Put it like that it's just a mission impossible! However, in this chapter we will show that by carefully selecting the system to be studied and the type of computational protocol very useful insights can be obtained.

The subject of our investigation will be "matter" starting from molecules up to larger and larger aggregates: clusters, nano-particles, phases, membranes. The length scale thus covers a range of about 10 log units, from 10^{-10} m of molecules up to real world-devices. The time scale varies somehow accordingly, from 10^{-15} s typical of electronic motion up to real time functions. How is it possible with computational chemistry to cover such ranges?

For the description of the smaller and faster objects, that is molecules, we have quantum chemistry (QM). This is a well established set of theoretical protocols and models that are able to describe, with a high level of accuracy, the structure of a molecule, that is a system composed by nuclei and electrons, or, in a more chemically appealing view, atoms covalently bound together. QM theory has a high predicting power: molecular structures and spectroscopic properties can be predicted often with an accuracy higher than the experimental results [1, 2]. However, since it explicitly describes the electronic structure, its demand for computational power increases as we select larger molecules, that is multi-electron systems. Soon the possibility to use high-level QM methods fades out and molecules as large as fullerene can be indeed treated by QM methods but only after several approximations have been introduced; this, in turn, lowers the accuracy of the results obtained by the calculation. Moreover, as we will see, the QM methods that can be applied to study large systems, that is those based on the Density Functional Theory (DFT) suffer from a specific drawback of not being able to properly treat weak intermolecular interactions, although significant improvements have been obtained recently. Finally, we note that QM approaches usually consider the nuclei at their relaxed position thus offering a static description of the molecules.

On the other extreme we have materials and/or large organized, functional aggregates of molecules, possibly of large size, whose structure is dominated by weak supra-molecular interactions and whose functions occurs over relatively long times comprising the motion of several parts of the assembly. This is the realm of computer simulations such as Molecular Dynamics (MD) and Monte Carlo (MC). To describe such systems we have often to coarse-grain our description, thus loosing the details, e.g. the electronic structure of the molecules in first place, but gaining a better overall view. Moreover, the time scale explored by MD simulations can, nowadays range from ns to seconds, that is it is possible to simulate real devices.

4.2 Non-covalent Interactions: A Physical and a Chemical View

For the sake of simplicity, in order to study molecules and objects larger than molecules, we can consider matter as composed by nuclei and electrons, that is we treat both particles as elementary particles. This is, in fact, true only for the electron, which has a well defined mass, *m*, spin, *S*, and charge $-e$. Nonetheless, we will ignore the internal structure of nuclei thus regarding them as simple particles with a well defined mass, *M*, spin, *I*, and charge $+Ze$, where *Z* is the atomic number. Therefore, being a molecule composed of nuclei and electrons, intermolecular interactions, which are often called non-covalent interactions, will be produced by the charge and/or the spin of the particles constituting the molecules, thus they all have an electromagnetic origin.

From a more physical viewpoint, based on perturbation theory, [3] supra-molecular interactions can be divided into four groups: electrostatic, inductive, dispersive and exchange. The source of the first three interactions will be easily identified in the charge of the particles while the source of the fourth interaction is more subtle: it is also an electrostatic contribution but appearing only at the quantum level because of the Pauli principle which requires the overall wave function to be antisymmetric.

On the other hand, the chemical literature is crowded with a plethora of supra-molecular interactions whose definition is based, often, on the type of atoms and/or molecular moieties involved or other chemical features of the system: hydrogen bond, halogen bond, CH-π interaction, charge-π, π-π stacking, van der Waals, dipole-dipole, electrostatic, hydrophobic, steric, ... to mention but the most common ones. In most cases the four physical terms defined above (electrostatic, inductive, dispersive and exchange) all contribute, to varying extent, to the particular chemical supra-molecular interaction; though it might seem a contradiction, also a covalent contribution can be, in some cases, found in a supra-molecular interaction. It should be stressed, as noted already by several authors [4, 5] that the chemical classification of non-covalent interactions is more linked to the structural features of the interacting partners rather than to fundamental differences in the physical mechanism underlying the type of interaction itself.

4.3 Physical Interactions

For a more detailed description of the four physical contributions we consider a system made of two interacting molecules, A and B, with a total

Hamiltonian $H_0 = H_A + H_B$, and a total wave-function (correct at large distances) given by the product of the two wave-functions, $\Psi_0 = \Psi_A \cdot \Psi_B$. The purely electrostatic term, which is the dominant contribution at large separations, is obtained from first order perturbation theory as the expectation value of the total wave-function of the interacting system with the perturbative Hamiltonian $H^{(1)}$ this is given by the sum, over all "cross" particles pairs (electrons and nuclei not belonging to the same molecule: this latter one is the contribution already included in H_A and H_B) of the charge interaction ($-e$ for the electrons and Ze for the nuclei) separated by a distance r, taking the nuclei at fixed positions (Born-Oppenheimer approximation).

$$E_{el}^{(1)} = \langle \Psi_A \Psi_B | H^{(1)} | \Psi_A \Psi_B \rangle \tag{4.1}$$

where

$$H^{(1)} = -\sum_{\alpha}\sum_{n} \frac{Z_\alpha e^2}{r_{\alpha n}} - \sum_{\beta}\sum_{m} \frac{Z_\beta e^2}{r_{\beta m}} + \sum\sum \frac{e^2}{r_{nm}} + \sum\sum \frac{Z_\alpha Z_\beta e^2}{r_{\alpha\beta}} \tag{4.2}$$

where α and β run over the nuclei of molecules A and B, respectively, while n and m run over the electrons of molecules B and A, respectively.

The interaction energy of Eq. 4.1 is obtained considering the two molecules in their ground states with their original charge distribution, not perturbed by the presence of the other molecule. The *electrostatic interaction* of two systems of charges, each separated by a large distance compared to their dimensions, can be expanded in a multipole series. Each charge distribution, of molecules A and B, can be written as the sum of a net charge (monopole), a dipole, a quadrupole, an octupole, an hexadecapole, etc. Then the total electrostatic interaction can be expressed as a sum of all interaction terms: charge-charge, charge-dipole, charge-quadrupole, charge-octupole, charge-hexadecapole... dipole-dipole, dipole-quadrupole, dipole-octupole, dipole-hexadecapole... quadrupole-quadrupole, quadrupole-octupole, quadrupole-hexadecapole... etc. A detailed derivation of the various terms can be found in Ref. [3].

Obviously, for ions the charge-charge term is the dominant one. For neutral molecules the dipole-dipole term is the most important. However, molecules lacking a dipole are sometimes termed non-polar, which is, in fact, a misleading term since in the absence of a dipole a quadrupole-quadrupole interaction (if the quadrupole is present, of course) is the

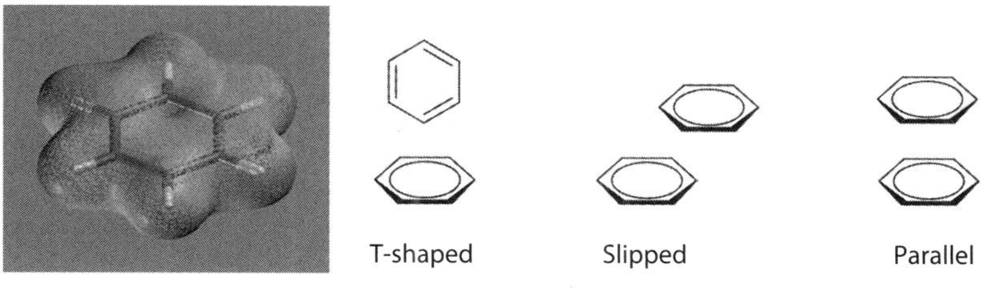

Figure 4.1 (Left) Isodensity surface of benzene colour-coded with the electrostatic potential showing the presence of a quadrupole. (Right) Three possible structures of the benzene dimer discussed in the text.

leading contribution. For example, the benzene dimer is a well known example of a system where the geometry of the most stable conformations has a significant contribution from the quadrupole-quadrupole interaction: [6, 7] this disfavours the parallel face-to-face arrangement while it favours the T-shaped and slipped arrangements. In Figure 4.1 we show an isodensity surface colour-coded with the electrostatic potential (B3LYP/6–31G** level of theory): the blue colour of the protons indicates a more positive charge and the green colour of the electron-rich central region indicates an excess of negative charge. The T-shaped and slipped configurations (see Figure 4.1) are stabilized, while the parallel configuration is destabilized, by the quadrupole-quadrupole interaction term.

This issue is of relevance in many areas of materials science (nano-carbon structures, liquid crystals) and biology. For example Hunter *et al.* [8], following a work by Burley and Petsko [9] nicely highlighted the low probability of parallel, face-to-face, arrangement of phenyl rings of phenylalanine residues in proteins, see Figure 4.2.

Another interesting example is that of chemically-induced liquid crystalline phases: for simple models of liquid crystals, like those based on Gay-Berne particles (soft ellipsoidal particles exhibiting a rich polymorphism [10, 11]) it was found that the presence of a point quadrupole on the centre of the discotic molecule strongly influenced the type of mesophase formed. [12]

The contribution of Eq. 4.1, as already mentioned, is obtained considering the unperturbed ground-state charge distribution of the two interacting molecules. However, the charge density of a molecule is also modified by the presence of a another molecule nearby having its own charge density. Additional multipoles are "induced" in molecule A by molecule B and vice versa. Therefore we have, for example, dipole-induced dipole interactions and similar terms known as *inductive interactions*.

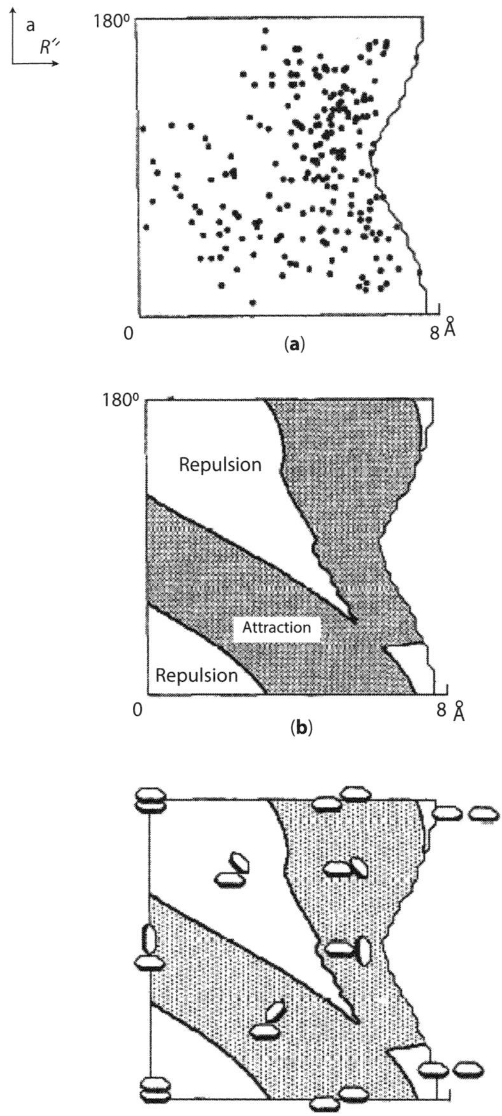

Figure 4.2 Distribution of the phenyl-phenyl structures (distance and orientation) of phenylalanine residues in proteins. Reproduced from Ref. 8 with permission from Elsevier.

$$E_i^{(2)} = -\sum_{n\neq 0} \frac{\left|\left\langle \Psi_A^{(n)} \Psi_B \middle| H^{(1)} \middle| \Psi_A \Psi_B \right\rangle\right|^2}{E_A - E_A^{(n)}} - \sum_{m\neq 0} \frac{\left|\left\langle \Psi_A \Psi_B^{(m)} \middle| H^{(1)} \middle| \Psi_A \Psi_B \right\rangle\right|^2}{E_B - E_B^{(m)}} \tag{4.3}$$

where n and m are excited electronic states of molecules A and B, respectively.

The expression of the inductive interaction energy in Eq. 3 is obtained by second order perturbation theory and, for molecules in their electronic ground state, it is always negative, thus corresponding to an attractive interaction [3].

A clear example of the existence of long-range attractive interactions in molecules is provided by the equation of state of a real gas, $(P+a/V^2)(V-b)=RT$. While the b term is easily understood as due to the finite size of molecules, the a coefficient describes the reduced effective pressure, compared to an ideal gas, for a system made of interacting particles.

The induced multipole contribution, however, does not explain the condensation of noble gases since none of the atoms have a static multipole to induce another multipole in a nearby atom. However the electronic charge distribution is not static: electrons move around the nucleus and, in doing so, they produce instantaneous multipoles. Thus, the instantaneous multipole in atom A can induce a multipole in the neighbouring atom. The same mechanism, obviously, is present also in molecules together with the other ones already mentioned.

$$E_d^{(2)} = -\sum_{n\neq 0}\sum_{m\neq 0} \frac{\left|\left\langle \Psi_A^{(n)}\Psi_B^{(m)} \middle| H^{(1)} \middle| \Psi_A \Psi_B \right\rangle\right|^2}{E_A - E_A^{(n)} + E_B - E_B^{(m)}} \tag{4.4}$$

This contribution is termed *dispersion interaction* and is always attractive. From second order perturbation theory we obtain Eq. 4.4 [3].

Finally, the *exchange interaction* is rooted in the Pauli exclusion principle: when the overlap between the electronic clouds of the two molecules becomes not negligible the system needs to be treated as a single "supermolecule". Thus the total wave function cannot be simply written as the product of the two unperturbed wave functions of molecule A and B; rather the product needs to be antisymmetrized for the exchange of electrons between the two molecules. This is achieved with the antisymmetrization operator $\hat{A}$ where

$$\hat{A}\Psi_A\Psi_B = N_{AB}\sum_{p}(-1)^p P\Psi_A\Psi_B \tag{4.5}$$

and P represents all possible permutations of electrons between the molecules.

The antisymmetrization operation brings in a new term in the first order perturbation theory, together with the known electrostatic contribution, Eq. 4.2, that is the exchange contribution shown in Eq. 4.6:

$$E_{ex}^{(1)} = \left\langle \Psi_A \Psi_B \middle| H^{(1)} \middle| \sum_{p \neq 1} (-1)^p P \Psi_A \Psi_B \right\rangle \tag{4.6}$$

This term is strongly repulsive for closed-shell molecules, for which the overlap of the electronic clouds would lead to a violation of the Pauli principle (in contrast it is largely responsible for the stability of covalent bonds where opposite spins are paired in the bonding orbitals).

4.4 Chemical Interactions

We will now briefly survey the "composition", in terms of physical contributions, of some of the most common supra-molecular interactions as encountered in the chemical literature.

Electrostatic interactions: the electrostatic interactions found in the chemistry literature are described above in Eq. 4.1. However they are often decomposed in their multipole-multipole contributions and each of these terms is often discussed separately. Thus it is common to encounter ion-ion, ion-dipole, dipole-dipole interactions, etc. The rationale behind such distinction is two-fold: first, ions are very different chemical objects from non-charged dipolar molecules which, in turn, are different from non-dipolar molecules. Second, the dependence of the interaction energy from the distance follows diverse power laws, therefore the strength and the range of action of the forces involved are different. In particular ion-ion interaction energies can be of the same order of a covalent bond, that is about 100 kcal/mol (attractive for ions of opposite charge), with distance dependence as the inverse of the distance, $E \approx q_1 q_2 / r$, where q_1 and q_2 are the two ion charges. In contrast, dipole-dipole interactions are usually of the order of about 10 kcal/mol and decay as $\mu_1 \mu_2 / r^3$ where μ_1 and μ_2 are the molecular dipoles. Also, the former ones have an orientational dependence which is missing in the ion-ion term. By definition, electrostatic interactions do not contain the other terms described in Equations 4.3 — 4.5 and they have a clear classical analogue.

Hydrogen bond (HB): this is a very well known type of chemical bond which has a fundamental role in the supra-molecular aggregation of biomolecules such as proteins and DNA as well as being responsible of the extraordinary properties of water and protic solvents. In the hydrogen bond a donor group, X-H, interacts with an acceptor group Y, through the hydrogen atom, X-H···Y. Both X and Y are electronegative atoms such as nitrogen, oxygen, sulphur, halogens, and Y must have a lone pair. It is largely due to an electrostatic interaction with important contribution from dispersive and inductive effects but also a partial covalent character [13]. The latter contribution is mainly responsible for the well-defined (at least with respect to what we would expect for a purely electrostatic interaction) distance and geometry of the bond. For the same reason, and because the HB donor and acceptors are relatively close in space, the exchange interaction is very significant and cannot be neglected in any quantum chemical treatment. Interaction energies are usually in the range of few kcal/mol, though they may vary strongly depending on the nature of the X-H and Y groups. A limiting case of hydrogen bond is that between the C-H group and an aromatic ring, often referred in the literature as the C-H/π interaction [14]. Rather than being a really different kind of interaction, compared to HB, it is the relative weight of dispersive and electrostatic contribution that differentiates between them. An experimental determination of the relative weight of electrostatic *vs* other quantum mechanical contributions in X-H/π interactions is reported in Ref. [15]. Since all terms in Equations 4.1 — 4.6 contribute to the HB, its accurate description by computational methods is not an easy task.

Van der Waals: the Van der Waals interaction is usually associated with the presence of a polarisable electronic cloud. It is an attractive interaction with a distance dependence of approximately $1/r^6$. One contribution to the van der Waals interaction, of inductive type, is the so-called Keesom term, which is due to rapidly rotating dipoles. This is expected to play a role particularly for polar molecules in the gas phase. It can be shown that for fast rotating dipole the (attractive) energy contribution goes like $E \approx \mu_1^2 \cdot \mu_2^2 / r^6$ (where, again, μ_1 and μ_2 are the dipoles of molecule 1 and 2, respectively). A second term, still of inductive type, is the so-called Debye term, due to the polarization of a molecule (e.g. molecule 2) by a rotating dipole (e.g of molecule 1): again the attractive energy contribution goes like $E \approx \mu_1^2 \cdot \alpha_2 / r^6$, where α_2 is the polarizability of molecule 2. Finally there is a dispersive term, called the London contribution, related to the polarizability of both molecules for which the interaction energy goes like $E \approx \alpha_1 \cdot \alpha_2 / r^6$. The last term is often the most important and the only one responsible for the condensation of noble gases where there are no static dipoles. Van der Waals interactions are described by a r^{-6} dependence in the Lennard-Jones potential in Eq. 4.7, see below.

Halogen bond: the halogen bond, XB, is an intermolecular interaction D···X-Y where an electrophilic halogen atom X (Lewis acid, with Y = C, N, halogen…) in a X-Y moiety, interacts with a donor of electron density, D (Lewis base such as amines etc.) [16]. Though a halogen in an organic compound usually bears a negative partial charge, the charge distribution is highly anisotropic and along the X-Y bond direction, that is facing the halogen, there is a region of positive electrostatic potential. The main contribution comes from a strong quadrupole in addition to the well-known X-Y dipole [17]. The D···X-Y interaction is, in many respect, similar to the HB interaction, sharing significant contribution from dispersive and inductive terms as well as partial covalent character. The strength of the XB donor increases as Cl < Br < I and the interaction energy ranges from about 2 to 50 kcal/mol.

π-π stacking: with this name we usually indicate the interactions responsible for the stabilization of aggregates of aromatic moieties. These can be found in biological macromolecules, materials science, catalysis etc. A significant work in this area has been done by Hunter and Sanders [18]; subsequent experimental work by Cozzi *et al.* [19] and theoretical investigations by Sinnokrot and Sherrill [20] addressed the importance of substituent effects. There is ample literature concerning π–π interactions and it is not possible to give an exhaustive list of contributors here. We mention a recent paper by Lima *et al.* [21] where the relative weight of electrostatic contribution vs dispersive interaction was analyzed in detail through a combination of thermochemical data and computational protocols. The Authors considered 1,8-diphenylnaphtalenes as model systems, as shown in Figure 4.3.

They observed a very good linear correlation between the π–π interaction energy and the dispersive contribution, while the elelctrostatic terms resulted less important than previously thought. However dispersive interactions are not orientational dependent so, while the energetics is dominated by them, the geometrical arrangement is influenced by the electrostatic terms. Thus, the importance of the various (physical) contributions to the π–π stacking is still an area of active investigation.

In a recent paper Kuchenbecker and Jansen have highlighted the importance of the various contributions in π-π interactions between donor and

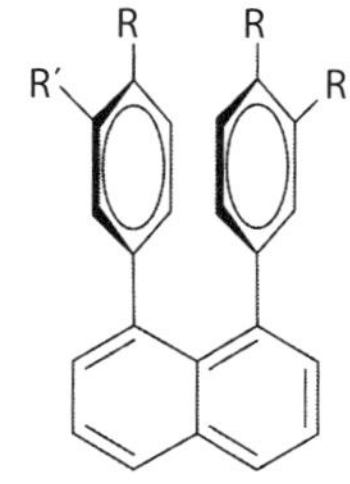

Figure 4.3 Structure of the compounds investigated by Lima *et al.* in Ref. [21]. R = H, OCH_3, CHO; R' = H, OCH_3, CHO.

acceptor π-systems, where charge-transfer between donor and acceptor is believed to occur [22]. The Authors have studied tetracyanoethylene–benzene and tetracyanoethylene–p-xylene as examples using a variety of methods, including MP2, DFT-SAPT, CCSD(T) extrapolating the results to the complete basis set limit. It is found that the most important contribution to the attractive interaction comes from second-order dispersion terms, while the second most important is the electrostatic term.

Steric: steric interactions are essentially due to the impossibility for two closed-shell electronic clouds to overlap, because of the Pauli exclusion principle. Thus the exchange term in Eq. 6 is the most important contribution. Clearly electrostatic repulsion is also important, but its distance dependence cannot explain the very steep increase of the interaction energy that occurs when molecules are in contact. In simple model pair potentials, such as the Lennard-Jones potential, this contribution is modeled as a repulsive $1/r^{12}$ term, while the attractive part (mostly due to van der Waals contributions) has a $1/r^6$ dependence, see Eq. 4.7:

$$V(r) = 4\varepsilon\left(\frac{\sigma}{r^{12}} - \frac{\sigma}{r^{6}}\right) \tag{4.7}$$

σ, the distance where the interaction potential is zero, is taken as the contact distance, while ε is the interaction energy at the equilibrium position (well depth).

4.5 Computational Methods for Supramolecular Interactions

Quantum chemistry. Quantum chemical methods can be divided in to those based on the Wave Function Theory (WFT), usually referred to as the true *ab-initio* methods, and those based on the Density Functional Theory (DFT). WFT methods are based on the Hartree-Fock (HF) theory: the Schroedinger equation for a system of *m* nuclei and *N* electrons is solved, in the Born-Oppenheimer approximation, assuming that each electron is moving the mean field of the remaining electrons [23]. The energy obtained for the molecules lacks the instantaneous electron-electron interaction, also called "correlation energy". As can be understood this term contributes largely to the weak dispersive and inductive terms of Eq. 4.3 and Eq. 4.4. Therefore, basically all non-covalent interactions, except the purely electrostatic ones, are very poorly described at the Hartree-Fock level of theory. Post-HF methods are used to describe weak non-covalent interactions. Among them, the reference method, which is often used to calibrate, parameterize and/or judge the performance of other QM methods is CCSD(T), that is Coupled Cluster method with inclusion of single and double excitations plus the perturbative treatment of the triple excitations.

Accurate investigation of the geometry and interaction energy of the benzene dimer were reported by Sinnokrot and Sherril [4] who also investigated the effect of the substituents. By using high level ab-intio methods they found that simple electrostatic arguments fail in predicting the geometry and structure of substituted aromatic systems. A proper description can only be achieved by using the computationally demanding coupled cluster method CCSD(T). Unfortunately it can only be applied to rather small systems.

Recently accurate interaction energies have been reported for a dataset intended to be used as reference, the S66 data set, in the complete bases set limit (CBS) at the CCSD(T) level [24]. The pairs, in various geometries, and their interaction energies, are included in Table 4.1.

It is noteworthy that the size of the systems investigated is of the order of 10–15 atoms (excluded hydrogens). The computational cost of CCSD(T) scales as $O(N^7)$ so extension to larger molecules is impractical. Other *ab-initio* methods are computationally cheaper, such as 2nd order Møller-Plesset perturbation theory (MP2), which scales as $O(N^5)$. Its performance is also much better than HF, since single and double excitations are included up to second order. However it is known to overestimate dispersive interactions. For example the interaction energy of the benzene dimer at the MP2 level, CBS limit, in a face-to-face arrangement, is -4.7 kcal/mol, about 75% more than the value obtained at the CCSD(T)/CBS level. However, MP2 allows the treatment of relatively large systems including electron correlation to a good degree. Systems as large as two-layer nanographene sheets $(C_{96}H_{24})_2$ have been considered by Katouda and Nagase [25] using MP2 and the resolution of identity (RI) approximation. However, the errors, with respect to the reference CCSD(T) method, can be significant, especially when pure π–π interactions are considered (while H-bonded systems are well described) thus, not surprisingly, several "hybrid" methods where corrections to the MP2 energy are introduced in some *ad-hoc* way have been proposed. One of this is the MP2.5 method. It has been found that the errors obtained from third order MP perturbation theory have usually opposite sign and similar magnitude compared to the errors obtained at the MP2 level. Thus, a version of the perturbation theory where the MP2 and MP3 correlation energies are weighted each 50%, or some other combination depending on the basis set used, has been proposed

Table 4.1 Interaction energies at the CCSD(T)/CBS (Complete Basis Set) level, in kcal/mol, for the dimers of the S66 dataset. Reproduced from Ref. [24] with permission from American Chemical Society.

H Bonds		**Dispersive**		**Others**	
H_2O···H_2O	−5.01	benzene···benzene (π-π)	−2.72	benzene···benzene (TS)	−2.83
H_2O···MeOH	−5.70	pyridine···pyridine (π-π)	−3.80	pyridine···pyridine (TS)	−3.51
H_2O···$MeNH_2$	−7.04	uracil···uracil (π-π)	−9.75	benzene···pyridine (TS)	−3.29
H_2O···peptide	−8.22	benzene···pyridine (π-π)	−3.34	benzene···ethyne (CH-π)	−2.86
MeOH···MeOH	−5.85	benzene···uracil (π-π)	−5.59	ethyne···ethyne (TS)	−1.54
MeOH···$MeNH_2$	−7.67	pyridine···uracil (π-π)	−6.70	benzene···AcOH (OH-π)	−4.73
MeOH···peptide	−8.34	benzene···ethene	−1.36	benzene···$AcNH_2$ (NH-π)	−4.40
MeOH···H_2O	−5.09	uracil···ethene	−3.33	benzene···H_2O (OH-π)	−3.29
$MeNH_2$···MeOH	−3.11	uracil···ethyne	−3.69	benzene···MeOH (OH-π)	−4.17
$MeNH_2$···$MeNH_2$	−4.22	pyridine···ethene	−1.80	benzene···$MeNH_2$ (NH-π)	−3.20
$MeNH_2$···peptide	−5.48	pentane···pentane	−3.76	benzene···peptide (NH-π)	−5.26
$MeNH_2$···H_2O	−7.40	neopentane···pentane	−2.60	pyridine···pyridine (CH-N)	−4.24
peptide···MeOH	−6.28	neopentane···neopentane	−1.76	ethyne···H_2O (CH-O)	−2.93

(*Continued*)

Table 4.1 *(Cont.)*

H Bonds		**Dispersive**		**Others**	
peptide···$MeNH_2$	−7.56	cyclopentane···neopentane	−2.40	ethyne···AcOH (OH-π)	−4.97
peptide···peptide	−8.72	cyclopentane···cyclopentane	−2.99	pentane···AcOH	−2.91
Peptide···H_2O	−5.20	benzene···cyclopentane	−3.51	pentane···$AcNH_2$	−3.53
uracil···uracil	−17.45	benzene···neopentane	−2.85	benzene···AcOH	−3.75
H_2O···pyridine	−6.97	uracil···pentane	−4.81	peptide···ethene	−3.00
MeOH···pyridine	−7.51	uracil···cyclopentane	−4.09	pyridine···ethyne	−4.10
AcOH···AcOH	−19.41	uracil···neopentane	−3.69	$MeNH_2$···pyridine	−3.97
$AcNH_2$···$AcNH_2$	−16.52	ethene···pentane	−1.99		
AcOH···uracil	−19.78	ethyne···pentane	−1.72		
AcNH···Uracil	−19.47	peptide···pentane	−4.26		

[26, 27]. Other empirical modifications of the MP2 method has been proposed by Grimme[28] by scaling separately the parallel- and antiparallel spin components. A similar approach has been used also to improve, at a lower computational cost, the coupled cluster method limited to single and double excitation, CCSD. [29] Hesselmann, instead, combined MP2 with TDDFT to improve the interaction energies in pairs of interacting molecules. [30] Another method is DFT-SAPT theory, presented by Misquitta *et al.* [31] and by Hesselmann and Jensen [32]. In this method the Symmetry Adapted Perturbation Theory (SAPT) is combined with time dependent DFT to obtain accurate interaction energies also for systems dominated by weak dispersive interactions. In ref. [33] DFT-SAPT theory has been applied to the study of *cis-syn* cyclobutane pyrimidine dimer (CPD), one of the photoproduct (potentially leading to skin cancer) formed from adjacent pyrimidine bases in DNA under solar irradiation.

The various terms contributing to the stability of the photodamaged CPD moiety and the repaired pyrimidine pairs were carefully analyzed and a successful description was obtained by DFT-SAPT. These last methods drive us directly into the realm of Density Functional Theory (DFT), the only quantum chemical method which is really capable of treating large systems made of hundreds of heavy (non-hydrogen) atoms.

In contrast to wave function theory, in DFT the focus is on the electron density, $\rho(r)$. The energy is written as a functional of the electron density as in Eq. 4.8:

$$E(\rho) = T_s(\rho) + V_{ne}(\rho) + J(\rho) + E_{ex}(\rho) \tag{4.8}$$

where the density can be expressed in terms of the set of one electron Kohn-Sham orbitals,

$$\rho(\mathbf{r}) = \sum_i |\Phi_i(\mathbf{r})|^2 \tag{4.9}$$

and the various terms in Eq. 4.8 are the kinetic energy, T_s, the nucleus-electron potential energy, V_{ne}, the classical electron-electron repulsion

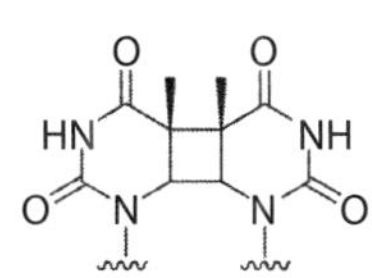

Figure 4.4 Cyclobutane pyrimidine dimer photoproduct from two adjacent pyrimidine bases in DNA, studied by DFT-SAPT. [33]

energy, J [these terms are explicitly described in Eqs. 410 — 4.12], and the unknown remaining term, the exchange-correlation functional, E_{xc}.

$$T_s(\rho) = \sum_i \langle \Phi_i | -\frac{1}{2}\nabla^2 | \Phi_i \rangle \tag{4.10}$$

$$V_{ne}(\rho) = -\int \rho(\mathbf{r}) \sum_a \frac{Z_a}{\mathbf{r} - \mathbf{R}_a} d\mathbf{r} \tag{4.11}$$

$$J(\rho) = \frac{1}{2} \iint \frac{\rho(\mathbf{r})\rho(\mathbf{r}')}{|\mathbf{r} - \mathbf{r}'|} d\mathbf{r} d\mathbf{r}' \tag{4.12}$$

where $\mathbf{R}_\alpha$ represent the position of the nuclei.

Several approximations of E_{xc} have been proposed in the literature. DFT became a very popular QM method probably starting from 1993 after the inclusion of some Hartree-Fock exchange into the exchange-correlation functional, thus leading to the so called hybrid functionals. Of these, the most popular and widely used is indeed B3LYP. The success of DFT was, in fact, the success in correctly describing covalent bonds, thus the structure and energy of molecules. However, it became soon clear that weak non-covalent interactions were not well described by DFT, particularly dispersive interactions which dominate π–π stacking and van der Waals interactions, since DFT functionals do not exhibit the correct long-range behaviour. Thus a lot of effort has been put recently into the development of new functionals having a good performance also for the description of weak interactions.

Truhlar and co-workers have developed several new functionals, the most recent ones are those of the M06 family: M06, M06L, M06–2X [34–38]. These have been found to have a rather good performance for the calculation of the interaction energies of several systems dominated by weak interactions. One interesting applications was the evaluation of the differential free energy of solvation of oxonium vs ammonium ions in water [39]. This subject has several implications, for example in the determination of the preferential site of interaction of ethanolamines for the extraction of CO_2 and H_2S from fossil fuels or for the choice of the site of protonation of polyfunctional bases such as carboxylic and noncarboxylic amides, see Figure 4.5.

It is clear that the free energy of solvation is strongly dependent on a proper description of the hydrogen-bond network that such systems establish with the water solvent. A very good agreement with accurate

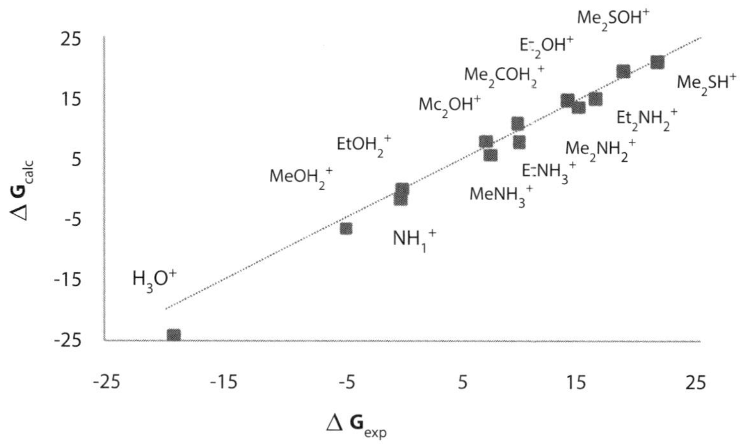

Figure 4.5 Correlation between calculated and experimental differential solvation free energies of onium ions [39]. Reproduced with permission from American Chemical Society.

experimental data was found only after inclusion of an explicit water molecule hydrogen bonded to the onium ions. Another interesting application was the prediction of the pk_a of several amines in a wide temperature range [40]. Thus, the calculations of hydrogen-bonded interactions is indeed feasible within the framework of DFT, if proper functionals are selected. Indeed, for a correct investigation of solvation energies by quantum methods the choice and parameterization of the continuum model is of paramount importance. We direct the reader to the appropriate literature [41].

An alternative approach is to use DFT theory where the dispersion contribution has been introduced as a correction, as in the B97-D method by Grimme [42]. For example, using this method Nagy and Erhardt have investigated the association constant of acetate anions with protonated amines as model systems for the study of the most stable form (ion-pair or hydrogen bonded complex) in the binding cavity of a receptor in contact with the aqueous environment [43].

Hydrogen bonded complexes were also investigated by Arey *et al* [44]. The Authors investigated a set of complexes including simple molecules such as H_2O, NH_3, DMSO, CH_3OH and others, as well as guanine-cytosine and thymine-adenine pairs. Inclusion of dispersion-corrected atom-centered potentials were found to significantly improve the description of H-bonded systems: with the BLYP functional a mean signed error of 0.010 Å in the hydrogen bond distance and a mean relative error of 5.1% in the interaction energy was found, compared to benchmark high level ab-initio calculations.

Much larger discrepancies of 0.036 Å and 15.9%, respectively, were found for the BLYP functional without the inclusion of dispersion terms.

However, the real benchmark for computational methods is, indeed, in the so-called π–π stacking interactions. Thus DFT have been tested rigorously in recent years concerning this issue. π–π interactions of dimers of linear acenes and the corresponding all-trans all-anti saturated compounds have been investigated by Grimme [5] using the dispersion-corrected B2PLYP-D [45] with energy-minimized structures at the B97-D level, with very large basis sets. Interestingly, it was pointed out that very similar interactions exists between dimers, either aromatic or aliphatic, providing they have a small number of phenyl or cyclohexyl units (1 or 2, as in the benzene and naphthalene dimers compared with the cyclohexane and bicyclohexane dimers). In contrast, special non-local electron correlations between the π electrons in the two monomers exist for larger systems and they are responsible for the different stacking properties of aromatic and aliphatic systems, see Figure 4.6.

Mackie and DiLabio have recently presented a protocol for the calculation of the interaction energy of H-bonded and van der Waals dimers based on standard DFT and small basis sets, [46] thus very appealing for the calculation of the structure and energy of large systems. Dispersion correcting potentials, in conjunction with counterpoise corrections, can be easily implemented in standard software packages in the same way as atomic pseudo-potentials are used. Mean absolute deviations of about 14%

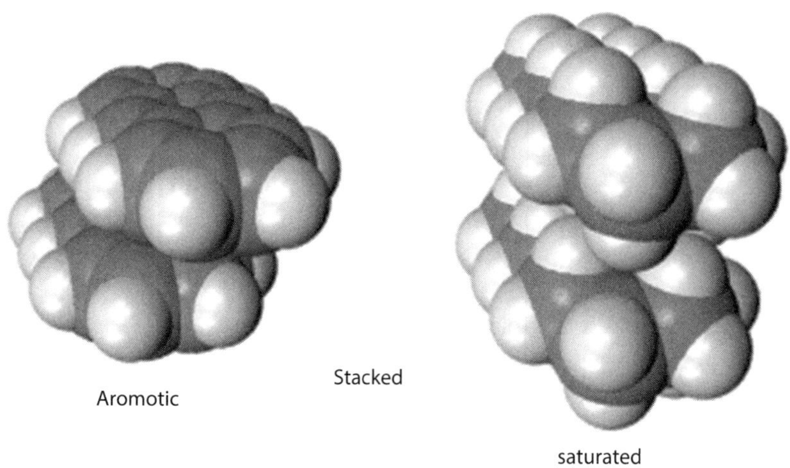

Figure 4.6 Aromatic and aliphatic dimers investigated in Ref. [5] Reproduced from Ref. [5] with permission from Wiley.

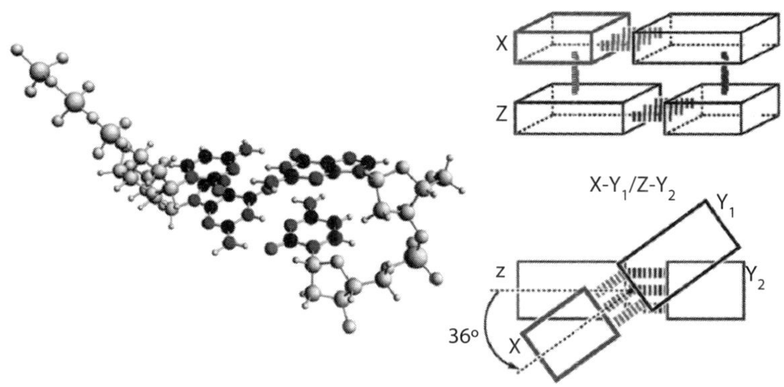

Figure 4.7 Model complex of DNA investigated in Ref. [48] Reproduced from Ref. [48] with permission from the Royal Society of Chemistry.

were observed at the B971/6–31G(d)-DCP with respect to the energies calculated with high-level ab-initio methods in the S22 reference set developed by Hobza and co-workers [47].

Bickelhaupt and co-workers investigated, by dispersive-corrected DFT (BP86-D/TZ2P), the many factors responsible for the high fidelity in DNA replication, see Figure 4.7 [48].

The computational study revealed the important role of hydrogen-bonding, as already known, but also the cooperative interplay between hydrogen-bonding and solvent effects. Stacking interactions have been found to be less important in determining the base pairing matching, although they play a fundamental role in the overall stability of the system.

An interesting application of dispersion-corrected DFT protocols to large systems has been recently presented by Grimme [49].

Several host-guest and supramolecular complexes in various solvents were investigated (some of them are reported in Figure 4.8) ranging from cases dominated by dispersive interactions to charged complexes in water. Solvation effects were treated within a continuum formalism. The free energies of complexation were found to be in very good agreement with experimental data, within 2 kcal/mol, on average. It should be noted that the final ΔG values generally resulted from a large compensation of opposite gas-phase ΔE and entropic and solvation effects. Therefore accurate evaluation of dispersion terms is necessary.

A significant work has been done recently using the M06 family of functionals developed by Zhao and Truhlar [34–38]. These functionals have been specifically tailored for the description of non-covalent interactions.

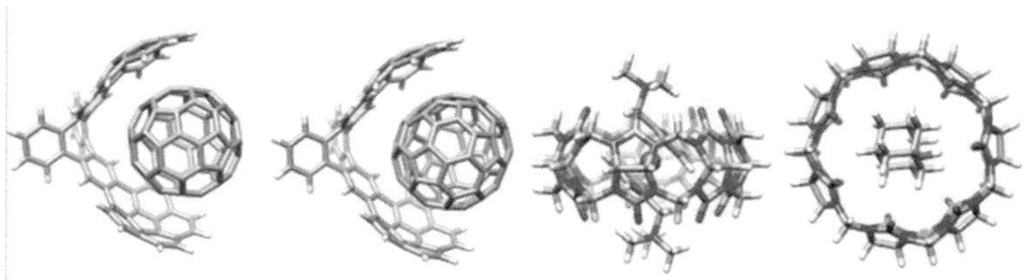

Figure 4.8 Some of the host-guest supramolecular systems investigated by Grimme in Ref. [49]. Adapted from Ref. [49] with permission from Wiley.

The performance of such DFT methods has been tested for real systems and a very good agreement with experimental data has been found. The geometry and binding energies of buckyball tweezers $C_{60}H_{28}$ and supramolecular complexes with C_{60} fullerene have been investigated at the M06-2X/6-31+G(d,p)//M06L/Midi! [50]. The same level of theory was used by Casella and Saielli to study the supramolecular complexes of C_{60} and C_{70} fullerenes with buckybowls; after a proper inclusion of solvent effects and enthalpic and entropic terms a very good agreement with the experiments was found [51]. In the same paper the Authors also explored the complexation of and C_{60} fullerene with the reduced and oxidized form viologen dimers, see Figure 4.9:

It was found that the reduced form of the complex is composed of a reduced viologen dimer and a neutral C_{60} fullerene with a negative complexation free energy, ΔG, of -7.7 kcal/mol, while the free energy of complexation of C_{60} by the oxidised viologen dimer is positive (+12.5 kcal/mol). This suggests the possibility of using viologen dimers as redox switches for fullerene complexation/release in solution.

Similar complexes between corannulene, sumanene, and pentaindenocorannulene and C_{60} and C_{70} fullerenes were studied at the M06-2X/6-311G level [52] while in Ref. [53] Josa *et al.* investigated substituent effects in corannulene dimers by several DFT methods including M06-2X, B97-D and ωB97X-D. Though electronwithdrawing and electrondonor substituents had little impact on the structure and curvature of the monomers a significant effect was observed for the interaction energies. The effect of the curvature on the interaction energy of buckyplates and buckybowls has been investigated in Ref. [54] using several DFT functionals.

Finally, double hybrid density functionals (DHDFs) have been applied to the calculation of interaction energies in system of biological relevance obtaining good results [55].

The overall picture that emerges from these data, which are by no means exhaustive of the literature on the subject but serve simply to establish a

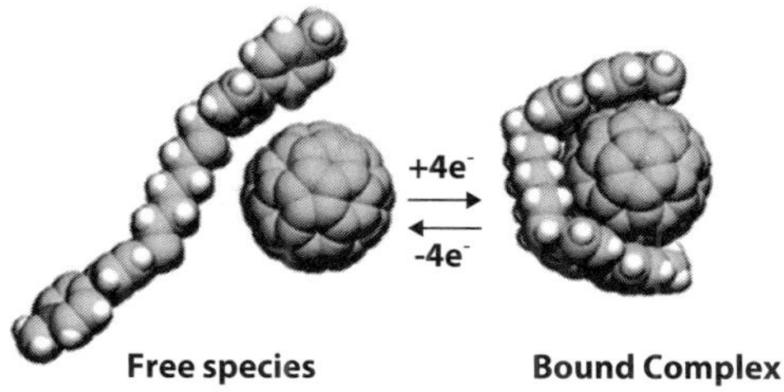

Figure 4.9 Complexes of (left) oxidized and (right) reduced viologen dimer with C_{60} fullerene. Reproduced from Ref. [51] with permission from the Royal Society.

proof of principle, is that nowadays very effective computational protocols, based on DFT methods, are available for the description of large systems interacting through non-covalent interactions. Both hydrogen-bonded systems as well as those dominated by π–π stacking can be well described at a reasonable computational cost even if containing few hundreds of atoms.

What for larger systems? What for fluid phases like liquid crystals, biological membranes, zeolites, proteins, cellulose? This will be the subject of the next Chapter.

4.6 Classical Force Fields

Up to this point we have discussed methods that explicitly include the electronic degrees of freedom. The description of the electrons requires the use of quantum chemistry, either in the form of *ab-initio* methods or as DFT protocols. A first step towards a higher level of coarse-graining is provided by classical molecular dynamics (MD) Force Fields (FF) [56]. A FF is a set of parameters that describes a molecule after the electronic degrees of freedom have been averaged out. What remains is a set of nuclei each with a given mass. The total energy is decomposed in a pair-wise additive fashion: the role of the electrons is now represented in different ways, depending whether the two atoms we consider are separated by one bond (bonds), two bonds (angles), three bonds (dihedrals) or more. For bonds and angles usually a harmonic potential is used, while torsional angles are described by periodic functions. For interactions between atoms separated by more than three bonds, thus also including interactions between atoms of different molecules, an interaction potential of Lennard-Jones type and the electrostatic contributions (if the atoms bear a charge) are considered. The general expression of the interaction energy is as in Eq. 4.13

$$U = \sum k_b (r - r_0)^2 + \sum k_a (\theta - \theta_0)^2 + \sum V_n \cos(2n\pi\phi) + \sum\sum V_{LJ} + \sum\sum V_{EL} \tag{4.13}$$

The first term represents the harmonic interaction between bound particles, having an equilibrium distance r_0 and a harmonic force constant k_b; the second is the similar term for the bending of the angles, with the equilibrium value θ_0 and the harmonic constant k_a. The third one is the periodic potential of torsional angles φ, with amplitudes V_n (n=1–3). Finally there is a sum over all pairs describing the van der Waals interaction (by a LJ potential as in Eq. 4.7) and the electrostatic contribution.

Since the atoms have a much larger mass than the electrons their dynamics can be described, at a reasonable level of accuracy, by classical mechanics, without the need to recur to quantum chemistry. This is a huge saving in computational time, thus not only molecules with tens to hundreds of atoms can be considered: instead, after applying proper periodic boundary conditions to avoid surface effects, systems of the order of 10^5–10^6 particles can be simulated for time scales ranging from ns to μs. This allows the description of the phase behavior of complex fluids such as ionic liquids, liquid crystals, biological membranes, micelles, etc. as well as the investigation of the mechanism of action of large biopolymers, such as proteins and DNA and the morphology and conduction properties of polymeric membranes. The system dynamics is solved, obeying to the Newton equations of motions, by using several algorithms after discretization of the time. The step size used for the integration, δt, must be chosen the longest possible that still allows a correct integration of the fastest degrees of freedom of the system [57].

If all the atoms are explicitly described by the FF, including the hydrogen atoms, the FF is said to be fully atomistic. Starting from here a large number of FFs can be generated by coarse-graining the description of the molecule even more, that is removing some degrees of freedom which are not significantly contributing to the property under consideration. Essentially, the "mapping" of the fully atomistic FF into a Coarse-Grained one (CG) consists in replacing groups of atoms by a single "superatom" or "bead". The CG particle is often spherical, but other shapes can be used, e.g. ellipsoids of spherocylinders. In the United Atom (UA) approach all methyl and methylene hydrogens are combined in a single particle, as in the Optimized Potential for Liquid Simulation (OPLS-UA) [58], though an All-Atomistic version fo the same FF has been presented later (OPLS-AA); [59] higher level of coarse-graining can be reach by merging together more atoms, such as entire phenyl rings, imidazolium rings, or even by considering single anisotropic particles, such as in the Gay-Berne model, as representative of complex liquid crystal molecules. Several procedures to develop a CGFF can be used such as Iterative Boltzmann Inversion (IBI), [60], fitting to some relevant experimental data as in the MARTINI FF, [61] or Force Matching [62]). A step further is the coarse graining of more than one molecule into a single particle. The latter approach is particularly useful in describing the solvent. Integration time steps, therefore the total length of the simulation, can vary significantly, from less than a *fs* for fully atomistic FFs up to tens of *fs* for coarse-grained FFs [56].

In this chapter we will focus in some recent applications of several types of FFs for the description of membranes and similar systems.

Fully atomistic FFs for the simulation of biological membranes have been developed recently. A consistent FF for several types of phosphatidylcholine and phosphatidylethanolamine lipids, called Slipids (Stockholm lipids) has been reported by Jämbeck and Lyubartsev [63]. The Slipid FF overcomes some flaws that have been found in several previous FFs used for the simulation of biological membranes. Of particular importance is the good performance of the Slipid FF concerning the reproduction of experimental NMR deuterium order parameters and X-ray and neutron scattering form factors. As pointed out by the Authors, such experimental data are very sensitive probes to test the goodness of a given FF. Order parameters of the C-D bonds obtained from NMR experiments (C-H bonds in the simulations) are very well reproduced over the entire chain for both *sn*-1 and *sn*-2 chains of DOPC (1,2-dioleoyl-*sn*-glycero-3-phosphocholine), POPC (1-palmitoyl-2-oleoyl-*sn*-glycero-3-phosphocholine) and POPE (1-palmitoyl-2-oleoyl-*sn*-glycero-3-phosphoethanolamine) lipid bilayers. The experimental observation that the *sn*-1 chain lies in a more perpendicular orientation with respect to the bilayer normal has been correctly reproduced by the simulations.

It is of interest to study not only the structure and dynamics of membranes but also the way nanosized objects penetrate the bilayer. In particular carbon nanotubes have received recently a great deal of attention both as drug-delivery systems as well as for their potential toxicity. In Ref. [64] Kraszewski *et al.* have investigated the permeation of carbon nanotubes using atomistic MD simulations. Their results show that the functionalized nanotubes penetrate the membrane by a passive diffusive mechanism (nanoneedle mechanism). A snapshot is shown in Figure 4.10.

Another important issue concerning biological membranes is the mechanism of action of anesthetics, which is still not completely understood. One hypothesis is connected with the structure and order of cell

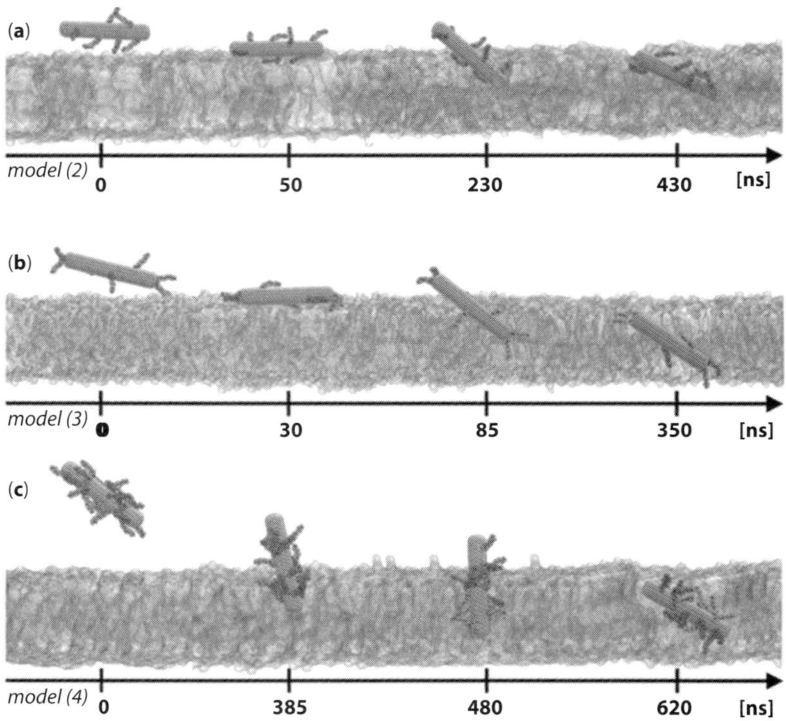

Figure 4.10 Nanoneedle penetration of carbon nanotubes into a lipid bilayer studied by MD simulations. Reprint with from Ref. [64] with permission. Copyright © 2012 Kraszewski *et al.* This is an open-access article distributed under the terms of the Creative Commons Attribution License, which permits unrestricted use, distribution, and reproduction in any medium, provided the original author and source are credited.

membranes when an anesthetic is dissolved in the lipid bilayer, since, as a rule of thumb, the solubility of a compound in oil is directly correlated with its strength as an anesthetic [65]. Another possibility considers the direct interaction of the general anesthetic with membrane proteins. Moreover, a "pressure reversal" occurs in general anesthesia, that is the anesthetic effect is lost under high pressure. Many details of the mechanism need to be addressed: how the thickness and surface area of a membrane change under the presence of general anesthetic; how the order changes; what is the distribution of anesthetic molecules along the bilayer normal; how all these properties depend on the pressure.

Darvas *et al.* [66] have investigated the effect of four different anesthetics (chloroform, halothane, diehtylether and enflurane) on the structural properties of a lipid bilayer of DPPC (dipalmitoylphosphatidylcholine) using a United Atom FF where methyl and methylene groups are considered as a

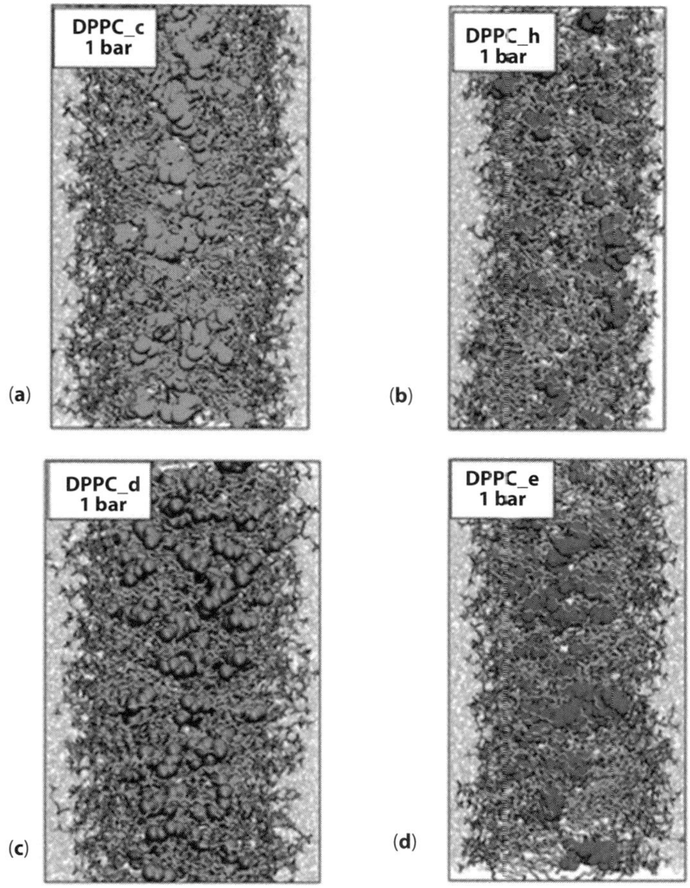

Figure 4.11 Distribution of four common anesthetics in a lipid bilayer studied by MD simulations. (a) chloroform; (b) halothane; (c) diethyl ether; (d) enflurane. Reproduced with from Ref. [66] with permission from the Royal Society.

single particle, that is a light coarse-graining compared to a fully atomistic model. The Authors performed long simulations of the order of 35 ns for each of the five system (a hydrated bilayer of 256 DPPC molecules with one of the four anesthetics plus the free DPPC bilayer for reference). As can be observed in Figure 4.11 a significantly different distribution of the four anesthetics in the DPPC membrane was observed. Moreover, order parameters were calculated and found to be in good agreement with experiments.

In Ref. [65] Yamamoto *et al.* have studied the effect of Xenon, another well known anesthetic, dissolved in a lipid bilayer of POPE using MD simulations.

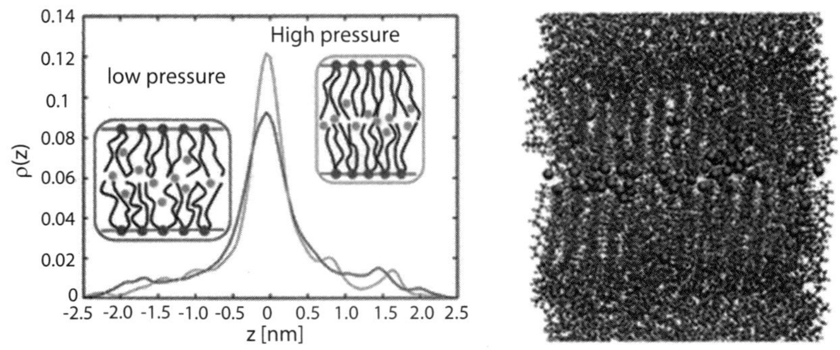

Figure 4.12 (Left) Density profile of xenon in a lipid bilayer and (right) snapshot of the MD simulation. Reproduced from Ref. [65] with permission of The American Chemical Society.

Again the distribution of xenon atoms in the bilayer was investigated at ambient and high pressure to highlight the differences in membrane structure and the pressure reversal phenomenon, see Figure 4.12.

The results of these simulations are helping to further our understanding of the molecular mechanism of general anesthesia since, apart from the coarse-graining of the hydrogen atoms of methyl and methyelene groups, all the other relevant degrees of freedom of the system are maintained.

Although atomistic simulation are extremely useful they cannot be employed to study large systems for very long times. Therefore coarse-graining procedures have been introduced, as mentioned already, to remove, or average out, some fast degrees of freedom which are deemed not essential for the description of the specific problem at hand. These methods usually consist in replacing several atoms with a single site representing the average size, interaction and dynamics of the original particles.

Izvekov and Voth have proposed a coarse-graining procedure based on a "force matching" approach [62]. This ensures the possibility of extending the FF to multiscale simulations. The level of CG is significantly higher than in a United Atom approach, as can be seen in Figure 4.13. Several heavy atoms, in fact, are grouped into a single site. The proposed FF has been applied to the simulation of structural properties of dimyristoylphosphatidylcholine (DMPC) lipid bilayer obtaining a good agreement with experimental data of structural properties. In Figure 4.13 we can clearly see the mapping procedure used to replace three-four heavy atoms with a single interaction site.

A similar level of coarse-graining is also used in a popular CGFF for simulation of biological membranes, the MARTINI Force Field [61].

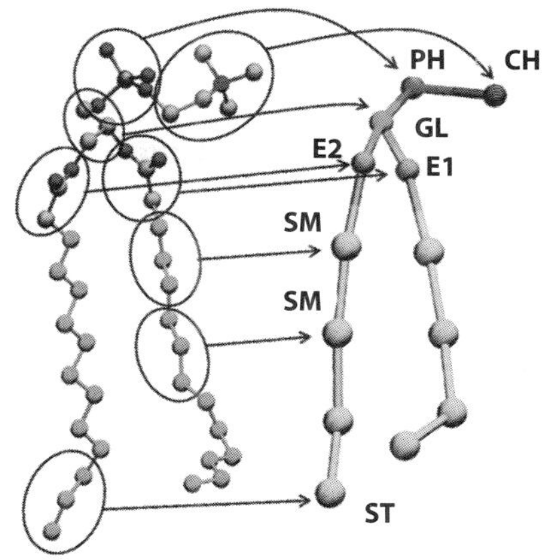

Figure 4.13 Mapping scheme of an atomistic Force Field onto the Coarse-Grained Force Field for a lipid molecule. Reproduced from Ref. [62] with permission from the American Chemical Society.

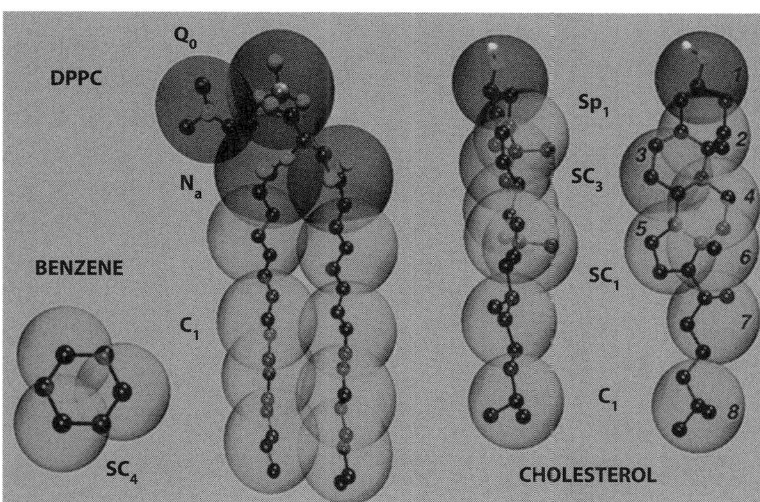

Figure 4.14 Schematic representation of the MARTINI FF. Reproduced from Ref. [61] with permission from the American Chemical Society.

As can be seen in Figure 4.14, on average four heavy atoms are replaced by a single particle. The MARTINI procedure of coarse-graining is mostly based on a parameterization of the FF in order to reproduce thermodynamic quantities with a special emphasis to the partition coefficient between oil and water. Four main interaction sites are defined: polar (P), nonpolar (N), apolar (C) and charged (Q). The FF has been used for several studies concerning membranes and surfactant molecules. It has also been extended considerably to include new types of systems. For example in Ref. [67] Sergi *et al.* have developed a suitable parameterization of graphene in order to study the wetting properties of graphitic surfaces with mixtures of water with several surfactants. The Authors found that small concentrations of surfactants, of about 1%, are responsible for sizable reduction of the contact angle of the droplet.

Another interesting application of the MARTINI CGFF has been reported in Ref. [68] Sangwai and Sureshkumar have investigated the shape transition from spherical to rod-like micelles, see Figure 4.15, for a system composed of cetyltrimethylammonium chloride in water containing varying concentrations of other salts, such as sodium salicylate and sodium chloride. The simulations offered a clear view at the microscopic level of the effect of amphiphilic salicylate anions: the sphere-to-rod transition was induced after the micelle corona was saturated with salicylate thus resulting in a decrease of the micelle-water interfacial tension.

Nafion is one of the most popular polyelectrolite commonly used in proton exchange membranes (PEM) in fuel cells. It is a random copolymer composed by two monomers: the non-polar tetrafluoroethylene unit [-CF_2-CF_2-] and the polar pefluorosulfonic vinylether [-CF_2-CF(O-CF_2-CF(CF_3)-O-CF_2-CF_2-SO_3H)] unit. Therefore the membrane consists in a Teflon backbone with randomly attached hydrophilic chains and has a significant proton conductance, as needed in fuel cell membranes. However the nano-sized structure of this macroscopically amorphous system, in particular hydrated Nafion, is still debated and several different morphological models have been proposed in the literature. Atomistic MD

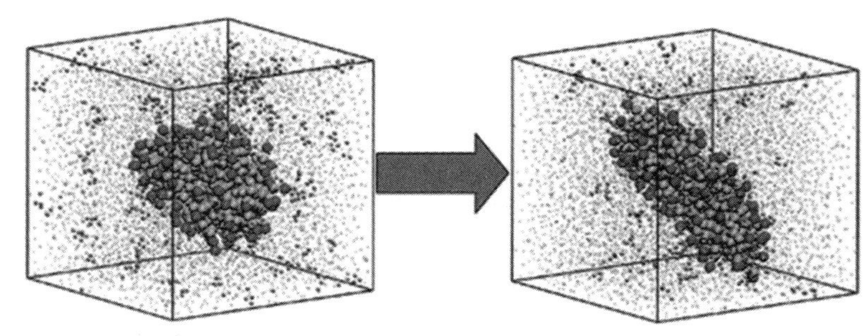

Figure 4.15 Sphere to rod transition of surfactant micelles studied by MD simulations. Reproduced from Ref. [68] with permission from the American Chemical Society.

simulations have been reported by Knox and Voth [69]. Very large systems of about 2 million atoms in a box of about 30 nm side have been studied for long times, of the order of tens of ns. The systems were prepared in a given nano-sized aggregation state and relaxed, though a full equilibration of the system was not possible.

The large size of the system investigated allowed a meaningful description of the long-range nano-aggregates that can, in principle, be present in Nafion membranes and the comparison of calculated vs experimental scattering spectra. These latter experimental observations have been used to support morphological models of Nafion; however all non-random aggregation motifs investigated by the MD simulations, see Figure 4.16 for an example, revealed the characteristic scattering peaks found experimentally thus suggesting that such techniques are not sensitive to the details of the Nafion structure.

The study was then extended to a multiscale approach, thus removing some of the limitations of the atomistic simulation [70]. A mesoscale level simulation was obtained by combining together a coarse-grained description of the membrane with a dissipative particle dynamics and smoothed particles hydrodynamics to capture the proton transport properties. The method allows to use as input parameters those derived from atomistic simulations, thus offering a bridge between a detailed molecular description and the long time scales and large sizes required to model the membrane behaviour.

A different PEM has been investigated in Ref. [71], that is a hydrated crossed-linked and sulfonated poly(1,3-cyclohexadiene), having an improved proton conductivity, compared to Nafion, at high temperatures.

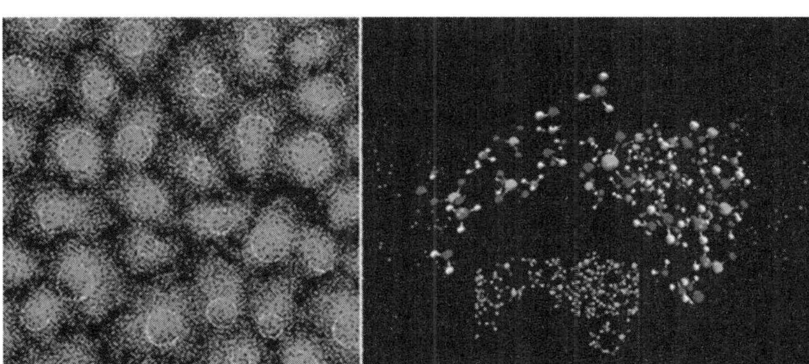

Figure 4.16 (Left) inter-cluster connecting bridge formation and (right) representative bridge between clusters in a coarse-grained model Nafion membrane. Reproduced from Ref. [69] with permission from the American Chemical Society.

Also in this case a multiscale approach was followed: atomistic simulations (where hydrogens were merged with carbons in CH_n groups) were first run to derive a CG potential through the Iterative Boltzmann Inversion method. The transferability of the CG model potential was also tested by varying the water content of the membrane, but poor transferability was observed.

Another common polymeric membrane is polystyrene. A comparison of various CGFF for polystyrene, developed following different mapping schemes and having a different level of coarse-graining, see Figure 4.17, has been recently reported [72]. The Authors have compared the performance of the FFs and their ability to reproduce several properties, such as

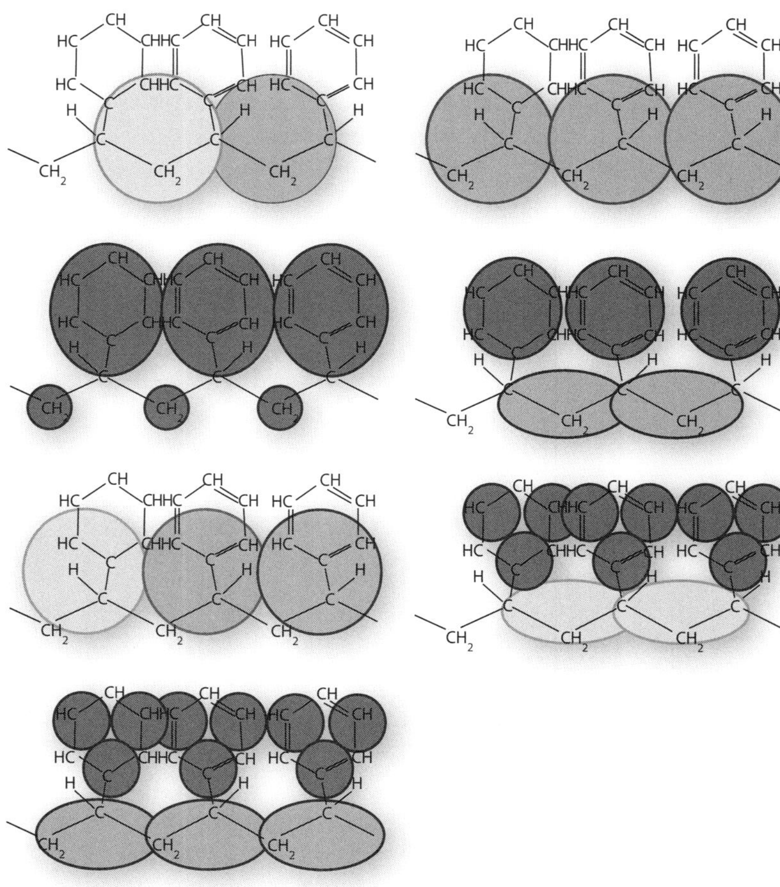

Figure 4.17 Schematic representation of the various coarse-grained FF for polystyrene. Reproduced from Ref.[72] with permission of John Wiley and Sons.

structural properties of the melt, tacticity, properties under shear, properties in solution and dynamical properties.

As highlighted by the authors, the choice of the CG mapping scheme have an impact on the accuracy of the properties that one wishes to reproduce. In particular, the transferability to state points (that is temperature, pressure, composition) different from that one used for the coarse-graining was found to be poor and difficult to predict.

The issue of transferability was also discussed in Ref. [73]. Mukherjee *et al.* have investigated the transferability of a CGFF for a liquid crystal compound, dioctyloxyazabenzene. If the CGFF is obtained by a mapping in a given phase, e.g., isotropic, then the formation of a mesophase is not observed on cooling the temperature. Therefore the Authors selected, as state point for the coarse-graining procedure, a super-cooled liquid, with no long-range orientational order but with local structures arranged as in the LC phase, see Figure 18. The derived CGFF was observed to be capable of exhibiting both isotropic and smectic A phases as a function of the temperature. Due to the presence of an azo group in the molecule this compound may be used in a photo-switchable device: the presence of an amount of *cis* isomer destabilize the LC phase in favour of the isotropic phase.

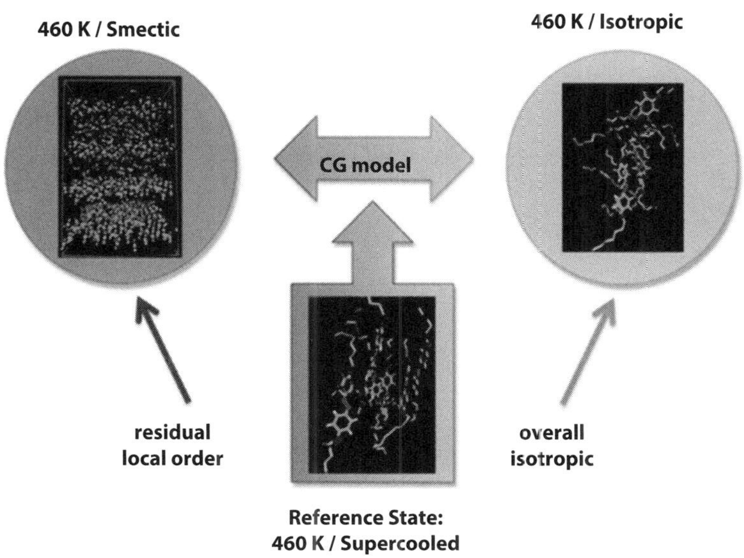

Figure 4.18 Developing of a transferable coarse-grained FF for the simulations of phase transitions in liquid crystals. Reproduced from Ref. [73] with permission from The American Chemical Society.

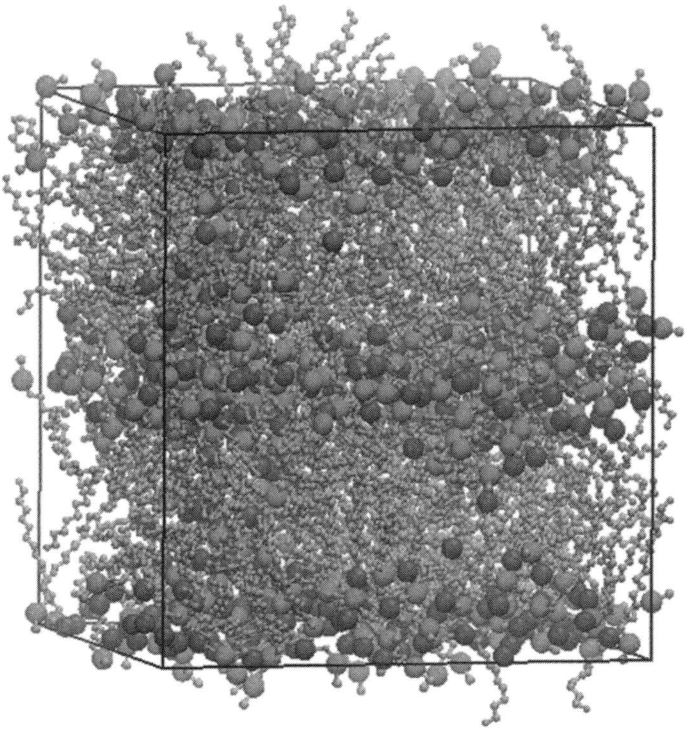

Figure 4.19 Ionic SmA phase exhibited by a coarse-grained FF developed for short chain ionic liquids. Reproduced from Ref. [76] with permission from the Royal Society.

In contrast, for ionic liquid crystals, where smectic phases are formed essentially by micro-segregation between alkyl chains and ionic layers, it was found that an CGFF developed for the description of short chain imidazolium ionic liquids [74, 75] was capable of forming a stable ionic mesophase, see Figure 4.19 [76].

A much higher level of coarse-graining was proposed by Voth and coworkers to study protein-protein interactions (PPI) [77]. The electrostatic induced self-association phenomena at high protein concentrations are responsible for a different dynamical behavior of two therapeutic monoclonal antibody. The systems, however, are extremely large, of the order of 10^5 atoms, and at least a thousand particles should be simulated, in a large box of the order of 10^3 A, to have a reasonable representation of the structure of the aggregates.

To be able to simulate such large scales, not only the description of the atomic details needs to be dropped down, but also the primary and even

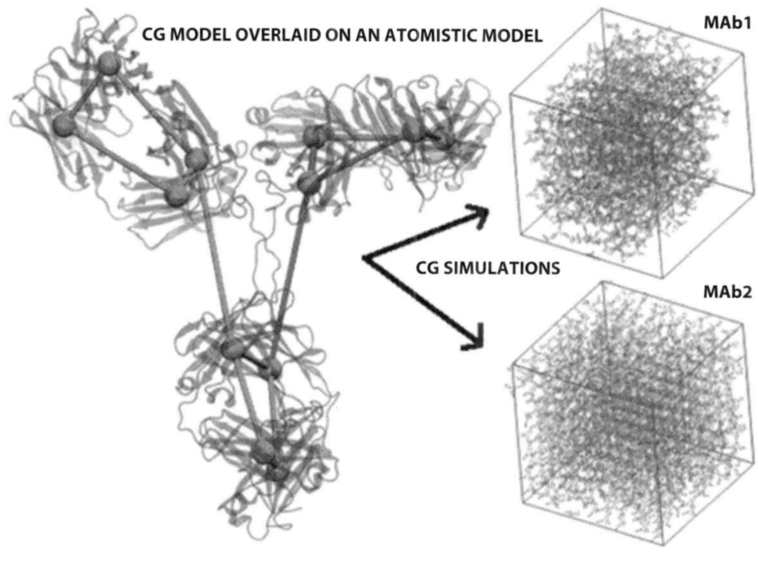

Figure 4.20 Schematic representation of the mapping scheme of a coarse-grained model of a monoclonal antibody. Reproduced from Ref. [77] with permission from the American Chemical Society.

the secondary structure of the polypeptide cannot be accounted for, see Figure 4.20. Nevertheless, the careful choice of the relevant degrees of freedom, here low modes of vibration connected to the dynamics of each subunit, allowed a meaningful description of the aggregation behavior of two monoclonal antibody, providing an explanation of their different behavior, in spite, or better say, thanks, to the simplicity of the representation.

4.7 Conclusions

In this Chapter we have reviewed and described the recent methods available to investigate, by means of computational protocols, weak interactions driving self-assembly and nano-aggregation in membranes and polymeric systems. The overview has covered a large time and space scale: from accurate description by ab-initio methods of small dimers dominated by dispersive interactions, to larger aggregates where DFT can give a very valuable results at a relatively low computational cost, up to MD simulations, first atomistic, then coarse-grained, used to investigate protein motion and aggregation and structure and function of membranes and polymers.

Acknowledgement

The Author thanks MIUR (PRIN 2010N3T9M4, FIRB RBAP11C58Y) and Fondazione CARIPARO (Progetti di Eccellenza "Nano-Mode" 2010) for financial support.

References

1. B. M. Cossairt, C. C. Cummins, A. R. Head, D. L. Lichtenberger, R. J. F. Berger, S. A. Hayes, N. W. Mitzel and G. Wu, *Journal of the American Chemical Society,* Vol. 132, p. 8459–8465, 2010.
2. B. Persson, P. Taylor and T. Lee, *Journal of Chemical Physics,* Vol. 107, p. 5051–5057, 1997.
3. I. G. Kaplan, *Intermolecular Interactions*, Wiley, Chichester, 2006.
4. M. O. Sinnokrot and C. D. Sherrill, Vol. 110, p. 10656–10668, 2006.
5. S. Grimme, *Angewandte Chemie International Edition,* Vol. 47, p. 3430–3434, 2008.
6. M. Pitonak, P. Neogrady, J. Rezac, P. Jurecka, M. Urban and P. Hobza, *Journal of Chemical Theory and Computation,* Vol. 4, p. 1829–1834, 2008.
7. T. Janowski and P. Pulay, *Chemical Physics Letters,* Vol. 447, p. 27–32, 2007.
8. C. A. Hunter, J. Singh and J. M. Thornton, *Journal of Molecular Biology,* Vol. 218, p. 837–846, 1991.
9. S. Burley and G. Petsko, *Science,* Vol. 229, p. 23–28, 1985.
10. M. Bates and G. Luckhurst, *Journal of Chemical Physics,* Vol. 110, p. 7087–7108, 1999.
11. R. Berardi, A. P. J. Emerson and C. Zannoni, *Journal of the Chemical Society, Faraday Transactions,* Vol. 89, p. 4069–4078, 1993.
12. M. A. Bates and G. R. Luckhurst, *Liquid Crystals,* Vol. 24, p. 229–241, 1998.
13. G. A. Jeffrey, *An Introduction to Hydrogen Bonding (Topics in Physical Chemistry)*, Oxford University Press, USA, 1997.
14. M. Nishio, M. Hirota and Y. Umezawa, *The CH/π Interaction*, Wiley, Chichester, 1998.
15. M. Saggu, N. M. Levinson and S. G. Boxer, *Journal of the American Chemical Society,* Vol. 134, p. 18986–18997, 2012.
16. P. Metrangolo, F. Meyer, T. Pilati, G. Resnati and G. Terraneo, *Angewandte Chemie International Edition,* Vol. 47, p. 6114–6127, 2008.
17. H. Torii and M. Yoshida, *Journal of Computational Chemistry,* Vol. 31, p. 107–116, 2010.
18. C. A. Hunter and J. K. M. Sanders, *Journal of the American Chemical Society,* Vol. 112, p. 5525–5534, 1990.
19. F. Cozzi, M. Cinquini, R. Annuziata and J. S. Siegel, *Journal of the American Chemical Society,* Vol. 115, p. 5330–5331, 1993.

20. M. O. Sinnokrot and C. D. Sherrill, *Journal of the American Chemical Society*, Vol. 126, p. 7690–7697, 2004.
21. C. F. R. A. C. Lima, M. A. A. Rocha, L. R. Gomes, J. N. Low, A. M. S. Silva and L. M. N. B. F. Santos, *Chemistry ? A European Journal*, Vol. 18, p. 8934–8943, 2012.
22. D. Kuchenbecker and G. Jansen, *ChemPhysChem*, Vol. 13, p. 2769–2776, 2012.
23. A. Szabo and N. S. Ostlund, *Modern Quantum Chemistry: Introduction to Advanced Electronic Structure Theory*, Dover Publication, 1996.
24. P. Hobza, *Accounts of Chemical Research*, Vol. 45, p. 663–672, 2012.
25. M. Katouda and S. Nagase, *International Journal of Quantum Chemistry*, Vol. 109, p. 2121–2130, 2009.
26. M. Pitonák, P. Neogrády, J. Cerný, S. Grimme and P. Hobza, *ChemPhysChem*, Vol. 10, p. 282–289, 2009.
27. K. E. Riley, J. Rezac and P. Hobza, *Physical Chemistry Chemical Physics*, Vol. 13, p. 21121–21125, 2011.
28. S. Grimme, *Journal of Chemical Physics*, Vol. 118, p. 9095–9102, 2003.
29. T. Takatani, E. G. Hohenstein and C. D. Sherrill, *Journal of Chemical Physics*, Vol. 128, p. 124111, 2008.
30. A. Hesselmann, *Journal of Chemical Physics*, Vol. 128, p. 144112, 2008.
31. A. Misquitta, B. Jeziorski and K. Szalewicz, *Physical Review Letters*, Vol. 91, p. 033201, 2003.
32. A. Hesselmann and G. Jansen, *Chemical Physics Letters*, Vol. 367, p. 778–784, 2003.
33. K. Sadeghian, M. Bocola and M. Schütz, *ChemPhysChem*, Vol. 12, p. 1251–1254, 2011.
34. Y. Zhao and D. Truhlar, *Journal of Chemical Theory and Computation*, Vol. 1, p. 415–432, 2005.
35. Y. Zhao and D. G. Truhlar, *Journal of Chemical Theory and Computation*, Vol. 3, p. 289–300, 2007.
36. Y. Zhao and D. G. Truhlar, *Theoretical Chemistry Accounts*, Vol. 120, p. 215–241, 2008.
37. Y. Zhao and D. G. Truhlar, *Journal of Chemical Physics*, Vol. 125, p. 194101, 2006.
38. Y. Zhao and D. G. Truhlar, *Journal of Chemical Theory and Computation*, Vol. 3, p. 289–300, 2007.
39. G. Saielli, *The Journal of Physical Chemistry A*, Vol. 114, p. 7261–7265, 2010.
40. M. Gupta, E. F. da Silva and H. F. Svendsen, *Journal of Physical Chemistry B*, Vol. 116, p. 1865–1875, 2012.
41. C. Curutchet, M. Orozco, F. J. Luque, B. Mennucci and J. Tomasi, *Journal of Computational Chemistry*, Vol. 27, p. 1769–1780, 2006.
42. S. Grimme, *Journal of Computational Chemistry*, Vol. 27, p. 1787–1799, 2006.
43. P. I. Nagy and P. W. Erhardt, Vol. 116, p. 5425–5436, 2012.
44. J. S. Arey, P. C. Aeberhard, I. Lin and U. Rothlisberger, *The Journal of Physical Chemistry B*, Vol. 113, p. 4726–4732, 2009.

45. T. Schwabe and S. Grimme, *Physical Chemistry Chemical Physics,* Vol. 9, p. 3397–3406, 2007.
46. I. D. Mackie and G. A. DiLabio, *Physical Chemistry Chemical Physics,* Vol. 12, p. 6092–6098, 2010.
47. P. Jurecka, J. Sponer, J. Cerny and P. Hobza, *Physical Chemistry Chemical Physics,* Vol. 8, p. 1985–1993, 2006.
48. J. Poater, M. Swart, C. F. Guerra and F. M. Bickelhaupt, *Chemical Communications,* Vol. 47, p. 7326–7328, 2011.
49. S. Grimme, *Chemistry-a European Journal,* Vol. 18, p. 9955–9964, 2012.
50. Y. Zhao and D. G. Truhlar, *Physical Chemistry Chemical Physics,* Vol. 10, p. 2813–2818, 2008.
51. G. Casella and G. Saielli, *New Journal of Chemistry,* Vol. 35, p. 1453–1459, 2011.
52. P. A. Denis, *Chemical Physics Letters,* Vol. 516, p. 82–87, 2011.
53. D. Josa, J. R. Otero and E. M. Cabaleiro Lago, *Physical Chemistry Chemical Physics,* Vol. 13, p. 21139–21145, 2011.
54. M. R. Kennedy, L. A. Burns and C. D. Sherrill, *The Journal of Physical Chemistry A,* Vol. 116, p. 11920–11926, 2012.
55. I. Y. Zhang and X. Xu, *Physical Chemistry Chemical Physics,* Vol. 14, p. 12554–12570, 2012.
56. S. Riniker, J. R. Allison and W. F. van Gunsteren, *Physical Chemistry Chemical Physics,* Vol. 14, p. 12423–12430, 2012.
57. M. P. Allen and D. J. Tildesley, *Computer Simulation of Liquids*, Clarendon Press, Oxford (UK), 1991.
58. W. L. Jorgensen and J. Tirado-Rives, *Journal of the American Chemical Society,* Vol. 110, p. 1657–1666, 1988.
59. W. L. Jorgensen, D. S. Maxwell and J. Tirado-Rives, *Journal of the American Chemical Society,* Vol. 118, p. 11225–11236, 1996.
60. D. Reith, M. Pütz and F. Müller-Plathe, *Journal of Computational Chemistry,* Vol. 24, p. 1624–1636, 2003.
61. S. J. Marrink, H. J. Risselada, S. Yefimov, D. P. Tieleman and A. H. de Vries, *The Journal of Physical Chemistry B,* Vol. 111, p. 7812–7824, 2007.
62. S. Izvekov and G. A. Voth, *The Journal of Physical Chemistry B,* Vol. 109, p. 2469–2473, 2005.
63. J. P. M. Jämbeck and A. P. Lyubartsev, *Journal of Chemical Theory and Computation,* Vol. 8, p. 2938–2948, 2012.
64. S. Kraszewski, A. Bianco, M. Tarek and C. Ramseyer, *PLoS ONE,* Vol. 7, p. e40703, 2012.
65. E. Yamamoto, T. Akimoto, H. Shimizu, Y. Hirano, M. Yasui and K. Yasuoka, *The Journal of Physical Chemistry B,* Vol. 116, p. 8989–8995, 2012.
66. M. Darvas, P. N. M. Hoang, S. Picaud, M. Sega and P. Jedlovszky, *Physical Chemistry Chemical Physics,* Vol. 14, p. 12956–12969, 2012.
67. D. Sergi, G. Scocchi and A. Ortona, *The Journal of chemical physics,* Vol. 137, p. 094904–10, 2012.
68. A. V. Sangwai and R. Sureshkumar, *Langmuir,* Vol. 27, p. 6628–6638, 2011.

69. C. K. Knox and G. A. Voth, *The Journal of Physical Chemistry B,* Vol. 114, p. 3205–3218, 2010.
70. R. Jorn and G. A. Voth, *The Journal of Physical Chemistry C,* Vol. 116, p. 10476–10489, 2012.
71. Q. Wang, N. S. Suraweera, D. J. Keffer, S. Deng and J. Mays, *Macromolecules,* Vol. 45, p. 6669–6685, 2012.
72. H. A. Karimi-Varzaneh, N. F. A. van der Vegt, F. Müller-Plathe and P. Carbone, *ChemPhysChem,* Vol. 13, p. 3428–3439, 2012.
73. B. Mukherjee, L. Delle Site, K. Kremer and C. Peter, *Journal of Physical Chemistry B,* Vol. 116, p. 8474–8484, 2012.
74. Y. Wang, S. Feng and G. A. Voth, *Journal of Chemical Theory and Computation,* Vol. 5, p. 1091–1098, 2009.
75. Y. Wang, W. G. Noid, P. Liu and G. A. Voth, *Physical Chemistry Chemical Physics,* Vol. 11, p. 2002–2015, 2009.
76. G. Saielli, *Soft Matter,* Vol. 8, p. 10279–10287, 2012.
77. A. Chaudhri, I. E. Zarraga, T. J. Kamerzell, J. P. Brandt, T. W. Patapoff, S. J. Shire and G. A. Voth, *The Journal of Physical Chemistry B,* Vol. 116, p. 8045–8057, 2012.

5

Sensing Techniques Involving Thin Films for Studying Biomolecular Interactions and Membrane Fouling Phenomena

Gabriela Diaconu[1,*] and Thomas Schäfer[1,2,*]

[1]*POLYMAT, NanoBioSeparations Group, University of the Basque Country UPV/EHU Donostia-San Sebastián, Spain*
[2]*Ikerbasque, Basque Foundation for Science, Bilbao, Spain*

Abstract
Interfacial phenomena are ubiquitous in sensing and separation applications involving membranes. In fact, their presence might strongly determine the efficiency of the overall sensing or separation system, if not even deteriorate its performance with time as is the case in membrane separations (fouling). Characterization of these phenomena at the earliest possible stage is therefore indispensable in order to minimize their detrimental impact. This chapter discusses two advanced surface characterization techniques for this purpose, quartz crystal microbalance with dissipation monitoring (QCM-D) and surface plasmon resonance (SPR), and how they can be employed to provide valuable and complementary information on molecular adsorption phenomena occurring at a very early stage on polymer membranes.

Keywords: Quartz crystal microbalance with dissipation monitoring, surface plasmon resonance, membrane fouling, interfacial phenomena

5.1 Introduction

Most separation processes are dominated by phenomena occurring at an interface rather than the macroscopic bulk conditions under which they operate. In order to characterize such phenomena, surface characterization techniques are required that are not only sufficiently sensitive to detect

**Corresponding author*: thomas.schafer@ehu.es, gabriela.diaconu@ehu.es

Annarosa Gugliuzza (ed.) Smart Membranes and Sensors, (145–160) 2014 © Scrivener Publishing LLC

or monitor the respective phenomena, but that also exhibit a versatility and robustness which allow applying different operating conditions. This chapter provides an overview of the quartz crystal microbalance with dissipation monitoring (**QCM-D**) and surface plasmon resonance (**SPR**) techniques from application and material characterization point of view including a brief description of both techniques, and identification of some areas of application such as their use for the characterization of membrane-related adsorption phenomena.

5.2 Quartz Crystal Microbalance with Dissipation Monitoring (QCM-D)

Quartz crystal microbalance with dissipation monitoring, **QCM-D** is a highly sensitive technique that utilizes acoustic waves generated by oscillating a piezoelectric, single crystal quartz plate. QCMs have been used for decades in order to measure the adsorption of compounds on the quartz sensor surface, possibly modified by a thin selective deposition layer, through mass effects. The heart of the QCM instrument is a piezoelectric AT-cut quartz crystal sandwiched between two metal electrodes. The quartz crystal oscillates at its resonant frequency when an alternating electric field is applied (Figure 1A). This frequency is monitored. Any adsorption of molecules onto the crystal surface as a thin film will increase the mass on the sensor surface, thus resulting in a damping of the frequency monitored. If the adsorbed film is thin, rigid and evenly distributed over the surface, then the conditions are given under which Sauerbrey [1] demonstrated the extremely sensitive nature of the piezoelectric crystal towards mass changes at the surface of the QCM electrodes. He correlated the changes of the oscillation frequency with the mass adsorbed by what is nowadays widely referred to as the Sauerbrey equation:

$$\Delta m = -\frac{C}{n}\Delta F \tag{5.1}$$

where Δm is the adsorbed mass, ng/cm^2,

C is the sensitivity factor for the quartz crystal, 17.7ng/cm^2·Hz for a 5MHz quartz crystal, and

n is the overtone number, n = 1, 3, 5,7, 9, 11 and 13.

The ease of using the QCM for adsorption studies, as well as the possibility to easily modify the sensor surface by thin deposition layers which can lend further selectivity have resulted in a plethora of applications and studies using QCM. However, strong limitations were encountered in these traditional QCM measurements when the mass deposited on the sensor was not rigid: visco-elastic changes of a film deposited could result as much in frequency changes as mass effects, thus rendering the interpretation of data measured extremely difficult, if not impossible. For a long time QCM was therefore known as a straightforward, facile experimental technique on one hand, but with an underlying theory that could be extremely intricate. This limitation was overcome to a great extent by coupling the traditional QCM-measurement with that of monitoring the energy dissipation of the deposition layers, resulting in what is known as "QCM-D". By repeatedly switching the electric field off for a very short time and subsequently measuring the oscillating energy dissipated from the crystal into the surrounding environment, valuable information could now be obtained on the nature of the adsorbed layer: a high dissipation means that the overall quartz crystal loses its energy quickly and suggests a thick and/or viscoelastic/soft film adsorbed on the sensor surface; a low dissipation suggests a flat and/or rigidly adsorbed film [2] [Figure 5.1(*b*)]. With the dissipation being seen as a measure of the rigidity or viscoelasticity of the adsorbed layer on the sensor surface, changes in frequency and dissipation could now provide information not only about the mass but also the structure of adsorbed layer: if an adsorbed layer does not change in mass but its dissipation changes, then a structural change must have taken place. Hence, in addition to the adsorbed mass measured as changes in frequency, ΔF, the QCM-D technique is able to also simultaneously detect and measure changes in viscoelastic properties of the adsorbed layer by monitoring the dissipation factor, D.

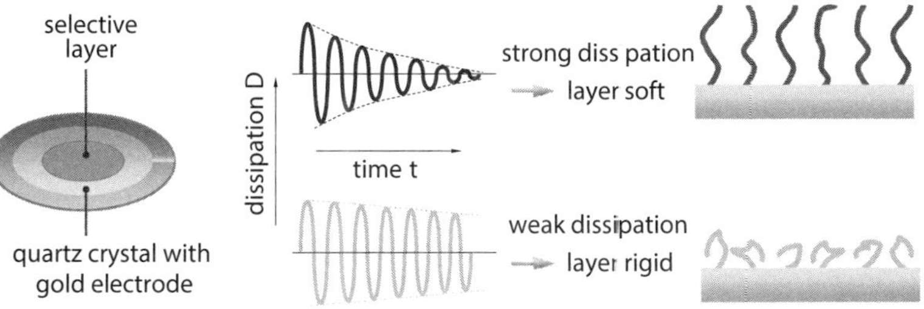

Figure 5.1 (*a*) Top figure illustrates the top, side and bottom view of a QCM-D quartz crystal and bottom figure shows a schematic representation of electric circuit applied to excite the crystal. (*b*) Schematic representation of the frequency and dissipation response of a rigid (dashed red line) and soft (solid green line) layer adsorbed on crystal surface, respectively.

5.3 Surface Plasmon Resonance (SPR)

Since the development of the first biosensor [3] based on surface plasmon resonance, the number of its applications have increased steadily. SPR is a surface sensitive analytical method based on a longitudinal charge-density propagating wave along the interface of two media, where one is a metal (typically gold and silver) and the other a dielectric [4]. It is used to detect and measure changes in the refractive index of the adjacent medium next to the metal surface. In a given system [Figure 5.2(*a*)] where a polarized light is shone through a prism on a sensor chip with a thin metal film on top, the light will be reflected by the metal film. By changing the angle of incidence of the light and monitoring the intensity of the reflected light it passes through a minimum. At this very angle of incidence, coupling between a photon from the laser beam and a surface plasmon at the metal/air interface causes excitation of the surface plasmon inducing surface plasmon resonance and subsequent reduction in the intensity of the reflected light. The angle at which the maximum loss of the reflected light intensity occurs is called *resonance angle* or *SPR angle,* and strongly depends on the thickness and dielectric constant of the dielectric layer present on the top of the metal layer [5]. Accordingly, the SPR response defined either as the shift in resonance angle ($\Delta\theta$) or in wavelength ($\Delta\lambda$) is associated with changes in the refractive index of the medium in contact with the metal

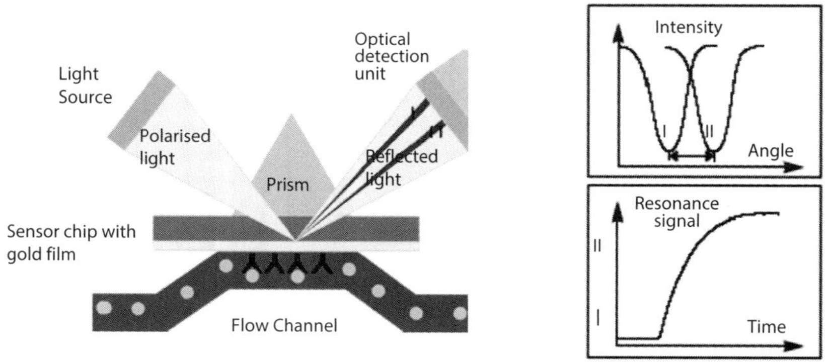

Figure 5.2 (*a*) Schematic representation of surface plasmon resonance set-up: a sensor chip with gold or silver coating is placed on a prism. The polarized light shines from a light-source onto the sensor chip, then the light is reflected by the metal surface acting as a mirror. The reflected light intensity is then measured in the detector. (*b*) Upper graph shows a *reflectivity curve* – the intensity profile of the light reflected by the metal surface before (I) and after adsorption (II); Lower graph displays a *sensorgram* – kinetics aspects of the adsorption process obtained by monitoring the time evolution of the SPR angle.

surface. Therefore, if the refractive index immediately above the metal surface changes by the adsorption of molecules from bulk solution, a change in the angle of incidence required to excite a surface plasmons will occur (Figure 2B). By monitoring the resonance angle during the adsorption process with time, an SPR adsorption profile, the so-called sensorgram, can be obtained. Jung *et al.* [6] introduced a simple but quantitative mathematical formalism to estimate the SPR response to an adsorbed film taking into account the decay length of the evanescent field. According to their work, the change in SPR angle, $\Delta\theta$, can be estimated from the following equation:

$$\Delta\theta = k\Delta n\left[1 - exp\left(\frac{-2d}{l_d}\right)\right] \quad (5.2)$$

where $\Delta\theta$ is the shift in SPR angle, k is the sensitivity factor of the system,
Δn is change in refractive index of the adsorbed film ($\Delta n = n_a - n_b$ with n_a and n_b are the refractive indexes of the adsorbed film and of the bulk solution, respectively),
d is the thickness of the adsorbed film, and
l_d is the decay length of evanescent field.

Equation 5.2 indicates that the change in SPR angle, $\Delta\theta$, is related to the thickness and the refractive index of the adsorbed layer. Thus, assuming $l_d \ll l_d$, from equation 5.2, one can calculate: *i)* the thickness of the adsorbed film, d

$$d = \frac{l_d}{2}\Delta\theta\frac{1}{k\Delta n} \quad (5.3)$$

and

ii) the optical mass of the adsorbed film, Γ, taking into account that it is approximated to be linearly related to the change in refractive index, $\Gamma = d \cdot \Delta n \cdot (dc/dn)$

$$\Gamma = \frac{l_d}{2}\frac{dc}{dn}k \cdot \Delta\theta \quad (5.4)$$

where dc/dn is the inverse of the refractive index increment with the bulk concentration.

The applications of both QCM-D and SPR as advanced surface characterization techniques are diverse and ever growing. One of the best-studied areas of application is the real-time monitoring of adsorption and desorption kinetics. A typical QCM-D and SPR kinetic profile is shown

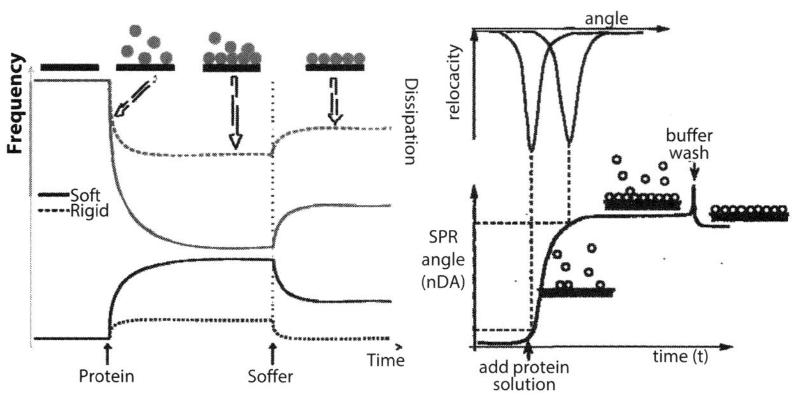

Figure 5.3 Schematic representation of a typical QCM-D (*a*) and SPR (*b*) adsorption profiles for the adsorption of a protein on the sensor surface. The change in the value of frequency and SPR angle is used to monitor the kinetics of the adsorption process.

in Figure 5.3 for the adsorption of a protein onto the sensor surface. In QCM-D, posterior to a baseline obtained under reference conditions, such as a buffer solution, the adsorption of protein molecules causes a decrease in frequency owing to a mass increase on the sensor surface. Concurrently, the dissipation factor increases. The extent of the QCM-D response depends on the rigidity of the adsorbed layer: in Figure 5.3(*a*), the solid line represents a soft layer, while the dashed line schematically illustrates a rigid layer. When the QCM-D is subsequently contacted again with the reference solution, some of the loosely adsorbed protein molecules are removed, resulting in both a relative increase in frequency and relative decrease in dissipation. Note that a decreasing dissipation might also occur owing to a rearrangement of the protein molecules leading to a more rigid layer.

Figure 5.3(*b*) depicts a schematic which represents the same adsorption kinetics for SPR. Here, contacting the SPR sensor with a protein solution results in an increase of the resonance angle. The adsorption occurs until saturation of the surface which results in a plateau in adsorption profile, followed by a "desorption"-step in which the protein feed solution is again replaced by buffer in order to remove the loosely bound material. The difference between the initial and final frequency or resonance angle indicates the extent of protein adsorption. As can be seen, no direct information on the rigidity or viscoelastic properties of the adsorbed layer can apparently be obtained during the SPR measurement since there exists only one independently measured parameter, namely the change in refractive index, as opposed to QCM-D, where dissipation D and frequency F as two independently measured parameters allow determining two variables. However, as will be illustrated later, both SPR and QCM-D can yield valuable information on the nature of adsorbed layers on sensor surfaces owing to their complementarity.

5.4 Applications of SPR and QCM-D

During adsorption, the typical output parameters obtained from QCM-D data analysis are adsorbed mass or thickness versus time, and viscoelastic properties versus time. The adsorbed mass (Δm) can be calculated using the Sauerbrey equation (Eq. 5.1) if the adsorbed layer is thin, rigid and evenly distributed over the surface, or the Voigt-based viscoelastic model when the adsorbed layer does not fulfil the Sauerbrey assumptions. Knowing or assuming the adsorbed layer density (ρ), it is possible to calculate its thickness (d). On the other hand, from SPR data analysis the output parameters obtained are adsorbed mass or thickness versus time. However, the full SPR curve obtained through measuring the reflectivity versus the angle of incidence holds all the necessary information to determine both thickness and refractive index of the adsorbed layer on the sensor surface. If using SPR with one wavelength and, hence, measuring one parameter (the angle shift), it is in principle not possible to determine both thickness and mass of the layer adsorbed. The same concept applies for alternative measurement techniques that rely on changes of the refractive index, such as ellipsometry. However, by simultaneously measuring SPR curves in two different media with known refractive indices, such as water and air, it is possible to gain the additional information needed to solve this problem [7]. Naturally, such an approach is only useful in case one aims at charactering solid thin films deposited on the sensor surface which can easily be exposed to different media.

Applications of SPR and QCM-D to sensing surface adsorption phenomena are diverse. Often than not, monitoring studies are related to biomolecules, be it with regard to boufouling, biocompatibility or biosensors. Ivanov *et al.* investigated the activity of antimicrobial peptides when as a function of different surface immobilization protocols [8]. Changing the sensor surface chemistry as well as the linker chemistry employed, and by combining both the frequency and the dissipation measurements, they learnt that physisorbed peptides resulted in a horizontal orientation whilst covalently attached peptides formed vertically positioned layers. This result was significant inasmuch as only the latter exhibited a high antibacterial activity. This example illustrates a more fundamental question, namely how macromolecules are arranged at surfaces and under which conditions and how they undergo conformational changes. Zhang and Wu investigated this aspect as for thermally sensitive PNIPAM polymer chains [9]. With PNIPAM possessing a lower critical solution temperature (LCST) of about 32 °C, they could show how shrinking and swelling of PNIPAM surfaces occurred as a function of temperature, going along with changes in the water adsorbed within the polymer chains. It should be noted that common SPR and QCM-D measurements occur in the range between 20 – 40 °C owing to the vast biological applications that these techniques face. Naturally, other temperature ranges can be covered but require and adequate flow cell design such as to warrant an efficient temperature control of the sensor surface. The advantage of both SPR and QCM-D, as well as naturally any other surface sensitive techniques such as ellipsometry or dual polarization interferometry, is the fact that surfaces can be tailor-made and their formation as well as the subsequent interaction with model adsorbates be monitored. A natural limitation of such sensitive techniques is the fact that bulk concentrations of potential adsorbates must not be excessively high as it otherwise may mask the measurement of adsorption phenomena. Many commercial SPR devices therefore come with an at least two-channel configuration which allows defining a measurement (layer for specific interaction) and a reference channel (layer with supposedly no specific interaction). The signal of the latter includes both bulk and non-specific binding phenomena and can be subtracted from the former in order to single out the contribution of the specific surface interaction sought for. The commercial QCM-D equipment comes with a single channel configuration, only, such that these phenomena need being studied either in parallel, independent flow cells or subsequently using one and the same device.

Protein adsorption is a typical experiment that demonstrates very well the complementarity of both QCM-D and SPR. The adsorption process of BSA on gold surface was investigated by means of QCM-D and SPR techniques. In a typical adsorption experiment using SPR, buffer solution (PBS, pH 7.4) was first injected until a stable SPR angle baseline were achieved. When a significant response is anticipated during an experiment, the baseline stability is of minor importance. However, if small molecule binding is studied and responses are, hence, expected to be small, a stable baseline is indispensable in order to yield high-quality adsorption data. A protein solution (BSA, 1mg/ml) in PBS-buffer was injected for 4 minutes over a gold surface at a flow rate of 20ml/min, followed by rinsing with PBS-buffer solution to remove loosely bound protein molecules. The measurements were performed at a constant temperature of 23°C.

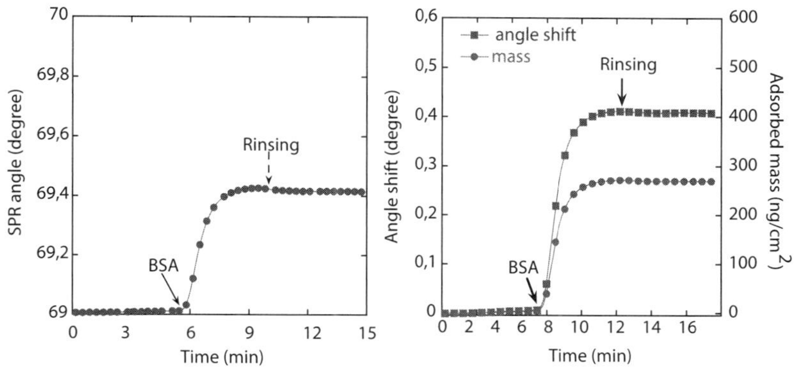

Figure 5.4 Adsorption of BSA (1mg/ml) on a gold sensor surface at pH 7.4 as monitored by means of SPR.

Figure 5.4 shows that within contacting time, the adsorption kinetics reaches an equilibrium. When subsequent rinsing is initiated, hardly any protein is removed from the surface, indicating irreversible protein binding under the experimental conditions. The angle shift measured between the baseline and the irreversibly bound protein layer is about 0.4 degrees, which translates into approximately 270 $ng{\cdot}cm^{-2}$. In a complementary QCM-D experiment, the gold-coated QCM-D sensor was mounted in QCM-D system and equilibrated in PBS-buffer at a constant temperature of 23°C until a baseline was obtained. A BSA-solution (1mg/ml BSA in PBS) was introduced to the flow cell with a flow rate of 100mg/ml for 15min, and then rinsed with PBS buffer solution for at least 20min to remove loosely bound BSA molecules. Figure 5.5 depicts the respective results for QCM-D.

The frequency shift caused by the bound protein is about 35 Hz while the dissipation shift is $1{\cdot}10^{-6}$, indicating a soft adsorption layer as to the criterion mentioned before. It is pointed out that with the sensor surface being of gold in both SPR and QCM-D, the adsorption kinetics of BSA and the adsorbed mass should in principle be observed to be similar in both techniques. However, flow channel geometries need to be taken into consideration as identical flow rates can result in very different hydrodynamic flow regimes which in turn can affect the adsorption transient. Furthermore, gold surface roughness may also affect equilibrium adsorption. Employing Eq. 1, a protein equilibrium adsorption of about 610 $ng{\cdot}cm^{-2}$ can be calculated and it can be seen that this value is more than doubling the one measured with SPR. The reason for this apparent discrepancy is the water bound to the protein molecules, as depicted in Figure 5.6.

SPR relies on measuring differences in the refractive index such that it is not capable of distinguishing between bulk water and water bound

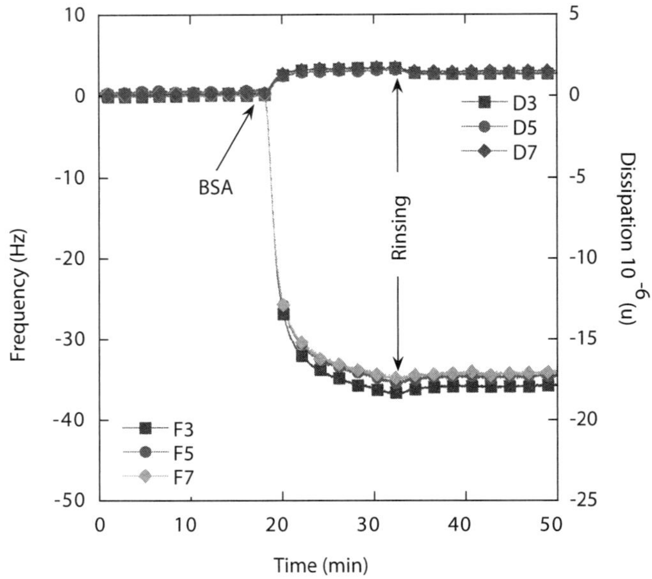

Figure 5.5 Adsorption of BSA (1mg/ml) on gold sensor surface at pH 7.4 as monitored by means of QCM-D.

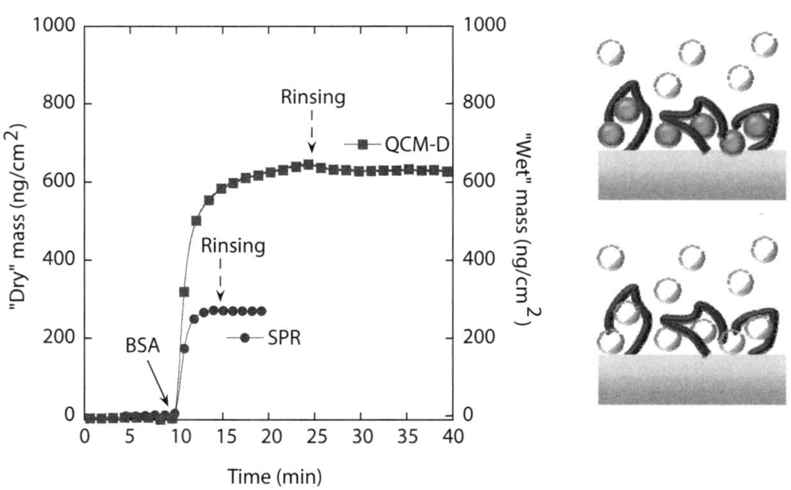

Figure 5.6 The concept of "dry" and "wet" mass: comparing adsorption data of BSA at pH 7.4 as measured by both SPR and QCM-D. As opposed to SPR, QCM-D is sensitive also to the water bound to the protein layer adsorbed on the sensor surface.

to adsorbed proteins. On the contrary, QCM-D as an acoustic technique measures the damping of the quartz oscillating frequency that occurs upon binding of molecules: protein-bound water molecules contribute to this damping in a different way than bulk water which can freely rearrange, and are therefore detected as additional mass. As a consequence, combining both SPR and QCM-D allows differentiating between the "dry" absorbed protein mass and the "wet" adsorbed protein mass. In some instances observing how the respective ratio changes can also indicate conformational changes which the protein is undergoing upon adsorption, as this may be correlated with a more or less exposure of hydrophilic domains. Bittrich *et al.* combined QCM-D and ellipsometry simultaneously for understanding in more detail the adsorption of BSA and the swelling behaviour of polyacrylic acid brushes adsorbed on the sensor surface [10]. A simultaneous rather than independent measurement of two complementary techniques bears the tremendous advantage of diminishing the variance in the experimental data obtained as one and the same surface modification is being investigated. Using this coupled set-up, they could elucidate adsorption phenomena below and above the isoelectric point of BSA (pH=5.6), differentiating between the uptake of BSA and buffer in the polymer film.

In this sense, QCM-D can also contribute to better characterize sensing applications involving DNA which can be understood as yet another type of polymer brush. Layer-by-layer DNA films [11], DNA hybridization [12], Holliday junctions [13], and aptamer-protein interactions [14] have been reported where the dissipation served as a major parameter for this film characterization. Using a small molecule binding aptamer as a model system, Serrano Santos *et al.* not only determined the interaction of adenosine-5′-monophosphate (AMP) with AMP-binding aptamer films using QCM-D, but also succeeded in quantifying in situ and in real-time the structural changes that occur in this particular DNA aptamer when binding to AMP [15]. Their objective was to elucidate in how far aptamers or aptamer hairpins differ in their conformational changes upon target binding, and in how far this feature could be used for stimuli-responsive nanoporous gating devices to be used in, for example, drug delivery. The difference between both aptamer forms was that the AMP-binding aptamer formed a helix structure with mismatched nucleotides and upon target binding underwent only a minor molecular conformation change. On the other hand, it was known that the aptamer hairpin structure based on the same aptamer sequence but extended by an additional nucleotide and linker sequence and would respond with a major molecular rearrangement [Figure 5.7(*a*) and (*b*)].

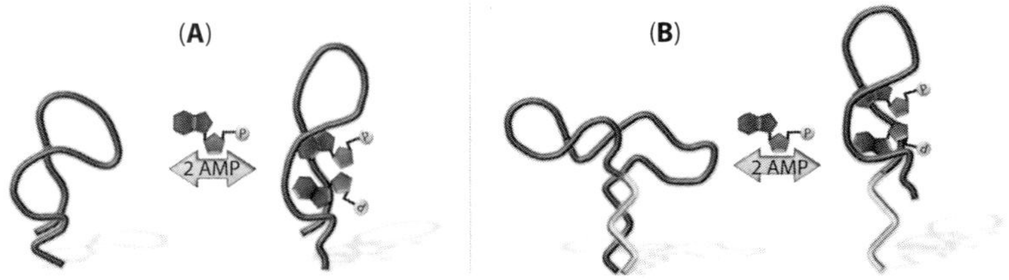

Figure 5.7 (*a*) Scheme of the AMP-binding aptamer (blue line), where the overall structure is not significantly disturbed by specific intercalation of the AMP ligand; (*b*) Scheme of the hairpin form of the AMP-binding aptamer (blue lines) which was created by adding 7 nucleotides at 3′end (yellow lines) forming a stem-loop structure.

They found that indeed the thickness of the aptamer film increased by only about 0.1 nm upon target binding whereas the thickness increase of the aptamer hairpin film reached almost 1.6 nm. The conformational changes were explained as a loss of the double stranded neck region of the aptamer hairpin, in which the single stranded loop region stabilized the mismatched double helix structure of the ligand-binding sequence. It was, hence, assumed that the aptamer hairpin film turned "softer" as the hairpin structires extended vertically away from the sensor surface. On the contrary, it was known from NMR studies that the aptamer did only undergo a minor rearrangement upon target binding with the internal binding pocket leading to a slightly more upright position of the molecule. It was this minor change of the film density which could indeed be observed by an only slight increase in dissipation.

These data obtained on the conformational changes were not inferred directly from frequency and dissipation measurements but through applying the Voigt viscoelastic model to the data measured. In this context, the work also revealed that caution must be taken when interpreting quantitative data obtained from such models: first, in absence of at least one other technique capable of quantifying conformational changes, such data should be regarded as a trend rather than be considered absolute; second, it must be emphasized that data measured are obtained as an average response of a film. Be it polymer brushes or DNA-molecules grafted onto the sensor surface, it often remains obscure how homogenously these molecules are oriented either vertically or horizontally, or both. As a consequence, quantitative data on conformational changes represent an average response to the external stimulus applied, rather than providing exact information such as a single-molecule measurement. Nevertheless, for many practical applications it is precisely the information that is required: many practical applications of polymer or biomolecule films suffer themselves from being intrinsically ill-defined on a molecular level.

A combination of both SPR and QCM-D can also be extremely powerful for elucidating surface phenomena occurring in membrane processes. For example, in membrane water treatment, matter dissolved or dispersed in water can adsorb to the membrane surface giving rise to the formation of a layer which is sufficiently thick to deteriorate dramatically the overall process performance. This phenomenon is well-known as membrane fouling. Provided that membranes are supposed to have a life-span of many years, often than not, membrane fouling therefore requires periodic membrane cleaning which not only means a process interruption but also contamination of the environment with detergents. It can easily be understood that what would be detected as a "rigid" film in QCM-D, could represent a more severe membrane fouling which will require a more thorough cleaning procedure. On the other hand, "soft" film may represent a situation in which pore clogging is more probable resulting in a faster deterioration of membrane filtration performance (decrease in transmembrane fluxes) although cleaning might be relatively easier. While membrane fouling has been under investigation for decades, most analytical methods employed have been at best microscopic. Considering that during the initial stage of the formation of the fouling layer its thickness does not exceed the lower nm-range, it becomes evident how crucial the information is which the SPR provides since it elucidates the very early-stage adsorption phenomena leading to membrane fouling.

Membrane barriers or interfaces usually consist of materials which have desirable bulk properties. These properties can either refer to easy processability and film formation, or be associated to desirable thermodynamic properties. Sometimes a compromise needs to be found between both. For example, thin polymeric films used in sensors ideally should have a high chemical affinity for the target compounds. Polyalkylmethylsiloxanes can be employed for this purpose. The alkyl groups can be chosen to be long-chains in order increase the hydophobicity, however, on the costs of the mechanical stability of the membrane polymer. For sensor applications this is not an issue since the thin polymer films are supported by substrates, but membrane stability can become an issue if self-standing thin films were required. In cases were bulk properties do or should not necessarily be the same as the thin film surface properties, surface modifications can be carried out. For example, the surface of intrinsically hydrophobic polyether sulfone filtration membranes can be made hydrophilic by plasma treatment in order to reduce membrane fouling. Similarly, the hydrophilicity of polyamide-based water filtration membranes can be further increased by grafting hydrophilic groups. Once such surface modifications have been conducted, screening of their efficiency would be desirable. Again, in this case QCM-D and SPR can serve as rapid screening methods and provide valuable information on the efficiency of the modification of membrane surfaces.

For example, Contreras *et al.* [16] employed QCM-D for studying the interaction of BSA and sodium alginate, which they considered model foulants, with different chemical sensor surface functionalities. Apart from confirming that in order to avoid membrane fouling surface functionalities should be employed that minimize the possibility of forming hydrogen bonds, they could show that the dissipation of alginate layers for incremental mass adsorbed was significantly higher than that of adsorbed layers of BSA. This indicated a "softer" adsorption layer in the case of alginate compared to BSA. As another example, the build-up of polyelectrolyte layers can be used as a matrix which fosters cell adhesion or provides a stimuli-responsive surface layer [17]. Polyetheretherketone (PKD) dissolved in chloroform was deposited onto SPR gold-slides and QCM AT-cut quartz crystals with bare gold electrodes by spin-coating. Subsequently, PDDA (polydiallyldimethylammonium chloride) and PAA (polyacrylic acid) were alternately spin-coated onto the modified sensor surfaces from aqueous solutions. Owing to the opposite charge of PDDA and PAA, a multiple poly-electrolyte layer was obtained whose step-wise formation was monitored by QCM-D and SPR and. The build-up of this layer is depicted in Figure 5.8 for SPR and in Figure 5.9 for QCM-D.

It can be seen that both techniques allow visualizing the multi-layer build-up. Comparing the mass adsorbed after the third layer (1: PDDA; 2: PAA; 3: PDDA) on both the SPR and the QCM-D sensor, there is a very good agreement between both techniques which confirms their

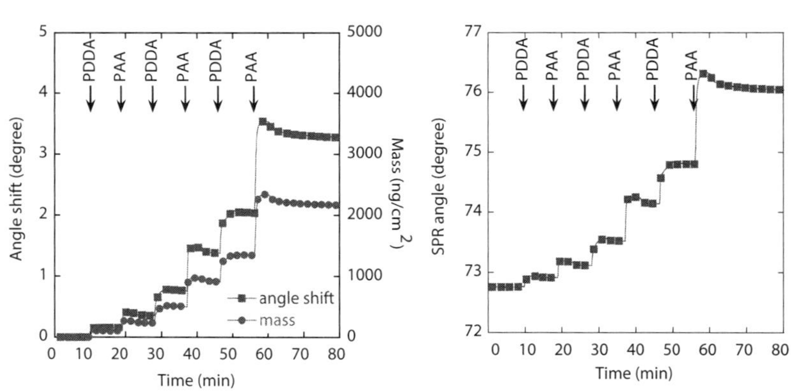

Figure 5.8 Build-up of a polyelectrolyte multi-layer consisting of consecutive PDDA-PAA depositions as measured by SPR. Left: angle shift and resulting calculated total mass of adsorbed layers; right: SPR angle.

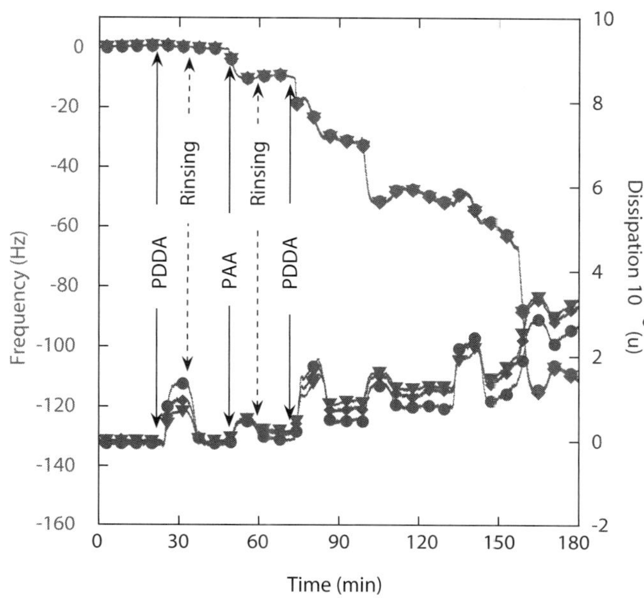

Figure 5.9 Build-up of a polyelectrolyte multi-layer consisting of consecutive PDDA-PAA depositions as measured by QCM-D.

complementary nature. In addition, QCM-D measurements reveal a rearrangement of the polyelectrolyte layer upon rinsing (Figure 5.9): while the frequency remains constant, dissipation values drop after each rinsing step. From the SPR measurements (Figure 5.8) we can deduce that no mass is lost from the sensor surface during single steps. Thus, only a rearrangement of the polyelectrolyte layer toward higher rigidity can explain the dissipation values measured at these points.

5.5 Conclusions

Adsorption phenomena occurring at the surface of membrane or thin films can be monitored in real-time and in a non-invasive manner using QCM-D and SPR. The information so obtained provides valuable insights on the early-stage formation of the very first molecular adsorption layers which often than not determine subsequent microscopic or macroscopic phenomena. While conformational changes of surface-adsorbed molecules can be investigated using a single technique such as QCM-D, it always is desirable to not rely on one technique only, mainly in order to minimize the risk of pitfalls and possibly interpreting experimental artefacts.

Acknowledgements

G.D. and T.S. would like to acknowledge financial support from the Basque Government under CTP10-P10 and the Diputación Foral de Gipuzkoa under grant Exp.30/12. T.S. would like to gratefully acknowledge ERC Starting Grant 209842-MATRIX.

References

1. Sauerbrey, G. *Zeitschrift fur Physik* 1959, *155* [2], 206–222.
2. Rodahl, M.; Höök, F.; Fredriksson, C.; Keller, C. A.; Krozer, A.; Brzezinski, P.; Voinava, M.; Kasemo, B. *Faraday Discuss.* 1997, *107*, 229–246.
3. Liedberg, B.; Nylander, C.; Lundstrom, I. *Sensors and Actuators* 1983, *4* [2], 299–304.
4. Lundstroem, I. *Biosens.Bioelectron.* 1994, *9* [9/10], 725–736.
5. De Bruijn, H. E.; Altenburg, B. S. F.; Kooyman, R. P. H.; Greve, J. *Opt.Commun.* 1991, *82* [5–6], 425–432.
6. Jung, L. S.; Campbell, C. T.; Chinowsky, T. M.; Mar, M. N.; Yee, S. S. *Langmuir* 1998, *14* [19], 5636–5648.
7. Liang, H.; Miranto, H.; Granqvist, N.; Sadowski, J. W.; Viitala, T.; Wang, B.; Yliperttula, M. *Sensors and Actuators B: Chemical* 2010, *149*, 212–220.
8. Ivanov, I.E.; Morrison, A.E.; Cobb, J.E.; Fahey, C.A.; Camesano, T.A. *Appl. Mater. Interfaces* 2012, *4*, 5991–5997.
9. Zhang, G.; Wu, C. *Macromol. Rapid Commun.* 2009, *30*, 328–335.
10. Bittrich, E.; Rodenhausen, K.B.; Eichhorn, K.-J.; Hofmann, T.; Schubert, M.; Stamm, M.; Uhlmann, P. *Biointerphases* 2010, *5(6)*, 159–167.
11. Singh, A.; Snyder, S.; Lee, L.; Johnston, A.P.R.; Caruso, F.; Yingling, Y.G. *Langmuir* 2010, *26*, 17339–17347
12. Larsson, C.; Rodahl, M.; Höök, F. *Anal. Chem.* 2003, *75*, 5080–5087.
13. Papadakis, G.; Tsortos, A.; Gizeli, E. *Nano Lett.* 2010, 10, 5093–5097.
14. Chen, Q.; Tang, W.; Wang, D.Z.; Wu, X.J.; Li, N.; Liu, F. *Biosens. & Bioelectron.* 2010, 26, 575–579
15. Serrano Santos, M.B.; Llobet, E.; Özalp, C.V.; Schäfer, T. *Chem. Commun.* 2012, 48, 10087–10089
16. Contreras, A.E.; Steiner, Z.; Miao, J.: , Kasher, R.; Li, Q. *Environmental Science and Technology* 2011, *45*, 6309–6315.
17. Diaconu, G.; Gugliuzza, A. Schäfer, T. *Procedia Eng.* 2012, *44*, 1260 1261.

6

Smart Membrane Surfaces: Wettability Amplification and Self-Healing

Annarosa Gugliuzza

Institute of Membrane Technology ITM-CNR, University of Calabria, Rende (CS), Italy

Abstract

This chapter is dedicated to deal with membrane surface sensitivity as a powerful tool to direct events such as fouling mitigation, self-cleaning, liquid flows and self-repairing. Wettability and self-healing concepts are discussed through the chapter with emphasis on the attractive opportunity to combine different strategies for switching and adjusting morphology and chemistry of the surfaces depending on desired target. Membrane ability to cause in situ-cleaning, liquid diffusion, self-restoring and auto-recovery is argued in relation to self-adjustment, adaptability and actuation mechanisms.

Keywords: Smart surfaces, wettability, self-powered liquid motion, self-cleaning, self-repairing, superhydrophobicity, antifouling

6.1 Introduction

The first stage of membrane separations occurs via surface, where the most of interactions takes place with materials. Abrupt changes in the surface chemistry and morphology can affect the properties of materials used, determining the final performance of a membrane process [1]. Fouling, fogging, cleaning, transport, transcription, catalysis, conversion, corrosion, abrasion, regeneration and healing are some of the most frequent events, which regard with the surface of many materials used in manufacturing industry, textiles, architecture, automotive, food-processing, environment, pharmaceutics, biomedicine and chemicals [2].

**Corresponding author*: a.gugliuzza@itm.cnr.it, annarosa.gugliuzza@cnr.it

Annarosa Gugliuzza (ed.) Smart Membranes and Sensors, (161–184) 2014 © Scrivener Publishing LLC

Adhesion and adsorption as well as wetting and diffusion can suddenly alter the equilibrium established at the solid-liquid-gas interfaces, causing interfacial forces to transform the surfaces thermodynamics; in this regard, membrane surface transformations can also be responsible for flux decline, selectivity loss, reduced resistance to dirty and harsh environments as well as mechanical stress [3, 4].

Surface phenomena can be controlled when physics and chemistry are fully understood and directed towards the identification of structural, electronic, biological and chemical factors involved in the rearrangement and confinement of functional components in predetermined geometries, layers and volumetric spaces. The design of advanced materials with associated smart functions becomes hence possible by directing the nature and the rate of surface phenomena towards durability, efficiency and quality of products and processes.

This chapter covers comprehensively basic aspects related to wettability and healing concepts, describing the attractive role played by the surfaces in many processes and events [5]. It provides also useful indication about the possibility to manipulate morphology, charge and chemical of the surfaces in order to actuate desired events and preserve original properties and integrity of the materials, including membranes. A special focus is dedicated to the wettability amplification for self-cleaning functions and self-powered liquid flows together self-repairing processes for restoring superhydrophobicity and antifouling properties.

6.2 Basics of surface wettability

The membrane surface is wettable when approaching liquids spread on it, preferring the solid phase to a fluid [6]. The degree of wetting or spreading (S) can be regarded as the result of interfacial forces, which control the adhesion processes at the solid/liquid/air interfaces. These controlling interactions are expressed and quantified by the surface free tension *(γ)*, which is, in turn, proportional to the cohesive energy necessary to minimize the free surface area at the interface. The mathematical relation used to quantify the wettability is the following:

$$S = \gamma_{SV} - \left(\gamma_{SL} + \gamma_{LV}\right) \tag{6.1}$$

Where $\gamma_{SV,}$ $\gamma_{SL,}$ and γ_{LV} are the free tensions at solid-vapor, solid-liquid and liquid-vapor interfaces, respectively. Positive values of adhesion are expression of hydrophilicity, whilst negative values give indication about the surface hydrophobicity. An empirical and direct measurement of the degree of hydrophilicity/hydrophobicity is obtained with the estimation of the apparent contact angle (θ_a). This is a thermodynamic parameter, which is expressed for ideal and almost flat surfaces by the Young's equation:

$$\cos(\theta_a) = \frac{\gamma_{SV} - \gamma_{SL}}{\gamma_{LV}} \tag{6.2}$$

A general convection establishes that values of θ_a<90° can be measured on hydrophilic membrane surfaces, whilst values of θ_a >90° are estimated for hydrophobic surfaces. Values of θ_a >150° are typical of ultraphobic surfaces. Another parameter considered for studying the liquid flow over the surfaces is the roll-off angle or sliding angle, which denotes the angle to which a surface may be tilted for the roll off of water drops (Figure 6.1).

Considering that the contact angle value is the result of combined morphological - i.e., tailed and loopy polymer segments, grooves and pores - and chemical - i.e., functional binding or non-binding sites - elements of the membrane surface, the Wenzel and Cassie–Baxter's equations are considered to be more suitable than Young's equation for a more accurate description of real surfaces.

The Wenzel equation is based on the assumption that the grooves through the surfaces are filled by the liquid. In this case, the contact angle is expressed as:

$$\cos(\theta_w) = R\left[\frac{\gamma_{SV} - \gamma_{SL}}{\gamma_{LV}}\right] \tag{6.3}$$

Combining this equation with the Young's one, the Wenzel contact angle (θ_w) is related to the apparent contact angle by the following relation:

$$\cos(\theta_w) = R \cdot \cos(\theta_a) \tag{6.4}$$

where R is the rugosity factor that quantifies the irregularities of the membrane surface, which are responsible for increased contact angle values [7].

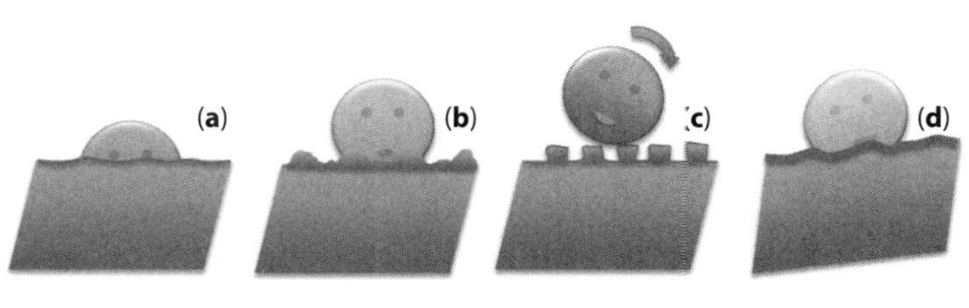

Figure 6.1 Schematic representation of surfaces with different degree of roughness and wettability according to (*a*) Young, (*b*) Wenzel, (*c*) Cassie-Baxter, (*d*) Roll-off angle models.

This phenomenon becomes an important element in the Cassie–Baxter's model:

$$\cos(\theta_{CB}) = f_s \cdot \cos(\theta_a) + f_v \cdot \cos(\theta_v) \tag{6.5}$$

where θ_a is the apparent contact angle measured on the flat surface, whereas f_s and f_v are the fractions of the solid and vapor on the surface (*i.e.*, $f_s + f_v = 1$), respectively. According to this model, the surface grooves are filled by the air so that the liquid-surface adhesion is significantly reduced, causing an increase in the contact angle values and, therefore, in the surface hydrophobicity. Indeed, the role of the chemistry on the surface wettability can be well-evaluated through the correction of the apparent contact angle value for the roughness factor according to the zig-zag model proposed by Taniguchi and Belfort [8].

Considering the effect that the rugosity factor can have in suspending small water droplets, the increase in the surface roughness is one of the most attractive approaches to make membrane unwettable. Despite that the direct measurement of the apparent contact angle by static or dynamic mode is one of the most used method to estimate the degree of hydrophobicity/hydrophilicity of a membrane surface, a more accurate estimation of the wettability can be done through the calculation of a semi-quantitative thermodynamic parameter such as the electron donor surface (γ) component of the overall surface free tension (γ) [9].

This parameter and other polar and non polar components can be estimated by the following set of mathematical equations:

$$\gamma_s^{LW} = \frac{\gamma_l^{LW} \cdot (1+\cos\theta)^2}{4} \tag{6.6}$$

$$\gamma_l \cdot (1+\cos\theta) = 2 \cdot \left(\sqrt{\gamma_s^{LW} \cdot \gamma_l^{LW}} + \sqrt{\gamma_s^{+} \cdot \gamma_l^{-}} + \sqrt{\gamma_s^{-} \cdot \gamma_l^{+}} \right) \tag{6.7}$$

$$\gamma_s^{AB} = 2 \cdot \sqrt{\gamma_s^{+} \cdot \gamma_s^{+}} \tag{6.8}$$

where the Lifshitz-van der Waals γ^{LW} [mJ/m^2], polar γ^{AB} [mJ/m^2], acid (electron acceptor) γ+ [mJ/m^2], base (electron donor) γ [mJ/m^2] are the components of the overall surface free energy γ [mJ/m^2].

6.3 Amplified Wettability

Animal wings and skins as well as vegetable leafs and mineral textures owe an amplified wettability to their multiscale surface roughness. The Figure 2 shows a typical example of rolling water drops across vegetable leafs.

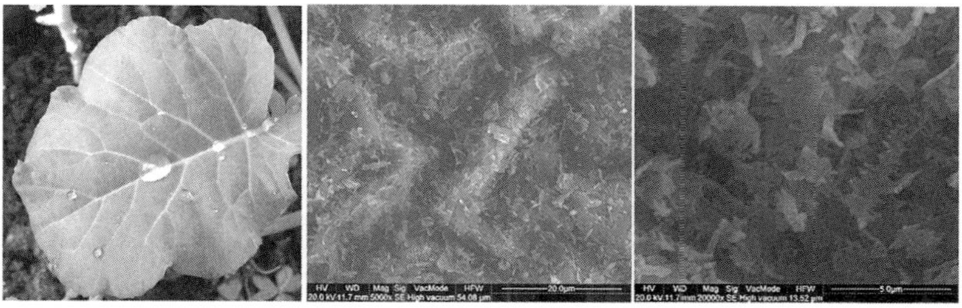

Figure 6.2 Picture of rolling water drops on broccoli leafs (*a*). SEM pictures (*b,c*) representing the multiscale roughness at 5000x and 20000 x, respectively.

The roll-off mechanism is due to the multiscale roughness, which characterizes the leaf texture, as shown in SEM micrographs.

The typical architectures and supra-molecular structures of these natural objects are often responsible for extraordinary events such as liquid motion, contaminant resistance, antifogging, antifouling and/or preferential surface interactions. So, the greatest aspiration of many scientists is to reproduce artificially some of these behaviors, thereby bringing revolutionary concepts to different areas of the biotechnology, biomedicine, textiles, architecture, microelectronics and membrane processes [9–14]. A general proof-concept for endorsing events such as liquid motion, self-cleaning, antifogging, and fouling mitigation consists of making the wettability adjustable, depending on local environmental conditions. Some recent attempts to manipulate the surface wettability through actuating mechanisms are herein discussed, analyzing the role of morphological and chemical features in the membrane surface events.

6.4 Actuation Mechanisms

The fabrication of complex textures showing switchable wetting properties represents an important target. Hydrophilic/hydrophobic transitions and *vice versa* can be actuated by using external triggers such as electrical/magnetic fields, light, temperature, pH, molecular stimulation or mechanical stress [15].

6.4.1 Electrical Switching

Electrowetting is frequently used to manipulate little amounts of liquids on surfaces under applied voltage [16]. Usually, droplets of salt solutions with a diameter less than 1 mm are put in contact with a planar surface in a medium that can be air or an immiscible oil. In absence of electrical field, the behavior of the droplet depends on the surface free tension forces. When a voltage is applied, a spreading of the droplet is induced with an increase in the droplet contact area. From a thermodynamics point of view, this behavior is ascribed to cooperative factors such as the variation of the surface free tension at the droplet/surface interface and the energy stored in the capacitor formed by the conductor and salt solution, which are separated by an insulator. The literature refers about the use of hydrophobic insulators and conductive materials for the fabrication of electrical pathways. Parylene has been recently used as an insulating layer and repulsion forces between similar electrical charges at the water-parylene interface have been identified as the main responsible for a decrease in contact angle values [17]. Carbon nanotube-based membranes have been used as anode to induce superhydrophobic-hydrophylic transitions, while a *Pt* wire has been used as cathode in order to generate a small electrical tension with strong polarity dependence [18].

Fowlkes et coworkers [19] have proposed a study on reversible behavior of Water/KCl/Glycerol droplets (saline) on vertically aligned superhydrophobic carbon nanofibers, showing a competitive and reversible two-liquid (dodecane/saline) electrowetting. Replacing air by dodecane, a wetting reversibility has been induced by measuring contact angle of 160° and 120° before and after application of an electrical tension. In absence of tension, a return to 160° has been observed only for systems where CNTs filaments were coated by dodecane. The effects of the oil on reversible wetting processes has been also examined by Bhat *et al.* [20] during electrowetting experiments carried out on textiles based on polymer and wood microfibers. After treatment with parylene/fluoropolymer solution, water contact angle values of 120° were achieved, while a transition from 120 to 70° was induced under voltage. The reversibility was observed in presence of oil, confirming a type-Cassie configuration of the droplet.

Recently, redox process of a smart dodecylbenzenesulfonate doped polypyrrole (PPy(DBS)) has been also proposed to manipulate wetting properties. Under a voltage of 0.9 to 0.6 V, a surface tension gradient across the contact line causes the spreading of a dichloromethane (DCM) droplet due to Marangoni-type stress. When the electrical tension is removed, a full recovery of the original wetting properties can be observed.

6.4.2 Light-Driven Switching

The light is regarded as a powerful trigger to stimulate membranes surface and induce changes in wettability through mechanisms, which can involve changes in both the surface polarity and texture [21]. Light-responsive inorganic and organic materials can be used to amplify the surface wettability flexibly. Metal oxides such as zinc or titanium oxide can be patterned in intricate structures combining their semiconductor properties with complex topographies. Indeed, the irradiation of ZnO-based surfaces at a wavelength less than 375 nm, which has an energy more than ZnO band gap, caused a reversible hydrophobic-hydrophilic transition from 124° to 5°, owing to competitive sorption/desorption events [22]. Treatments of smooth TiO_2 sol-gel films with CF_4 plasma and octadecylphosphonic acid (ODP) allowed getting surfaces in a superhydrophobic state as well [23]. Under UV irradiation (1 $mWcm^{-2}$), the wettability of these surfaces was switched towards a superhydrophilic state, due to the photocatalytic decomposition of the ODP monolayer. The superhydrophobic state was recovered after re-adsorption of ODP molecules.

Also, photo-induced responsive surfaces based on organic-compounds can adapt themselves to light through reversible conformational transitions. Typical classes of compounds used to make photo-switchable systems are azobenzenes, spiropyrans, dipyridylethylenes, stilbenes, and pyrimidines [24]. These molecular switches are usually immobilized onto the surface by self-assembly processes, ion beam techniques, chemical and/or physical crosslinking [25]. The related ability to switch reversibly from an isomer to another under UV/light irradiation makes them suitable to get controllable surface properties. In this respect, biodegradable polycarilactone (PLC) nanofibers were modified with azobenzene and trans-to-cis isomerization of the azobenze group was induced under UV irradiation, yielding a change in the static contact angle from 132.2 ± 2.8° to 53.1 ± 3.2°; full reversibility was observed under visible light [26]. Similarly, an increase in hydrophilicity (112° to 90) was obtained with micro-structurally rough polymer surfaces functionalized with spiropyran molecules (1′,3′ -dihydro-1′,3′,3′-trimethyl-6-nitro- spiro[2H-1-benzopyran-2,2′ -(2H)-indole]). The reversible change in the wettability was caused by a transition from non-polar spiropyran molecules to the polar merocyanine isomers [27]. Another example of photo-switchable system is that of a vinyl spiropyran ((1'-(2-propylcarbamylmetha-crylamide)ethyl)-3',3'-dimethyl-6-nitrospiro[2H-1]benzopyran-2,2'-indoline) UV-grafted on poly(ether sulphone) (PES) membranes, which allowed an optical control of wetting and protein adsorption under light irradiation [28]. Considering that the nonpolar closed form (spiropyran) is colorless, while the polar open form (merocyanine) is colored, the isomerization event can be easily investigated. Depending on the isomer induced, a different protein adsorption and permeation rates were detected. Higher protein adsorption was measured on the closed configuration of the vinyl spiropyran surface, while a reduction of the 26% was quantified on the open configuration of the merocyanine surface. Consequently, higher permeation rates were estimated for samples having lower amount of protein adsorbed onto the surface (open form). Wang *et al.* [29] demonstrated how the functionalization of surfaces with spiropyran molecules can be induced the fastest and highest change in wettability reducing the contact angle from 138.8±1.3° to 42.7±1.7° after 5 min under 365 nm UV irradiation and recovering to the original state after 20 min of visible light irradiation.

6.4.3 Thermal Switching

Thermal-responsive poly(N-isopropylacrylamide) (PNIPAM) hydrogels have been proposed to get responsive and reversible control of wettability and adhesion [30]. In particular, the hydrogel surface can be switched from a superoleophobic and low adhesive state, below the PNIPAM lower critical solution temperature (LCST, 32°C), to an oleophobic and high adhesive state, above its LCST, with a reversible variation of about 24° of the contact angle value. The behavior of PNIPAM was examined in relation to the effects of its contraction and expansion on the local surface topography induced by temperature.

Thermo-responsive polymer brushes anchored on porous anodic aluminum oxide have been demonstrated to cause a variation of the surface roughness due to a volume expansion, resulting in a significant change in wettability (158° to 40°C) [31].

Polybenzoxazine (PBZ) thin films on silica supports have been also proposed to amplify the degree of wettability by combining UV irradiation and thermal treatment. The purpose was to control the competition of intermolecular hydrogen bonding under UV treatment and intramolecular hydrogen bonding under thermal treatment, respectively [32]. In the first case, a superhydrophilic state was induced, while in the second case superhydrophobicity was recovered fully.

A surprising discovery has been recently done for non-wetting droplets on hot superhydrophilic surfaces [33]. Despite spontaneous spreading is expected on microstructured superhydrophilic surfaces, an evaporation-induced pressure prevents wetting ($\theta = 160°$) when heated over the saturation temperature.

6.4.4 pH-Driven Switching

Superoleophilicity-superoleophobicity switching in aqueous media was achieved by grafting pH-responsive poly(2-vinylpyridine) and oleophilic/hydrophobic polydimethylsiloxane (P2VP-b-PDMS). Controllable switching of oil wettability was obtained through protonation/deprotonation mechanisms, giving useful indication about the use of these smart surfaces in programmable oil/water separation [34–35]. In this respect, the literatures refers about different studies on wettability switching triggered by pH, including polyacrylic acid/polivinilpirrolidone (PAA/PVP) and polyacrylic acid/poly(diallyldimethylammonium chloride) (PAA/PDDA) [36–38], polyacrylic acid/poly(allylamine hydrochloride) (PAA/ PAH) [39] or polivinilpirrolidone/carboxyl-terminated polyether dendrimer (PVP/DEN-COOH) multilayer films [40].

Temperature-responsive materials have been also combined with pH-sensing compounds, yielding dual sensing smart surfaces based on P(NIPAAm-co-AAc) copolymers [41]. Under established pH conditions, these functional surfaces exhibited hydrophilic character at low temperature - expanded PNIPAAm conformation below the LCST -, whilst an increase in hydrophobicity was appreciated at higher temperature values - collapsed PNIPAAm conformation above the LCST -. Under fixed temperature conditions, the surface became hydrophilic when pH was lowered and hydrophobic when pH was increased. At pH 2 and 45°C the system exhibits very high contact angle value (≈149°), whereas quicker spreading was observed at temperature below LCST and under harsh basic conditions.

6.4.5 Molecular Switching

Solvent annealing is often used to switch the surface properties, including wettability. The basic concept is to promote a molecular rearrangement under attractive interactions between the vapor molecules and the portion of polymer and/or solid compounds with major affinity to them. With this purpose, poly(methyl acrylate)/polystyrene/ poly(methyl acrylate) (PMA-*block*-PS-*block*-PMA) were anchored on a Si/SiO_2 based surface and annealed with dichloromethane, thereby promoting an extended and smoother surface whereon contact angle of 69° were measured [42]. By using cyclohexane in place of dichloromethane, a rougher surface was obtained, due to the low affinity of PMA towards this solvent. As a result, a local collapse of the polymer structure was induced and contact angle values of 90° were measured. More recently, a compound such as 1,2,3,4,5-hexaphenylsilole (HPS), enabling to emit at 500 nm in the amorphous solid form and to exhibit 465 nm luminiscene in the crystal solid state, has been used to prepare a solvent fuming dual responsive system by combining photoluminescent behavior and wettability [43]. After exposition to ethanol or like-ethanol vapors, the HPS film became more hydrophobic due to the formation of crystalline structures. By annealing the film in toluene or like-toulene vapors a reversible amorphous structure was obtained along with a reduction of the water contact angle value.

Reversible wettability can be also obtained with ionic liquids (ILs). When these compounds contact solutions containing hydrophilic (Cl^-, Br^- and BF_4^-) or hydrophobic anions (PF_6^- and Tf_2N^-), variations of the spreading can be induced on the surfaces. In this respect, a poly(ionic liquid) such as ([1-(4-vinylbenzyl)-3-butylimidazolium hexafluorophosphate] (PVBIm][PF6]) was grafted on silicon wafer. Initially, contact angle values of 95° were measured on the surface. After immersion of the surface in 0.2 M NaCl solution, an exchange of $PF6^-$ with Cl^- was induced, causing a decrease in wettability (CA=41°) [44].

Another example of switchable wettability is given by Ag and CNTs-polyethylene composite surfaces, fabricated by using air-plasma treatments combined with surface fluorination. In this case, it was demonstrated that the water wettability could be switched from superhydrophobic (CA=156°) to superhydrophilic (CA=0°) state, depending on the alternation of the treatments [45].

6.4.6 Mechanical switching

Wetting properties have been also tuned under mechanical stress by using elastic rough poly(dimethysiloxane) films [46]. Specifically, the roughness factor was indicated as an energetic barrier for wettability tuning, the direction and the extent of the period being two crucial factors for spacing and/or geometry changes in the surface under stretching direction.

Adaptable wetting properties have been obtained with an elastic polyamide ($[-(CH_2)_{11}-$ $CO-NH-]_n$) film exhibiting a triangular net-like structure. Under biaxial extension and unloading changes in topography were achieved, causing a variation in the surface free tension. Contact angle values were varied from 151° to 0° under extension and reversibly recovered when the stress was released [47].

6.5 Self-Powered Liquid Motion

Self-powered liquid motion over solid surfaces is attractive for advanced research frontier in the area of biosensors, diffusive transport of biomolecules, labs-on-a-chip and sophisticated microfluidics. Despite actuation mechanisms based on thermocapillary [48] or electrocapillary [49] effects have been proposed in the recent past, the amplification of wettability is regarded as a powerful route for causing droplet liquid motion through local variations of the surface roughness and, therefore, of the surface free tensions (Figure 6.3).

Considered that an increase in the surface roughness causes an increase in hydrophobicity according to the Wenzel and Cassie-Baxter models, whilst flatter surfaces promote a larger spreading of the fluid, the fabrication of textures wherein flatter and rougher regions are adjacent is becoming of great interest for the concrete opportunity to drive the wettability. Literature refers about some attempts to make surface roughness locally adjustable in order to get liquid droplets to move over solid surfaces [50–51]. Mainly, two different approaches are preferred for realizing stimulus-triggered wettability, including the manipulation of adaptable surface chemistry and/or morphology.

Thin PDMS membranes suspended on a PDMS pillar substrate have been proposed for inducing liquid motion through membrane deflection mechanisms actuated by pneumatic tools [50]. During membrane deflection the suspended thin membrane is caused to contract and shape the underlying support, thereby modifying the surface roughness and inducing wettability switching. This allows the liquid to move when changing from rough to flat surfaces. The design of liquid-infused dynamic materials has been also proposed to solicit liquid droplet motion [52]. Elastic responsive porous substrates - a composite polydimethylsiloxane (PDMS) and Teflon

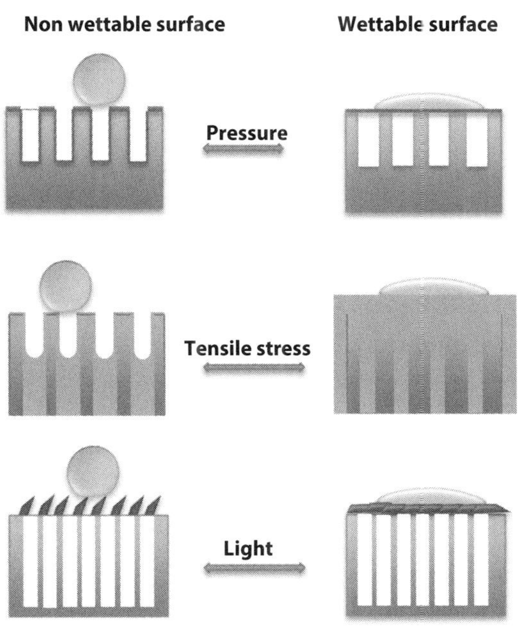

Figure 6.3 Scheme of local variation of the surface roughness for self-powered liquid motion.

membrane - have been proposed in place of a rigid one and combined with a lubricant, i.e. a perfluoropolyether enabling to wet and flow through the pores; this makes versatile, tunable and adaptive topographies through which elastic deformation, fluid flow, and solid-liquid-interactions can be actuated. The mechanism consists of a sequential expansion and contraction of the membrane pores filled by a lubricating liquid when a tensile stress is applied and removed (Figure). Under physical stress conditions, the pores work as mechanical actuators, their stretching causing the liquid to move inside the pores; this leads to a change in the surface tension until the equilibrium is reached at the interface of infused liquid, outside media and radius of the meniscus. Morphologically, this corresponds to a rough surface state. When the tensile stress is released, the liquid expands again outwards leading to a smooth surface. In this case, the fluid mobility, along with optical properties, can be successfully tuned, revealing the adaptability of the device and the efficacy of the approach to start, stop and start again the droplet mobility.

Another viable route for modifying the surface roughness is the photo-isomerization [21]. Membranes poly(vinyl alcohol)-co-ethylene blended with azobenzene polymers have been demonstrated to obtain a variation of the contact angle of 26° due to isomerization from E to isomer Z under irradiation [53]. The behavior of azobenze derivate anchored on silica surface has been studied in relation to the motion of oil droplets under induction of a gradient in the surface tension [54]. Indeed, light-driven motion was effective when the gradient of wettability was established between the one-half of the droplet irradiated with UV light and the other half irradiated with visible light by interfacial forces exceeding droplet pinning on the smooth surface. In this case, 7-[(trifluoromethoxyphenylazo)phenoxy] pentanoic acid was assembled on poly(allylaminehydrochloride)/SiO_2 films by using layer-by-layer techniques. Therefore, changes in dipole and surface roughness were induced under selective UV irradiation, thereby switching the surface wettability from a hydrophilic to superhydrophilic state. [55].

Films containing self-assembled photochromic molecules coordinated to a metal atom and exhibiting light-responsive wettability have been also disclosed in a US patent for applications in micro- and nano-manipulation of fluid motion and fabrication of micro- and nanofluidic devices [56]. Under irradiation changes until values of 1–15° in wettability were measured.

6.6 Self-Cleaning Mechanisms

Self-cleaning behavior against any sort of contamination is often observed for many natural systems, including vegetable leafs, i.e. lotus effects, animal

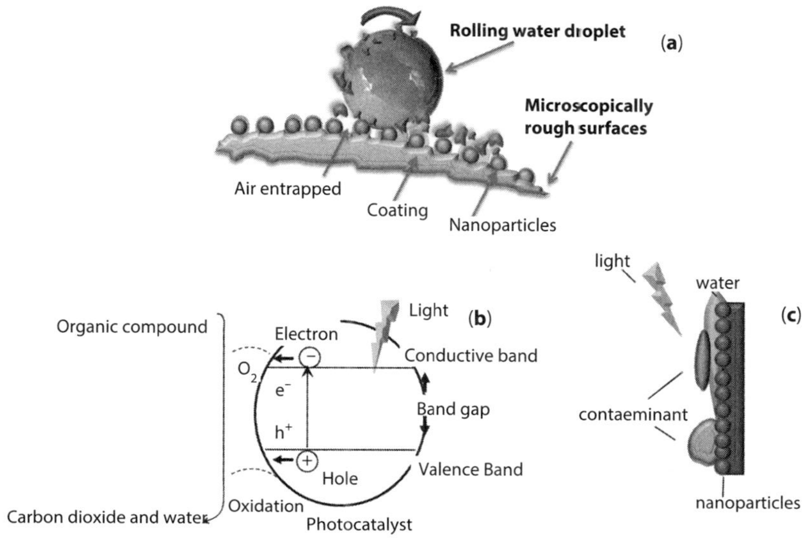

Figure 6.4 Scheme of roll-off mechanism (*a*), photocatalysis (*b*) and pick up surface decomposed contaminants (*c*) adapted from [57].

skins and/or wings/shells. It is, hence, instructive to examine how these natural organisms can remove or prevent organic or inorganic contaminant adhesion through two kinds of actuating mechanisms: droplet roll-off and photocatalysis [57] (Figure 4).

In this respect, there is a lot of research about the fabrication of membranes and/or thin films with complex textures or chemically active surfaces [57–63]; the major ambition is the development of surfaces with very high roughness factor in order to actuate rolling water droplet and/or photocatalytic functions to decompose dirty sticky under irradiation.

6.6.1 Droplet Roll-Off On Superhydrophobic Surfaces

Different manufacturing strategies, including nanolithography, etching, breath figure, nano-molding chemical vapor deposition, electrospinning, atom transfer radical polymerization (ATRP) and grafting approaches [51], have been selected together with classes of compounds such as metals [58], carbon nanotubes [59], silica [60] and metal oxides [61], polymers [62, 64] and peptides [63] to realize hierarchical lotus/cauliflower/gecko-inspired self-cleaning architectures.

Super-amphiphobic silicon wafer have been proposed by forming a fluorinated self-assembled monolayer onto the surface, thereby allowing the removal of dirty by glycerin droplets [65].

Lotus-inspired textures with very high degree of hydrophobicity have been fabricated by assembling modified carbon nanotubes (CNTs) onto cotton surface [58], whereas flexible and superhydrophobic carbon films have been prepared by combining nanocasting, electroplating and physical vapor deposition techniques, achieving lotus-like textures with a good balance of hardness, toughness and superhydrophobicity (160°) [66].

Lotus-like structures with very high superhydrophobic character (160±1°) have been formed with porous microspheres and nanofibers by electrospinning [67]. More recently, superhydrophobic micro and nano-SiO_2 and epoxy resin based structures have been fabricated by multicoating processes, leading to a maximum of contact angle values of 152±1° and a sliding angles of 6° [68].

Polyaniline (PANI) nanowires with binary micro and nanoscale structure have been synthetized onto a Ti/Si substrate in order to form pillar architectures. The polymer has been electrodeposited into the pores of aluminum oxide template, which has been removed in a successive step. After dipping into a solution of perfluorooctanoic acid and *N,N′*- dicyclohexylcarbodiimide (DCCD) in methanol for 2 h, the surface exhibited water CA of 160° [69].

6.6.2 Photocatalysis For Self-Cleaning Surfaces

Organic contaminants, (bio)fouling, microbes and soils can be decomposed in water and carbon dioxide through photocatalysis processes. Metals oxides including TiO_2 or ZnO, Fe_2O_3, CdS and CuO are often used to promote the formation of super-oxide anions and hydroxyl radicals intermediates [70–72], which cause the chemical decomposition making often the surfaces photo-inductively hydrophilic and, then, easily washable. TiO_2 is one of the most used photocatalytic materials for its large availability, non-toxicity, chemical stability, biocompatibility and recyclability [73–78].

Recently, the catalytic activity of poly(allyl amine hydrochloride) (PAH) and poly(styrene sulfonate sodium salt) (PSS) multilayers containing TiO_2 has been evaluated for the degradation of Rhodamine B dye under UV irradiation time of 4 h [79].

Composite membranes prepared from mixtures of carbon nanotubes and TiO_2/Al_2O_3 have been proposed for pollutant degradation during water treatment, achieving fluxes three times higher than traditional filtration and a humic acid removal rate of 10% higher than TiO_2/Al_2O_3 membranes [80]. This can be ascribed to the ability of CNTs to work as electrons acceptors, thereby suppressing the charges recombination under

UV irradiation. Particulate PTFE materials were doped in TiO_2 solutions in order to impart bactericidal and self-cleaning activity, while PVDF membranes were loaded with different amounts of TiO_2 yielding antibacterical against Escherichia Coliform (E. Coli), photoactive property against Reactive Black 5 (RB5) dye and self cleaning (antifouling) properties against BSA solution [81].

6.7 Self-Healing Concepts And Strategies

Thermal, mechanical or flashed events can seriously damage materials forming the membranes during their utilization, causing a loss of process efficiency. A major ambition is to design materials enabling to self-repair and restore the original structure and properties. There are many materials, such as polymers, metals ceramics and composites, which can have ability to self-repair themselves under specific conditions according to autonomic and nonautonomic mechanisms, including thermally activated solid phase healing, compartmentalized healing agents, photoinduced crosslinking, reversible bond formation, conduction and many others [82–85].

Polymers can undergo local stresses, crazes and microvoids due to disentanglement of segment chains or bond rupture. Spontaneous repair can be obtained by heating the polymer above its glass transition temperature and applying a pressure, thereby promoting molecular inter-diffusion through the chains and complete welding at the crack planes [86].

Addition of solvent can induce a lowering of the glass transition temperature and, hence, promote healing behaviors. Also, healing agents can be encapsulated inside microcapsules. After rupture of the capsules, these agents can be released and migrate towards the regions where the cracks is occurred. In presence of a catalyst and/or chemical activator uniformly dispersed through the matrix a chemical reaction is induced and the damaged material is sealed (Figure 5).

This approach needs to meet some criteria, including the ability to encapsulate and store healing agents in microspheres uniformly dispersed through the matrixes, chemical inertia of the agents during storage, long-term stability, proximity to the catalyst, robustness when processed but also weakness enough to break when a cracking is induced in the matrix. In addition, the healing agent, i.e. a monomer, must have low viscosity in order to reach the site of damage under capillary forces as well as the polymerization must be fast, thereby preventing shrinking events and/or stress relaxation. In some cases, the encapsulation of the catalyst in microcapsules and the dispersion of the healing agent in a phase separated

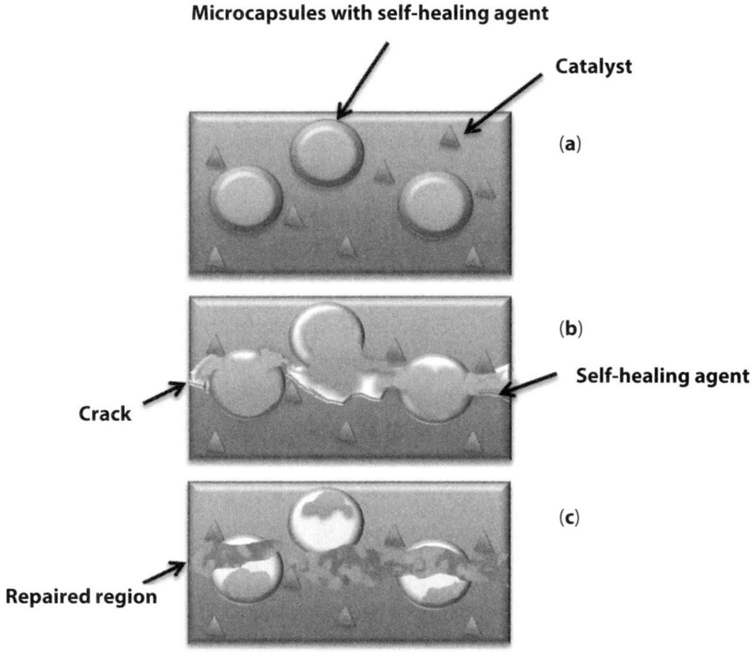

Figure 6.5 Scheme of microcapsules containing self-healing agents: *a)* dispersion of self-healing capsules and catalyst through the matrix; *b)* rupture of self-healing capsules in proximity to the crack; *c)* repairing action with welding of the matrix.

in the matrix are preferred. In both the cases, the systems are called as an autonomic healing. In this respect, high healing efficiency has been obtained when epoxy materials were used in a ring-opening metathesis polymerization (ROMP) reaction of microencapsulated dicyclopentadiene (DCPD) with solid particles of Grubbs catalyst [87].

Because the use of microcapsules allows a unique self-repairing event, hollow fibers [88] or bio-inspired microvascular networks [89] are preferred for multiple healing (Figure 6.6).

In the first case, larger amount of healing agents can be delivered to the overall crack plane closing the crack; in the second case, chemicals and building blocks of healing can be continuously transported to the damaged sites by using vascular structure-like devices.

Nanoparticles have been also demonstrated to have capability to migrate towards and fill cracked regions in flexible polymer domains [90]. Another approach is the reversible cross-linking, which shows self-healing ability when an external trigger such as temperature, light or chemical agent is applied. In this case, the system is called as a nonautonomic healing. Diels-Alder and retro Diels-Alder reactions, involving furanic polymers

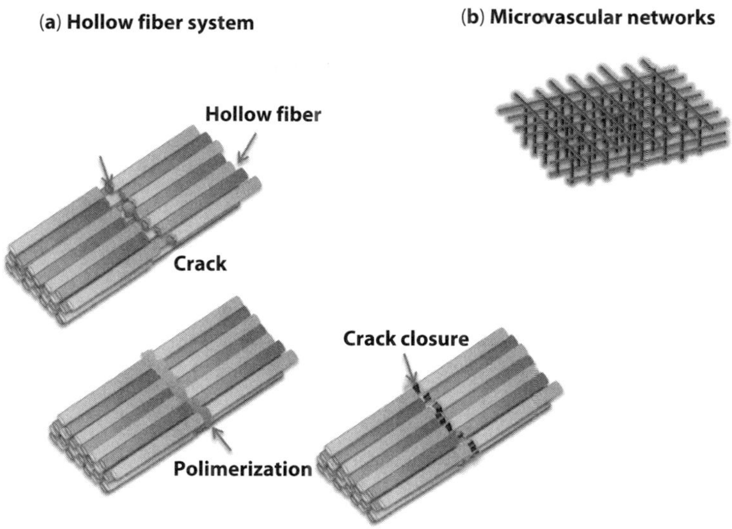

Figure 6.6 Representation of two selfhealing systems for multiple reparation: *a*) hollow fiber system; and *b*) microvascular network

with maleimide, are considered powerful tools for inducing reversible thermal cross-linking [91]. Photocycloaddition and recycloaddition of cinnamoyl groups have been also proposed to repair samples without using catalysts, additives or heat [92].

The use of conductive materials is becoming attractive for sensing and self-repairing systems under electrical field as well. Considering that micro-cracks, defects cause a decrease in conductivity, measurements of the resistance can be used to monitor and quantify the extent of damages through the materials. In addition, the application of a voltage bias can be exploited to generate local heat at the micro-crack forcing the system to recover the original state and selfheal. In this respect, organometallic polymers based on carbenes, metals and other conductive materials can be used to fabricate self-repairing devices. Shape memory alloys (SMAs) have been also demonstrated to exhibit healing functionality under heating [93].

6.8 Repairable Surface Properties

Loss of wettability properties as well as occurrence of fouling is a common event, which membranes suffer during separation processes. Attractively, self-healing approaches can be regarded as useful tools for recovering and restoring some original membrane functionalities when damaged.

6.8.1 Restoring Surface Superhydrophobicity

The loss of superhydrophobicity is due to chemical oxidation, exposition to harsh chemical environment, strong light, or physical friction. Generally, two strategies are proposed to restore surface repellency: a) encapsulation of hydrophobic components in nanoporous reservoirs; b) spontaneous self-organization of hydrophobic colloidal particles, which can form micro- and nanoscaled hierarchical topography at the damaged interface. Perfluoroalkyl silanes were used to fill the pores of a LBL film [94]. After loss of hydrophobicity due to mechanical stress and chemical reaction, the particles migrated towards the surface restoring the water repellency of the film. Recently, nanoporous anodized allumina [95] and silica particles [96] have been filled with low surface tension substances enabling to be released and restore superhydrophobicity of the surfaces. Mesoporous silica was used as a reservoir of octadecylamine (ODA). A polymdopamine layer was deposited onto ODA, thereby encapsulating and dispersing the silica particles. Contact angle values of 157.6° were estimated as well as self-healing properties.

Colloidal hydrophilic silica particles were immersed in wax and annealed in air above the wax melting point, resulting in superhydrophobic and self-repairing surfaces [97]. Considering the high tendency of the wax to crystallize, the particles caused rougher surface. After cracking, the surface was melted again thereby allowing the particles to self-organize themselves at the wax-air interface and restore superhydrophobic functions.

As a practical example, the use of self-healing compounds in textiles is in great demand due to the increasingly demand of durable self-cleaning properties. Fabrics treated with fluorinated-decyl polyhedral oligomeric silsesquioxane (FD–POSS) and a fluorinated alkyl silane (FAS) showed also self-healing properties and durable functions [98]. After plasma treatment, high degree of spreading was observed when water and organic liquids were put in contact with the coated fabric. Under heating, a quick recovery of superhydrophobicity and oleophobicity was estimated. The increased mobility and molecular rotation of the chemical moieties along with the tendency of the non-polar groups to orient themselves towards the substrate-air interface were proposed as the controlling factors for the self-healing event in the coated fabric.

6.8.2 Self-Healing for Durable Anti-Fouling Properties

Degradation and/or detachment of 2D grafted polymer chains and self-assembled monolayers (SAMs) cause frequent loss of antifouling properties. Long-lasting antifouling properties have been obtained with 3D grafted polymer structures, consisting of pH-responsive cross-linked poly(2-vinyl pyridine) (P2VP) films on the surface of Si-wafers [99]. In order to prevent this undesired event, polymer chains have been grafted onto the surface and inside the host substrate in proximity to the interface. The latter are able to replace the damaged or degraded segments under a gradient of chemical potential. A self-rearrangement of the grafted PEO chains residing in the network in proximity to the interfaces was indicated as a self-healing mechanism. Hydrogels prepared by UV-initiated copolymerisation of non-fouling zwitterionic carboxybetaine methacrylamide (CBMAA-3) and 2-hydroxyethyl methacrylate (HEMA) containing a uniform dispersion of clay nanoparticles (Laponite XLG) have been also proposed as self-healing materials with superior mechanical properties and resistance to fouling resistance to blood plasma [100]. In this case, the self-healing properties of the films were ascribed to the physical cross-linking of the hydrogels, which resulted to be better than chemical cross-linking. Some pH and sulfide ion sensitive nano-containers based on silica nanoparticles can be proposed to generate self-healing and antifouling multifunctional coatings. These systems exhibit pH-responsiveness, which regulates a durable self-healing function along with antifouling properties, owing to a biocide incorporated in the release system as well.

Hybrid antifouling coatings with a self-renewing topographical microgel surface have been fabricated by mixing small surface-functionalized microgels with a self-peeling hydrolysable terpolymer resin in presence of a poly-functional cross-linking agent (axiridine) [101]. When contacting water the microgels swell, while a slow hydrolysis of the top layer of the resin causes the formation of a brush-like soft/dynamic layer due to the rupture of crosslinking bonds. Successively, hydrolyzed resin chains together with microgels are released in water. This polishing mechanism along with the formation of microstructured topographies prevents and/or reduces biofouling events.

6.9 Conclusions and Perspectives

In membrane surfaces superior sensitivity and actuation functions can be accommodated in order to accomplish and manage important events, such as fouling mitigation, in-situ cleaning, liquid flows and self-healing of original features. All these phenomena occur mainly via surfaces and are responsible for durable, efficient, and proficient membrane processes performances. The surface manipulation reveals itself as a powerful tool to reproduce on lab-scale events, which daily occur in nature and seem to offer practical solutions to undesired events that affect frequently many production and separation processes. The possibility to associate smart functions to materials surfaces through combination and integration of structural and chemical components is a viable route for designing smart devices enabling to actuate sophisticated mechanisms of self-cleaning, self-repairing, storage as well as self-powered transport and auto-recovery. In this respect, membrane-based sensors represent a dynamic evolution of materials where chemical, physical, biological and engineering components work synergistically for making the devices adaptable to various environmental conditions and, hence, suitable for technologically sophisticated applications in the fields of microfluidics, ultra-smart textiles, drug delivery systems, (bio)separations.

References

1. A. Gugliuzza, R. Fabiani, M.G. Garavaglia, A. Spisso, E. Drioli, *J. Coll. Interf. Sci.*, Vol. 303, p. 388, 2006.
2. J. P. Youngblood, N. R. Sottos, *MRS Bulletin*, Vol. 33, p. 732, 2008
3. A. Gugliuzza and A. Basile, "Membrane contactor technology: fundamentals, interfaces and prospects" in A. Basile and A. Cassano, eds., *Membrane processes for a sustainable growth*, Nova Science Publishers, Inc., Hauppauge, NY, 2013
4. J. Genzer, K. Efimenko, *Biofouling*, Vol. 22, p. 339, 2006.
5. X. Gao, L. Jiang, *Nature*, Vol. 432, p. 36, 2004.
6. A. Gugliuzza, "Membrane Wettability" in E. Drioli and L. Giorno eds., *Encyclopedia of Membranes*, SpringerReferences, 2013.
7. T.R. Thomas, *Rough Surfaces, 2nd ed.*, Imperial College Press, London, 1999.
8. M. Taniguchi, G. Belfort, *Langmuir* , Vol.18, p. 6465, 2002.
9. M.E. Schrader and G.L. Loeb, *Modern Approaches to Wettability*, Plenum Press, New York, 1992.
10. F. Xia, Y. Zhu, L. Feng, L. Jiang, *Soft Matter*, Vol. 5, p. 275, 2009;
11. D. J. Beebe, J. S. Moore, Q. Yu, R. H. Liu, M. L. Kraft, B.-H. Jo, C. Devadoss, *Proc. Natl. Acad. Sci. U. S. A.*, Vol. 97, p.13488, 2000.
12. J. Lahann, S. Mitragotri, T. N. Tran, H. Kaido, J. Sundaram, I. S. Choi, S. Hoffer, G. A. Somorjai, R. Langer, *Science*, Vol. 299, p. 371, 2003.
13. T. L. Sun and G. Y. Qing, *Adv. Mater.*, Vol. 23, p. 55, 2011.
14. B. W. Xin and J. C. Hao, *Chem. Soc. Rev.*, Vol. 39, p. 769, 2010
15. N. Verplanck, Y. Coffinier, V. Thom, R. Boukherroub, *Nanoscale Res Lett*, Vol. 2, p. 577, 2007.
16. F. Mugele, J.-C. Baret, *J. Phys.: Condens. Matter*, Vol. 17, p. 705, 2005.
17. Z. Wang, Y. Ou, T.M. Lu, N. Koratkar, *J Phys Chem B*, Vol. 111, p. 4296, 2007.

18. Z. Wang, L. Ci, L. Chen, S. Nayak, P.M. Ajayan, N. Koratkar, *Nano Lett.*, Vol. 7, p. 697, 2007.
19. M.S. Dhindsa, N.R. Smith, J. Heikenfeld, P.D. Rack, J.D. Fowlkes, M.J. Doktycz, A.V. Melechko, M.L. Simpson, *Langmuir*, Vol. 22, p. 9030, 2006.
20. K. Bhat, J. Heikenfeld, M. Agarwal, Y. Lvov, K. Varahramyan, *Appl. Phys. Lett.*, Vol. 91, p. 024103, 2007.
21. F.P. Nicoletta, D. Cupelli, P. Formoso, G.D. Filpo, V. Colella, A. Gugliuzza, *Membranes*, Vol. 2, p. 134, 2012.
22. V. Khranovskyy, T. Ekblad, R. Yakimova, L. Hultman, *Applied Surface Science*, Vol. 258, p. 8146, 2012.
23. X. Zhang, M. Jin, Z. Liu, D.A. Tryk, S. Nishimoto, T. Murakami, A. Fujishima, J. *Phys. Chem. C.*, Vol. 111, p. 14521, 2007.
24. B. Xin, J. Hao, *Chem. Soc. Rev.*, Vol. 39, p. 769, 2010.
25. S. Wang, Y. Song, L. Jiang, J. *Photochemistry and Photobiology C: Photochemistry Reviews*, Vol. 8 p. 18, 2007.
26. M. Chen, F. Besenbacher, *ACS Nano*, Vol. 5, p 1549, 2011.
27. S.H. Anastasiadis, M. I. Lygeraki, A. Athanassiou, M. Farsari, D. Pisignano, *J. Adhesion Sci Tech*, Vol. 22, p. 1853, 2008.
28. A. Nayak, H. Liu, G. Belfort, *Angew. Chem. Int. Ed.* Vol. 45, p. 4094, 2006.
29. D. Wang, P.Jiao, J. Wang, Qi. Zhang, L. Feng, Z. Yang, J. *Appl. Polym. Sci.*, Vol. 125, p. 870, 2012.
30. L. Chen, M. Liu, L. Lin, T. Zhang, J. Ma, Y. Song, L. Jiang, *Soft Matter*, Vol. 6, p. 2708, 2010.
31. Q. Fu, G.V.R. Rao, S.B. Basame, D.J. Keller, K. Artyushkova, J.E. Fulghum, G.P. Lopez, *J. Am. Chem. Soc.*, Vol. 126, p. 8904, 2004.
32. W.-C. Su, S.-W. Kuo, J. Nanomat., Vol. 2013, p. 12, 2013.
33. S. Adera, R. Raj, R. Enright, E. N. Wang, *Nature Comm* Vol. 4, p. 1, 2013.
34. M. Zhang, L. Liu, C. Wu, G. Fu, H. Zhao and B. He, *Polymer*, Vol. 48, p. 1989, 2007.
35. L. Zhang, Z. Zhang, P. Wang, *NPG Asia Materials* Vol. 4, p. e8, 2012.
36. A. Synytska, M. S. Diez and L. Lonov, *Langmuir*, Vol. 23, p. 5205, 2007.
37. Y. Fu, S. L. Bai, S. Cui, D. Qiu, Z. Wang and X. Zhang, *Macromolecules*, Vol. 35, p. 9451, 2002
38. G. Diaconu, A. Gugliuzza, T. Schäfer, *Procedia Engineering*, Vol. 44, p.1260, 2012.
39. L. A. Connal, Q. Li, J. F. Quinn, E. Tjipto, F. Caruso and G. Qiao, *Macromolecules*, Vol. 41, p. 2620, 2008.
40. H. Zhang, Y. Fu, D. Wang, L. Wang, Z. Wang and X. Zhang, *Langmuir*, Vol. 19, p. 8497, 2003.
41. B. Xin, J. Hao, *Chem. Soc. Rev.*, Vol. 39, p. 769, 2010.
42. S. G. Boyes, W. J. Brittain, X. Wang and S. Z. D. Cheng, *Macromolecules*, Vol. 35, p. 4960,2002.
43. L. Heng, Y. Dong, J. Zhai, B. Tang, L. Jiang, *Langmuir*, Vol. 24, p. 2157, 2008.
44. X. He, W. Yang and X. Pei, *Macromolecules*, Vol. 41, p. 4615, 2008.

45. X. Zhu, Z. Zhang, K. Wang, J. Yang, X. Xu, X. Men, X. Zhou, *New J. Chem.*, Vol. 36, p. 1280, 2012.
46. S. Zhao, H. Xia, D. Wu, C. Lv, Q.-D. Chen, K. Ariga, L.-Q. Liue, H.-B. Sun, *Sof Matter*, Vol. 9, p. 4236, 2013.
47. J. Zhang, X. Lu, W. Huang, Y. Han, *Macromol. Rapid Commun.*, Vol. 26, p. 477, 2005.
48. V. Pratap, N. Moumen, R. S. Subramanian, *Langmuir*, Vol. 24, p.5185, 2008.
49. T.J. Lewis, J.P. Llewellyn, M.J. van der Sluijs, J. *Coll. Interf. Sci.*, Vol. 162, p. 381, 1994.
50. J. Lee, B. He, N.A. Patankar, *J. Micromech. Microeng.* Vol. 15, p.591, 2005.
51. K. Liu, Y. Tian, L. Jiang, *Progress in Materials Science*, Vol. 58, p.503, 2013.
52. : Yao, Y. Hu, A. Grinthal, T.S. Wong, L. Mahadevan, J. Aizenberg, *Nature Materials*, Vol. 12, p. 529, 2013.
53. B. Tylkowski, S.Peris, M. Giamberini, R. Garcia-Valls, J.A. Reina, J.C. Ronda, *Langmuir*, Vol. 26, p. 14821, 2010.
54. K. Ichimura, S.K. Oh, M. Nakagawa, *Science*, Vol. 288, p. 1624, 2000.
55. H.S. Lim, J.T. Han, D. Kwak, M. Jin, K. Cho, *J. Am. Chem. Soc.*, Vol. 128, p. 14458, 2006.
56. W. G. Mc Gimpsey, J.C. MacDonald, *7326526 B25*, 2008.
57. A. Gugliuzza, E. Drioli, *J. Membr. Sci.*, Vol. 446, p. 350, 2013.
58. K. Liu, L. Jiang, Bio-inspired self-cleaning surfaces, *Annu. Rev. Mater. Res.* , Vol.42, p. 231, 2012.
59. J.H. Zou, H. Chen, A. Chunder, Y.X. Yu, Q. Huo, L. Zhai, *Adv. Mater.*, Vol. 20 p. 3337, 2008
60. Q.F. Xu, J.N. Wang, K.D. Sanderson, *J. Mater. Chem.*, Vol. 20, p.5961, 2010.
61. KS Liu, J Zhai, L. Jiang, *Nanotechnology*, Vol. 19, p.165604, 2008.
62. H.S. Hwang, N.H. Kim, S.G. Lee, D.Y. Lee, K. Cho, I. Park, *ACS Appl. Mater. Interfaces*, Vol. 3, p. 2179, 2011.
63. J.S. Lee, J. Ryu, C.B. Park, *Soft Matter* Vol. 5, p. 2717, 2009.
64. A. Gugliuzza, E. Drioli, *Desalination*, Vol. 240, p.14, 2009.
65. T.I. Kim, D. Tahk, H.H. Lee, *Langmuir*, Vol. 25, p. 6576, 2009.
66. Y. Wang, L. Wang, S.C. Wang, R.J.K. Wood, Q. Xue, *Surf. Coat. Tech.*, Vol. 206, p. 2258, 2012.
67. L. Jiang, Y. Zhao, J. Zhai, Angew. *Chem. Int. Ed.* Vol. 43, 4338, 2004.
68. B. Zhang, H. Suo, H.D. Li, P. Shen, H. Tan, *Proceeding of the 6th PSU-UNS ICET-2013*, Vol. T.2-1.2, p. 1, 2013.
69. M. Qu, G. Zhao, X. Cao, J. Zhang, *Langmuir*, Vol. 24, p. 4185, 2008.
70. A.L. Costa, Self-cleaning textile: a basic approach for photocatalytic characterization, in: *Proceedings of Nanotechitaly 2012*, Vol. 31-Nanomat, 2012.
71. R.H. Wang, X.W. Wang, J.H. Xin, *ACS Appl. Mater. Interf.*, Vol. 2, p.82, 2010.
72. M. Miyauchi, A. Nakajima, K. Hashimoto, T. Watanabe, *Adv. Mater.*, Vol 12, p. 1923, 2000.

73. J.J. Park, J.G. Lee, D.Y. Kim, J.H. Hong, J.J. Kim, S. Hong, S.S. Yoon, *Environ. Sci. Tech.*, Vol. 46, p. 12510, 2012.
74. X. Zhang, M. Jin, Z. Liu, D.A. Tryk, S. Nishimoto, T. Murakami, A. Fujishima, *J. Phys. Chem. C,* Vol. 111, p. 14521, 2007.
75. S.D. Richardson, A.D. Thruston Jr, T.W. Collette, K.S. Patterson, B.W. Lykins Jr, J.C. Ireland, *Environ. Sci. Tech.,* Vol. 30, p. 3327, 1996.
76. B.R. Eggins, F.L. Palmer, J.A. Byrne, *Water Research,* Vol. 31, p. 1223, 1997.
77. L. Zhang, Y. Zhu, Y. He, W. Li, H. Sun, *Appl. Catal., B,* Vol. 40, p. 287, 2003.
78. E.I. Kapinus, T.I. Viktorova, T.A. Khalyavka, *Theor. Exp. Chem.,* Vol. 42, p. 282 2006.
79. D. Neela Priya, J.M. Modak, A.M. Raichur, *ACS Appl. Mat. Interf.*, Vol. 1, p. 2684, 2009.
80. H. Zhao, H. Li, H. Yu, H. Chang, X. Quan, S. Chen, *Sep. Purif. Tech.* Vol. 116, p. 360, 2013.
81. R.A. Damodar, S.-J. You, H.-H. Chou, . *J. Hazard. Mat.* Vol. 172, p.1321, 2009.
82. K. Jud and H. H. Kausch, *Polym. Bull.*, Vol. 1, p. 697,1979.
83. E.N. Brown, S.R. White, N.R. Sottos, *J. Mat. Sci.,* Vol. 39, p. 1703, 2004.
84. C. M. Chung, Y. S. Roh, S. Y. Cho and J. G. Kim, *Chem. Mater.*, Vol. 16, p. 3982, 2004.
85. J. H. Harreld, M. S. Wong, P. K. Hansma, D. E. Morse, G. D. Stucky, *US Pat., 2004007792-A1*, 2004.
86. K.A. Williams, A.J. Boydston, C.W. Bielawski, *J. Royal Society of Interface*, 4, 359–62. 2007.
87. J.D. Rule, E.N. Brown, N.R. Sottos, S.R. White, J.S. Moore, *Adv. Mater.* 17, 205 (2005).
88. C. M. Dry, W.McMillan, *Proc. SPIE - Int. Soc. Opt. Eng.*, Vol. 2718, p. 448, 1996.
89. J.A. Lewis, G.M. Gratson, *Mater. Today*, Vol.7, p. 32, 2004
90. S. Grupta. Q. L. Zhang, T. Emrick, A. C. Balazes, T. P. Russell, *Nat. Mater.*, Vol. 5, p. 229, 2006.
91. X. Chen, A. Dam, K. Ono, A. Mal, H. Shen, S.R. Nutt, K. Sheran, F. Wudl, *Science*, Vol. 295, p.1698, 2002.
92. C.M. Chung, Y.S: Roh, S.Y. Cho, J.G. Kim, *Chem. Mater.*, Vol. 16, p. 3982, 2004.
93. W. Nia, Y.T.v Cheng, D.S. Grummon, *Appl. Phys. Lett.*, Vol. 80, p. 3310, 2002.
94. Y. Li, L. Li and J. G. Sun, *Angew. Chem., Int. Ed.*, Vol. 49, p. 6129, 2010.
95. X. L. Wang, X. J. Liu, F. Zhou and W. M. Liu, *Chem. Commun.*, 2011, 47, 2324–2326.
96. Q. Liu, X. Wang, B. Yu, F. Zhou and Q. Xue, *Langmuir*, Vol. 28, p.5845, 2012.
97. N. Puretskiy, G. Stoychev, A. Synytska and L. Ionov, *Langmuir*, Vol. 28, p. 3679, 2012.
98. H. Wang, Y. Xue, J. Ding, L. Feng, X. Wang, T. Lin, *Angew. Chem. Int. Ed.*, Vol. 50, p. 11433, 2011.

99. H. Kuroki, I. Tokarev, D. Nykypanchuk, E. Zhulina, S. Minko, *Adv. Funct. Mater.*, p. 1, 2013.
100. N. Yu. Kostina, S. Sharifi, A. de los Santos Pereira, J. Michálek, D.W. Grijpm, C. Rodriguez-Emmenegger, *J. Mater. Chem. B*, Vol.1, p. 5644, 2013.
101. F. Hong, L. Xie,. C. He, J. Liu, G. Zhang, C. Wu, *J. Mater. Chem. B*, Vol. 1, p. 2048, 2013.

7

Model Bio-Membranes Investigated by AFM and AFS: A Suitable Tool to Unravel Lipid Organization and their Interaction with Proteins

Andrea Alessandrini[1,2] and Paolo Facci[3,*]

[1]*Centro S3, CNR-Istituto Nanoscienze, Modena, Italy*
[2]*Department of Physics, University of Modena and Reggio Emilia, Modena, Italy*
[3]*CNR-Istituto di Biofisica, Genova, Italy*

Abstract

The study of the biological membrane has largely benefitted from the exploitation of model bilayer systems. These simplified models of the complex biological membrane composed of thousands of different types of molecules allow both to understand basic physical principles underlying the membrane functioning and to test new techniques that will be subsequently applied to biological membranes. Here we concentrate on one of the most used model systems for this kind of investigations: the Supported Lipid Bilayer (SLB). In particular, we analyze the possibilities of investigation offered by Atomic Force Microscopy and Spectroscopy (AFM/AFS) on this model system. We discuss the information that these techniques are able to provide on the phase behavior of the lipid bilayers and on the partitioning of membrane proteins relative to the bilayer lateral heterogeneity. We discuss also the possibility to characterize the mechanical properties of lipid bilayers on the nanometer scale lateral resolution.

Keywords: Atomic force microscopy, atomic force spectroscopy, supported lipid bilayer, lipid phase transition, lipid/protein interactions

**Corresponding author*: paolo.facci@cnr.it

Annarosa Gugliuzza (ed.) Smart Membranes and Sensors, (185–226) 2014

7.1 Introduction

Biological membranes represent ubiquitous biological structures whose activity is central for cells to establish communications between different internal compartments and between the inside and outside [1]. They exemplify a typical complex biological system where the functional properties of the organization of many elements are more relevant with respect to the possible role due to the presence of a single molecule [2]. The Singer and Nicolson model for the biological membrane [3], proposed at the beginning of the '70s, assigned a secondary role to the lipid bilayer framework of the membrane. Essentially, the lipid bilayer represented a solvent for the membrane-related proteins allowing them to diffuse and to interact with each other while performing their functional activity. The possibility of domain formation in the lipid bilayer was contemplated, but no functional role was assigned to them. Besides the Singer and Nicolson model, other descriptions for the organization of the biological membranes were proposed. Nowadays, revitalized by the lipid raft idea [4], the heterogeneous structure of the biological membrane and its relevance for function is a concept largely shared in the biophysical and biological community. The lipid raft idea has continuously changed since the time of its introduction in the scientific community. Its origin is rooted in experiments on biological membranes showing the presence of Detergent Resistant Membrane domains (DRM). These experimental observations led to the introduction of the term "raft" to designate globally the presence of stable domains in the biological membranes enriched in cholesterol and in particular types of proteins. However, the technical aspects involved in the isolation of these domains raised some doubts on their existence in physiological conditions. Moreover, compared to the easiness of observation of cholesterol- and sphyngomyelin-rich domains in model membranes, lipid rafts have remained quite elusive in biological membranes under physiological conditions. Subsequently, it became clear that the existence of rafts in biological membranes had to be associated with small lateral extension (20–100 nm) and with a short aggregation lifetime for the constituent molecules [5–7]. From a biological point of view, several active roles were assigned to the raft structures. The up-to-date situation for the lipid raft idea is supported by the evidence of their presence in biological membranes obtained by super-resolution microscopies [8, 9] which highlighted the discontinuous diffusion of raft proteins in the membrane, in which periods of free diffusion alternate with periods of entrapment in nanometric domains. Moreover, protein diffusion has been found to be strongly related to the presence of cholesterol in the membrane. Very recently, another concept complemented the picture of the lipid rafts: their formation is strongly related to the involvement of transmembrane proteins linked to cytoskeleton underlying the plasma membrane [10]. In this picture, the two main structures assuring the stability of cells, the membrane and the cytoskeleton, are strongly related in performing relevant functional tasks in cells.

As already stated, the biological model membrane poses considerable challenges if a physical picture of the mechanisms underlying its function is to be implemented. Many different molecules are involved in the system and a thermodynamical description including concepts of phase transition, critical point, non-equilibrium fluctuations is required [11]. Many of the physical principles underlying the organization and function of lipid bilayer have been obtained starting from model systems of the biological membrane [12]. These models represent simplified but real systems where the number of involved molecular species is reduced (typically from 1 to 3 types of lipids and 1 or 2 different protein species) and where interactions with other structures of the cell are disregarded. Three classes of model systems are mainly used: liposomes, whose dimensions can range from 20–30 nm in diameter (Small Unilamellar Vesicles, SUV) to 100–200 nm (Large Unilamellar Vesicles, LUV) and up to microns (Giant Unilamellar Vesicles, GUV); Black Lipid Membranes (BLM), which are planar unsupported lipid bilayers especially useful for the studies of functional properties of transport proteins; Supported Lipid Bilayers (SLBs) which are planar lipid bilayers with one side facing a solid substrate. The role of lipid bilayer model systems is not limited to the possibility of understanding the basic physical principles ruling the organization of biological membranes. In fact, they are also useful as benchmarks for the assessment of new techniques that, before being applied to complex, *in-vivo* systems, need to be validated on samples of well-known composition. The final goal is that of increasing the complexity of the investigated system step by step, by adding structural elements which might be relevant for the functionality of biological membranes.

Giant vesicles (GUV), SLB and BLM structures can be used to study interaction of membrane proteins with lipids. Many of the studies performed with model systems incorporating membrane proteins established that the activity of membrane proteins is strongly related to the properties of the lipid bilayer hosting them [13]. Lipid environment can modulate the functional activity of membrane proteins in several ways: 1) through specific interactions between lipids and proteins; 2) through the mechanical properties of the lipid bilayer, considered as a structurally organized continuum medium which might affect the conformational transitions of membrane proteins; 3) mediating membrane protein aggregation by elastic interactions [14]. It is clear that the interactions between lipids and proteins can go both ways: from the lipids to the proteins and vice versa.

The possibility of formation of lipid rafts in biological membranes is equivalent to that of a new thermodynamic phase in model systems: the liquid ordered one [15]. According to the conformational order of the acyl chains or to the lateral organization of the lipid headgroups, different thermodynamic phases can be found in lipid bilayers: the solid ordered phase, in which the alkyl chains are in an all-trans configuration and the lipids show a reduced lateral mobility; the liquid disordered one, which is characterized by the presence of gauche defects in the chains and by a high degree of lateral mobility of the lipids; the liquid ordered one, in which the acyl chains are ordered but still allowing for a high degree of lateral mobility. In this scenario, lipid rafts might be the result of an ordering effect of cholesterol on the neighboring acyl chains establishing a coexistence of two liquid immiscible phases.

All the considered phases are characterized by specific mechanical properties related to different deformation modes of the lipid bilayer. The mechanical properties are identified by elastic constants which, considering small relative deformations, imply a square dependence of the deformation energy on the distortion. Given the remarkable connection between mechanical properties of lipid bilayers and membrane protein activity, the investigation of these mechanical aspects is highly desirable. Moreover, an investigation of the bilayer properties at the nanoscale would provide valuable information on a scale relevant to membrane protein functionality. At the same time, it would be interesting to access both microscopy and force spectroscopy information with a single technique. Among the considered model membrane systems, Supported Lipid Bilayers are prone to be studied by microscopy techniques such as Atomic Force Microscopy that exhibits, besides, force spectroscopy capabilities [16–20].

In the following we will discuss how Supported Lipid Bilayers studied by Atomic Force Microscopy can provide valuable information on the organization of lipid bilayers including their mechanical properties and the possible partitioning behavior of membrane proteins relatively to a heterogeneous lateral organization of the lipids. In Section 2 we will introduce the details of the lipid bilayer model system including its chemical-physical properties. In Section 3 we will provide examples of the phase behavior of Supported Lipid Bilayers as studied by Atomic Force Microscopy. In section 4 we will present the force spectroscopy technique concentrating on the information the technique is able to provide on the mechanical properties of the lipid bilayer. In Section 5 we will discuss the aspects related to protein/lipid interactions with some connections also to experiments aimed at correlating information obtained by Atomic Force Microscopy with functional properties of membrane proteins (ion channels).

7.2 Supported Lipid Bilayers

7.2.1 Preparation Techniques

Supported Lipid Bilayers consist of a lipid bilayer on a rigid, typically hydrophilic, substrate such as glass, silicon (di)oxide or mica (Figure 7.1*a*). Even if specific properties of supported bilayers might depend on the type of substrate, we will start with a general introduction to this lipid bilayer model system and we will come back later to aspects related to the type of substrate used. SLBs were initially developed by the McConnell's group to study the interaction of cells with lipid bilayers [21, 22]. They can be assembled by two different strategies: the Langmuir Blodgett/Schaefer approach [23] and the vesicle fusion technique [24]. The first technique is based on two consecutive transfers to a solid substrate of lipid monolayers assembled at the liquid/air interface in a Langmuir trough. The appealing feature of this approach is connected with the possibility of forming lipid bilayers characterized by a transbilayer lipid asymmetry, reproducing the actual situation found in biological membranes. However, it has been demonstrated that, due to rapid (as compared to measuring time) flip-flop transitions in these systems, an initial transmembrane lipid composition asymmetry is difficult to be preserved [25].

The vesicle fusion technique forms SLBs from unilamellar vesicles in solution [26]. Upon contact with surfaces, under specific conditions,

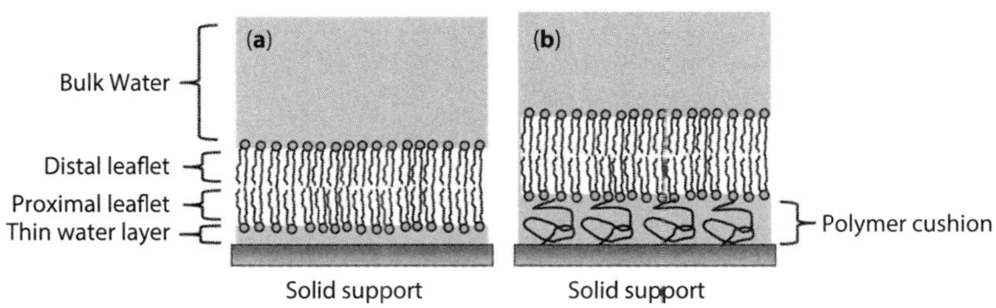

Figure 7.1 Scheme of a Supported Lipid Bilayer. (*a*) Scheme of a lipid bilayer directly supported on a solid substrate. The presence of a thin water layer between the bilayer and the substrate is highlighted together with the separation of the distal and proximal leaflets; (*b*) Scheme of a polymer-cushioned lipid bilayer. In this case a soft hydrophilic layer divides the lipid bilayer from the solid substrate.

unilamellar vesicles fracture, forming a planar bilayer [27]. In both preparation strategies, the presence of a thin water layer between the bilayer leaflet nearer to the substrate (proximal leaflet) and the substrate itself allows lipid diffusion to be preserved to a certain extent [28] (see discussion below). The thin water layer allows also to host transmembrane proteins even if a small portion protrudes from the bilayer towards the support side. The vesicle fusion technique allows the incorporation of transmembrane proteins in the SLB more easily than the Langmuir-Blodgett/Schaefer one. Even if strategies for the incorporation of detergent solubilized transmembrane proteins in already formed supported lipid bilayers have been developed [29–31], the direct fusion of proteoliposomes on surfaces appears as a more practical approach. Moreover, the drawback of the technique, which exploits the addition of detergent to slightly destabilize the lipid bilayer in order to favor the insertion of detergent solubilized membrane proteins, lies in the unknown amount of detergent left in the bilayer. Indeed, the residual detergent could affect the thermodynamics of the lipid bilayer, e.g. altering the compartmentalization of membrane proteins in different lipid domains.

It has to be stressed that in the vesicle fusion technique an undesired asymmetry in the lipid composition might be obtained. This asymmetry involves both lipid composition and lipid density. In case of mixtures, due to the presence of the substrate, lipid composition in the two leaflets could not be easily predicted on the basis of the composition used for SUV or LUV preparation and a preferential partitioning of some lipid species in a specific leaflet could be obtained [32].

Another largely exploited model system is the polymer-supported lipid bilayer (Figure 1b) [33]. In this case, the lipid bilayer is not directly facing the solid support but is separated from it by a soft polymer layer enabling an increased lipid lateral mobility. At the same time, the increased thickness of the hydrophilic region between the lipid bilayer and the substrate allows the incorporation in lipid bilayers of transmembrane protein with large extra-membraneous portions. The enhanced lateral mobility of both lipids and proteins makes this model system very interesting for the investigation of dynamic aspects of the membrane organization. However, dealing with the investigation of lipid bilayers by AFM, polymer-supported lipid bilayers do not assure the required flatness and stability for AFM imaging, so, very few AFM attempts have been performed on these systems [34, 35]. Typically, the effective presence of a lipid bilayer on the substrate must be confirmed by force spectroscopy measurements (see below), but, the presence of the polymer layer, could be misleading in the interpretation of the results.

Also lipid bilayers suspended over nanosized holes have been investigated by AFM [36]. This lipid bilayer configuration reproduces a situation very similar to that of freely suspended planar bilayers such as Black Lipid Membranes or GUV. In this case, the mechanical properties of the lipid bilayer could be investigated without the disturbing effect of the solid support and different compositions for the two aqueous compartments separated by the bilayer could be tested. Unfortunately, on this kind of samples the AFM high resolution imaging capabilities are strongly reduced.

In this work we will consider only lipid bilayers supported on solid substrates.

7.2.2 Chemical-Physical Properties of Supported Lipid Bilayers

To what extent the supported lipid bilayer model system can be compared with other model systems or even with a biological membrane? The answer to this question requires a detailed study of the chemical-physical properties of Supported Lipid Bilayers. In this analysis we will compare the chemical-physical properties as measured with AFM on SLBs with those of other model systems [37]. We will also make use of data obtained by other techniques, especially fluorescence techniques, on SLBs. The two most important and related topics are the behavior of lipid bilayers in the phase transition region and lipid and transmembrane protein lateral diffusion. In this section we will concentrate on the lipid behavior and in the next one we will discuss the reconstitution of transmembrane proteins in SLB and their diffusion properties.

The main concern about SLBs is the presence of a solid inorganic surface in the proximity of a bilayer that might affect its behavior. For example, the presence of a substrate might potentially have a strong influence on the vertical asymmetry of the physical and compositional properties of the two leaflets [38]. In this case, due to the lack of analytical capabilities of AFM, results from simulation studies could help. In particular, recent simulation studies demonstrated that in SLBs the surface density of lipid molecules in the two leaflets differs, being higher in the leaflet nearer to the substrate (proximal leaflet) than in the one in contact with the bulk solution (distal leaflet) [39, 40]. Moreover, the presence of an electrostatically charged substrate might induce a lipid composition asymmetry in the case of mixture of lipids with charged headgroups [41]. Valuable information about the physical properties of the bilayers can be obtained by monitoring the phase transition of a bilayer. In general, lipid bilayers display a reversible phase transition between a solid ordered (s_o) and a liquid disordered (l_d) phase. The transition is accompanied by changes in lipid chains (ordered or disordered) and lattice order (solid or liquid). This transition depends on parameters such as temperature, pH or ionic strength. Sterols induce a third phase, the so-called liquid ordered phase, with a loss in lattice ordering as for the l_d phase, but a higher lipid chains order as for the s_o phase. A phase similar to the l_d is likely to appear in biological membranes, where it is referred to as a lipid raft [6, 42]. Melting from the s_o to the l_d phase involves an increase in lipid bilayer area and a decrease in bilayer thickness. The decrease in thickness from the ordered to the disordered conformation of the lipid chains makes the transition easily recognizable by Atomic Force Microscopy. In fact, the thickness variation from the liquid to the solid phase of a lipid bilayer ranges from 0.7 to 1.5 nm, well above the noise level of AFM measurements. In model systems such as liposomes (LUV or GUV), the lipid phase transition involves a synchronous and in register behavior of the two leaflets. As we will see in a following section of this work, the study of temperature-induced phase transitions in SLBs can reveal the presence of a decoupling of the two leaflets in the bilayer. The decoupling is a result of the mutual weight of the interactions between the two leaflets and between the bilayer and the substrate and it results in two independent phase transitions of the leaflets (Figure 2). The independent behavior of the two leaflets might be caused also by a compositional asymmetry between the two layers. So it might be useful to discuss the origin of the interaction force between the substrate and the bilayer.

The interaction energy between a lipid bilayer and its substrate results from the balance of mainly three terms: the van der Waals interaction, the double layer interaction, and the hydration interaction [43]. The van der Waals interaction energy, in this specific case, results in an attractive force

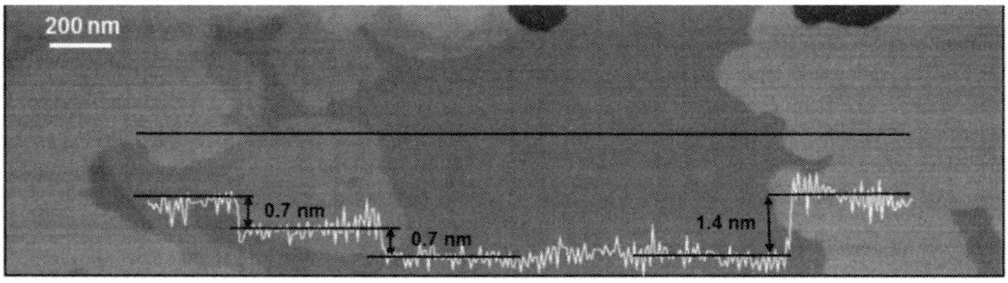

Figure 7.2 AFM image showing the presence of domains not in register between the two leaflets. The image shows both the presence of a domain characterized by both leaflets in the solid ordered configuration (on the right) and domains with only one leaflet in the solid ordered configuration. It can be seen that the solid domain in each leaflet corresponds to a height increase of about 0.7 nm.

at any distance. The double layer effect depends on the surface charge of the two interacting surfaces. The mica has a negative surface charge in water solution whereas the bilayer surface charge depends on the lipids. The decaying length of the double layer force is exponentially related to the presence of electrolytes in solution via the Debye screening length, whose magnitude decreases upon an increase of the ionic strength of the solution. In this context divalent ions will be much more efficient in screening surface charges than monovalent ions. The sum of the van der Waals and the double layer interactions is described by the DLVO (Derjaguin-Landau-Verwey-Overbeek) theory [44]. In solution, another short-range force, the hydration force, is present between two surfaces. It can be described as a consequence of stable structured water layers in the vicinity of a surface, especially in the case of a solid surface, or as a force due to entropic factors, namely thermal fluctuations. The second description applies mainly to soft interfaces such as SLBs. It can be broken into two motional contributions: protrusions and undulations. In the case of lipid bilayers the contributions come from the protrusion of the lipid headgroups and the collective undulations of the bilayers. In general, in the case of SLBs, in which the separation distance between the support and the lipid bilayer is very small (1–2 nm), the hydration force is repulsive. As a general comment, the interplay of all the forces acting between a lipid bilayer and its solid substrate is quite complicated. Moreover all the forces depend on the environmental conditions such as pH, electrolyte type and concentrations, temperature and hydrophilicity of the interacting surfaces. Nonetheless, the overall balance of the interaction energies between a lipid bilayer and a substrate might lead to an equilibrium distance between them. It is important to stress that all the involved forces have a strong dependence on distance. Thus, the substrate exerts its influence on the bilayer with an interaction stronger on the proximal leaflet than on the distal one. This differential interaction could affect the interaction at the interface between the two leaflets, which is usually referred to as the interleaflet coupling. If the interleaflet interaction is stronger than the interaction of the leaflets with the solid substrate, the bilayer will show a coupled behavior of the leaflets, otherwise, the two leaflets will behave independently of each other. The interleaflet coupling in lipid bilayers is at present investigated from both an experimental and theoretical point of view [45]. The relevance of this phenomenon stems from its possible role as a signaling mechanism between the two leaflets of a bilayer and, consequently, between the inner and outer regions of a cell. For example, the natural lipid composition asymmetry found in eukaryotic cell membranes between the inner and outer leaflet arouses the question if a liquid ordered (raft) domain that could form in the outer leaflet is able to induce a domain also in the inner one. This question is biologically relevant, because it has been shown that only the lipid composition of the outer leaflets is able to give rise to the liquid ordered phase [46, 47]. In general, it has been found that the interleaflet coupling is strongly dependent on the lipid composition of the leaflets, with an important difference between synthetic compositions and natural membrane compositions [48]. In particular situations of lipid composition asymmetry, the formation of a liquid ordered domain in one leaflet is however able to induce a liquid ordered domain in register with a second leaflet with a composition not prone to the formation of this kind of domain [47]. Apart from the specific case regarding interleaflet coupling, it has to be stressed that natural lipid compositions might have thermodynamic properties that cannot be completely superimposed to model systems. The dynamic interdigitation of the lipid chains is considered one of the main phenomena behind interleaflet coupling [45]. This interdigitation could be considered as a phenomenon that increases the entropy of the lipid chains and the entropy loss due to a restricted dynamic interdigitation could be the driving force for maintaining a coupling between the two monolayers. In the specific case of a SLB, Merkel *et al.* found that, by increasing the packing density of the proximal layer with respect to the distal one, exploiting hybrid bilayers in which the two layers are assembled progressively, an increase in the diffusion coefficient for the distal layer was obtained, probably due to a decreased interdigitation effect [49]. Accordingly, a strong asymmetry in the lipid density between the two leaflets could decrease the interleaflet coupling. Recalling that it has been suggested, on the basis of simulation studies, that the presence of a substrate increases the proximal leaflet lipid density, the overall situation can be rationalized by the following considerations. If, due to the presence of a substrate, the lipid density of the proximal leaflet largely increases with respect to the lipid density in the distal leaflet, the interdigitation of the facing portions of the alkyl chains will be impeded, inducing a decreased interleaflet coupling. Different strategies might be adopted to restore a similar lipid density in the two leaflets. The first one is that of using a substrate that does not interact strongly with the bilayer. Typically, a substrate with an extremely low roughness will have a smaller average distance from the bilayer and will consequently interact more strongly with the proximal leaflet. By choosing a substrate with a larger roughness, without preventing the possibility of clearly detecting different phases in the bilayer or the presence of transmembrane proteins by AFM, it is possible to reduce the substrate/leaflets interaction up to a point where the interleaflet coupling prevails (see below). Alternatively, it is possible to let the bilayer assemble on the substrate at a high temperature (above the phase transition temperature for the bilayer on the substrate), so to already present a low lipid density also in the proximal leaflet. The latter approach will prevent the increase of lipid density in the proximal leaflet allowing a stronger interdigitation effect with the distal leaflet and the coupling of the two leaflets.

The second important topic is connected to the lateral diffusion of both lipids and proteins in SLBs. Dealing with lipid diffusion, the problem could be divided in i) the comparison of lipid diffusion between unsupported and supported lipid bilayers and ii) the evaluation of possible difference in the diffusion between the two leaflets. The importance of the lateral diffusion issue stems from the fact that many important biological processes relies on lateral molecular diffusion [50].

Typically, the diffusion properties of lipid bilayers are measured by optical microscopy techniques involving fluorescence, such as Fluorescence Correlation Spectroscopy (FCS), Fluoresce Recovery After Photobleaching (FRAP) and Single Particle Tracking (SPT) [51, 52]. All these techniques could be also employed on biological membranes, enabling a direct comparison with model membranes. On the other hand, model membranes could be exploited to fully understand the information that these techniques can provide under well-established conditions. Generally speaking, from optical diffusion studies on lipid bilayers the structure of the membrane could be deduced. The different character of the observed diffusion can be interpreted on the basis of obstacles present in the bilayer. So, a combination of studies in which the results of optical diffusion experiments are compared to the nanometric structure obtained by AFM could be of high value in this context [53].

It has to be considered that the diffusion properties of SLBs depend on the chemical and physical properties of the substrate used to reconstitute the membrane [54]. The influence of the substrate on the diffusion properties could be also an expression of the substrate effect on the main transition temperature of the supported bilayer. In fact, the substrate, due to its interaction with the bilayer can shift the main transition to higher temperatures [55]. As a consequence, the fluidity of the bilayer measured at the same temperature on two different substrates might reflect the different phase condition of the bilayer. Moreover, due to the subtle dependence of the SLBs properties on their preparation and observation condition, it is very difficult to compare results obtained in different laboratories. Despite all these considerations, some general trends appear in the literature and these trends will be discussed here.

Concerning the comparison of the diffusion properties of SLBs with other lipid bilayer model systems, experimental results have shown that the lipid diffusion in free standing bilayers (GUV, Giant Unilamellar Vesicles) is more than two times faster than in supported lipid bilayers (the diffusion coefficient is D = 7.8 $\mu m^2 s^{-1}$ for GUVs and D = 3.1 $\mu m^2 s^{-1}$ for SLB on mica) [56–58]. The difference in the diffusion constant is usually attributed to the interaction of the substrate with the bilayer or to surface sticking of the lipids creating pinning points between the support and the bilayer. There are experimental evidences that the roughness of the substrate, chemical properties being equal, strongly affects the interaction between the bilayer and the support. For example, mica supports, which are quite useful for AFM studies, show a very low roughness value. On the one hand this property allows to better highlight the presence of protruding membrane proteins or small differences in height as a result of domain formation, but, on the other hand, the lipid diffusion is reduced with respect to rougher substrates. For example, a comparison of SLB formation on mica and silicon oxide established that, in spite of the similar chemical nature of the two surfaces, the interaction of the support with the bilayer is strongly reduced in the case of the SiO_2 substrate [59]. The reduced effect of the substrate can be established from the very low shift in the temperature of the main phase transition of the SLB with respect to SUV determined by Differential Scanning Calorimetry (DSC).

A different and more complicated issue is the possible difference in diffusion coefficients between the two leaflets. The asymmetric dynamic properties might also affect the strength of the interleaflet coupling. In the literature, different results can be found. Hetzer *et al.* [60] found that the outer leaflet has a diffusion constant which is two times faster than the inner leaflet, pointing to an independent behavior of the two leaflets. Recent results found the same translational diffusion coefficient for both leaflets within a 10% experimental uncertainty [61]. In the latter case a strong coupling between the two leaflets could be the reason for the same lateral mobility. However, many results show that the molecular behavior and the interleaflet coupling are strongly related to the experimental details of the sample preparation. So, it is not always possible to compare the obtained results even if on the same phospholipid and substrate system. The diffusion coefficient decreases as the surface density of the lipids increases in free bilayers. In the case of substrate supported bilayers we expect also a strong effect of the surface density and of each monolayer lateral pressure on the bilayer/surface interaction and on the interleaflet frictional effects. This means that the surface density of one of the two monolayers can influence the behavior of the other one.

Also the effect of surface treatment on the lipid diffusion coefficient has been studied and it has been found that surface processes that have the highest diffusion coefficient were the same that gave rise to the greatest probability of observing coexistence of different domains in the lipid bilayers [62, 63].

7.2.3 Transmembrane Protein Inclusion

To study the membrane protein/lipid interactions by means of an AFM it is mandatory to reconstitute the proteins in the SLB. If we consider the reconstitution of proteins by means of fusion of proteoliposomes it is important to stress that, according to the most probable pathway for supported lipid bilayers formation from vesicles, the external face of the liposomes will face the solid support in the final bilayer configuration, while the internal layer will face the bulk of the solution. However, contradictory results have been reported about the formation of SLBs from vesicles containing transmembrane proteins [64, 65]. Indeed, in some cases it has been found that proteins exposing the active site to the bulk solution in vesicles also exposed the active-site to the bulk solution in the SLB. In other cases, it has been found that a significant redistribution of protein orientation occurs during the SLB formation. Probably, the correct scenario depends on the specific case under investigation [66]. Among the parameters to consider there are the size of the vesicles, their lipid composition, deposition temperature, and the nature of the support. The orientation of the proteins in the lipid bilayer can be connected to the orientation in the proteoliposomes, but the final structure on the surface depends on the rupture pathway of the vesicles and on possible reorientations of the proteins. The accurate orientation of the proteins in the SLB can be established by functional tests or by measuring the distribution of the height of the inclusions [67].

When SLBs with reconstituted membrane proteins are studied by AFM, a homogeneous lipid-to-protein ratio in the vesicles, which are going to form the bilayer, is highly desirable. This situation would allow more reproducible results and a uniform distribution of the proteins in the supported lipid bilayer in the case of a homogenous lipid phase [68]. In fact, the possibility that vesicles with different lipid to protein ratio have a different affinity for the solid substrate could make the obtained SLB with reconstituted membrane proteins largely independent from the real vesicle composition in the solution. Even if the control over the lipid-to-protein ratio in the vesicles is difficult, it is important to take into consideration this parameter when interpreting the obtained AFM images. It is usually found that the density of proteins in the supported lipid bilayers is lower than the nominal concentration used to prepare the proteoliposome sample [69]. This fact could be the result of the presence of lipid vesicles without proteins in solution, especially in the case of an increased affinity of these vesicles for the solid support.

Many membrane proteins perform their tasks by forming dynamic assemblies with other proteins in the membrane. Thus, single molecule level studies of molecular interactions would increase our knowledge of biological processes. In order to study transmembrane proteins in SLBs in a functionally relevant case it is important to consider the possibility of lateral diffusion for the reconstituted proteins and their retained functionality. Dealing with the diffusion of membrane proteins in SLBs, it is usually found that proteins are able to diffuse, but the diffusion coefficient is orders of magnitude lower than expected from proteins embedded in free standing bilayers [70]. The reason for this behavior could be found in the thickness of the water layer between the membrane and the support that could increase frictional forces for membrane protein diffusion. Another explanation for the low diffusion coefficient is the presence of pinning points of the bilayer to the substrate that could produce an obstructed diffusion. Different types of motion, free diffusion and obstructed diffusion, have indeed been observed by AFM on SLBs [70]. It has to be noted that the diffusion of membrane proteins is related to the fluidity of the lipid bilayer. Indeed, the diffusion process is a thermally activated process and, by increasing sample temperature, it is possible to observe an increase of the displacements of the proteins in the membrane, eventually reaching a situation in which the proteins are no more visualized by AFM (personal observations), but are again visualized by lowering back the temperature. A major limitation in the observation of protein diffusion by AFM is the low time resolution of the technique. This limit allows only the observation of slow dynamics. In particular, depending on the area imaged and the time interval between two consecutive images, limits in the determination of diffusion coefficients are encountered [71, 72]. For example, it has been hypothesized that the prevalent observation of membrane proteins in association with more ordered lipid regions does not mean that the same proteins are not interacting with more liquid phases, especially in the case of small proteins or peptides [73]. In fact, highly mobile components in the lipid bilayer might not be detected and a complementary technique should be used to exclude the presence of small mobile proteins in the fluid regions of the bilayer. One of the most exciting area of development for the AFM technique in biological studies is the high-speed imaging which could allow the acquisition of time-lapse images with a very short time interval (down to tens of ms per frame) [74]. By using high-speed atomic force microscopy, the lateral and rotational diffusion of membrane proteins in supported lipid bilayer has been recently observed [75].

The study demonstrated that the behavior of the single proteins is very heterogeneous, with some molecules showing relatively high diffusion and other molecules fixed. Moreover, the frame period of about 400 ms was shown to be still too high to correctly characterize the rotational movements of the molecules. However, the reduced protein diffusion constant does not always imply an alteration of the functionality. For example, it has been demonstrated in a recent fluorescence study on the cooperativity of ion-channel subunits assembled in a solid SLB that, in spite of an apparent absence of mobility for all the molecules, only a small fraction showed no activity in response to a gating stimulus [76]. A method, which is potentially very interesting and useful is that of assembling membranes on supports in which holes have been produced [77]. In this case it would be possible to image by AFM membrane proteins separating two aqueous compartments, configuring thus a free-standing membrane.

7.3 Atomic Force Microscopy (AFM) and Phase Behavior of LBs

7.3.1 Transitions Induced by Temperature

In this section we will describe specific experiments in which the detailed features of the phase transition of SLBs (from the liquid disordered to the solid ordered phase) are studied. The main focus is on comparing the behavior of the phase transition in SLBs with the phase transition in other lipid bilayer model systems and to eventually develop particular strategies in order to make the different model systems thermodynamically equivalent.

The main phase transition in lipid bilayers is characterized by a variation in bilayer thickness, which can be easily tracked by AFM. Some studies on temperature-induced phase transitions, as observed by AFM, displayed features that raised some doubts on the equivalence of the SLBs and other unsupported model systems such as liposomes [55, 78]. In some cases, a clear decoupling in the behavior of the two membrane leaflets has been observed at the main phase transition. Two separate transitions, at variance with what observed in liposomes, where the two leaflets act together and the transition is characterized by in register domain between the two leaflets [79], have been observed. The two transitions have been attributed to the two leaflets undergoing separated phase transitions at different temperatures. The higher temperature transition has been assigned to the proximal leaflet. The transition occurring at lower temperature has been attributed to the lipid leaflet facing the bulk aqueous phase (distal leaflet). Concerning the comparison with other lipid bilayer model systems, it is remarkable that the lower temperature transition takes place in a temperature range similar to that of liposomes with the same lipid composition.

Figure 7.3 shows a series of images obtained on a POPE:POPG 3:1 supported lipid bilayer for different temperature values. The presence of the two transitions is clearly observed. In this case, the SLB was assembled at room temperature and the images were taken in pure water. In light of the previous discussion on the influence of the support on the lipid bilayer it is straightforward to attribute the particular observed behavior to the effect of the solid support on the properties of the assembled planar bilayer.

Figure 7.4 shows the evolution of the two phase transitions in two plots which allow to derive the van't Hoff enthalpy for each transition [37]. On the basis of this analysis and by comparing the AFM results with the thermodynamic enthalpy obtained from Differential Scanning Calorimetry (DSC) on liposomes of the same lipid composition it is possible to extract the average size, in terms of numbers of molecules, of the cooperative unit,

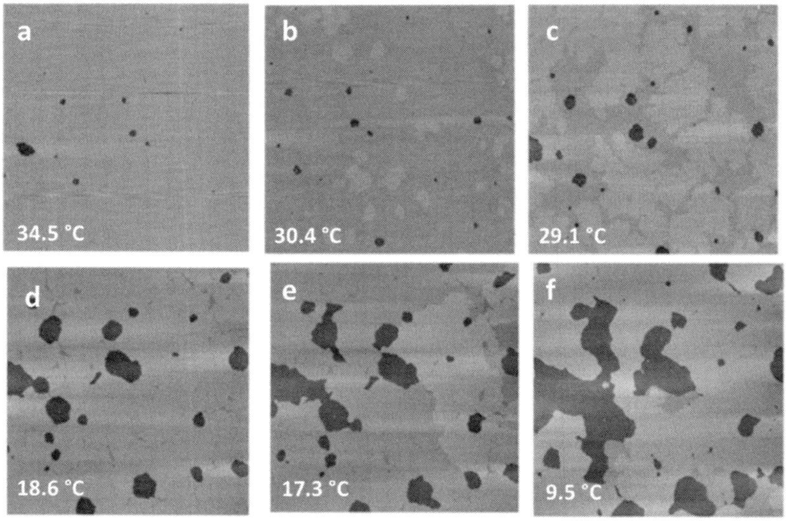

Figure 7.3 A temperature controlled AFM experiment performed on a supported lipid membrane of POPE:POPG 3:1 hydrated in pure water. (*a*)-(*f*): representative height images (9μm × 9μm) of the lateral membrane organization at different temperatures (from 34.5°C to 9.5°C). The lighter the color the thicker the membrane. The evolution of lipid domains and holes upon cooling is visible. Two phase transitions ascribable to independent phase transitions of the two leaflets composing the bilayer are clearly observed. Step sizes for the height increase of the domains are the same for the two transitions (0.7±0.2) nm. (Reprinted with permission from Biophys. J. Vol. 97, p. 1067, 2009).

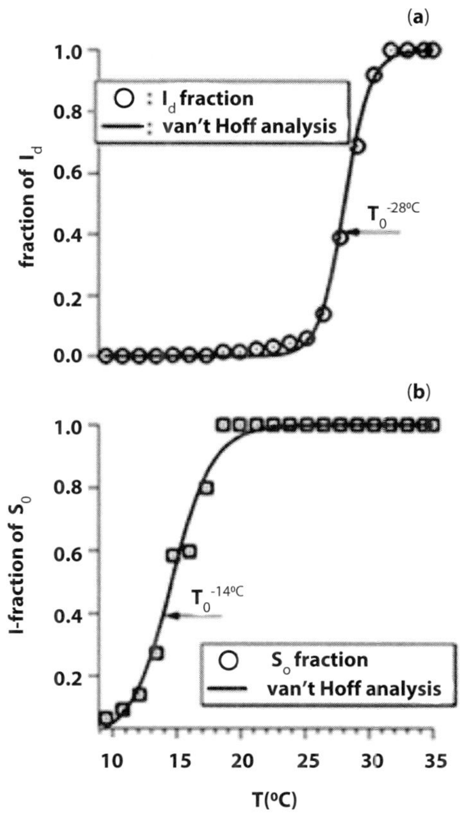

Figure 7.4 van't Hoff analysis of the two transitions observed in Figure 3. The high temperature phase transition of the proximal leaflet is reported in (a) whereas the low temperature phase transition of the distal leaflet is reported in (b). In (a) the fractional occupancy of the liquid disordered phase is reported as a function of temperature. In (b) the value of (1 – solid-ordered-fraction) is reported as a function of the temperature. Both solid ordered and liquid disordered phases are intended for the overall bilayer. (Reprinted with permission from *Biophys. J.* Vol. 97, p. 1067, 2009).

which represents the number of lipids in an intrinsic domain [80]. From the plots it appears that the transition of the distal leaflet is more cooperative than the transition involving the proximal leaflet and the presence of the substrate stabilizes the solid ordered phase of the proximal leaflet. In the bilayer shown in Figure 3, the two leaflets are clearly uncoupled, whereas, considering the same lipid system studied by DSC, only one transition is observed. So, it is possible to conclude that the SLB prepared and imaged according to the condition specified in Figure 3 behaves differently with respect to liposomes.

If an SLB of the same lipid composition is prepared at higher temperature (above the transition temperature of the bilayer on the support) and imaged at higher ionic strength, a different feature for the main phase transition is observed. Figure 5 shows a sequence of images performed at different temperatures on a POPE:POPG 3:1 bilayer prepared according to these new conditions. In Figure 5d a line section corresponding to the white line in the same image is reported. The section highligts that the formed solid domain has a height of about 1.4 nm, which is twice the height measured for domains observed in the case of uncoupled leaflets (Figure 2). It is evident from Figure 5 that the main phase transition develops with the two leaflets in a coupled configuration with domains in register. A possible explanation for the coupled behavior of the leaflets can be traced back to the lipid density conditions in each leaflet assured by preparing the lipid bilayer at high temperature. As explained in the chemical-physical properties section, the lipid density in the bilayer leaflets is affected by the presence of the support, inducing an increased density in the proximal leaflet with respect to the distal one. At the same time, a strong vertical asymmetry in the density can impede the interdigitation phenomenon in the midplane of the bilayer. So, it is plausible that high preparation temperature

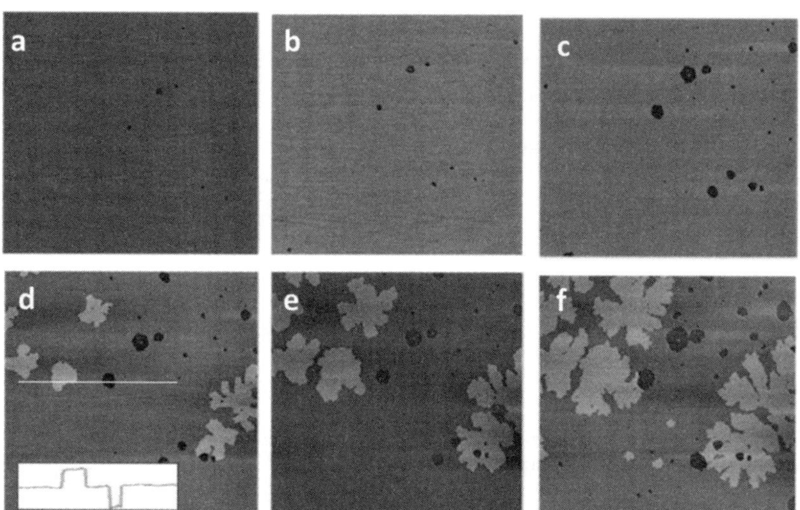

Figure 7.5 Sequence of AFM images on a POPE:POPG 3:1 SLB assembled at a temperature of 35°C and observed in high ionic strength conditions (150 mM KCl). The images have been at different temperatures (a: 33°C; b: 27°C; c: 20°C; d: 19°C; e: 13°C; f: 10°C). The apparent domains are characterized by an height of 1.4 nm with respect to the liquid disordered phase of the lipid bilayer as shown in the line section of (d). Scan size: 10 x 10 μm^2.

reduces the lipid density asymmetry between the two leaflets. The lipid density alteration in SLBs with respect to unsupported bilayers can also be the reason for the typically increased temperature for the main phase transition. In fact, in a way similar to the role of a pressure increase in three-dimensional cases, an increased lateral pressure can shift the phase transition region to higher temperature [81].

The specific features of the phase transition in SLBs are also affected by the chemical or physical properties of the support. For example, the same lipid system deposited on mica presented in Figures 7.3–7.5, when assembled on silicon dioxide presents always the coupled configuration for the leaflets, independently of the deposition temperature [59]. In the last situation, the silicon dioxide surface is not substantially different from the chemical point of view, but it is considerably different from the point of view of its roughness, with respect to mica. The larger roughness of silicon dioxide with respect to mica produces an increase in the average distance between the support and the lipid bilayer. Consequently the effect of the support on the proximal leaflet is largely reduced. The reduced effect of the support on the lipid bilayer manifests itself also with a main phase transition temperature which is very similar to that measured by DSC. Indeed, Figure 6 shows the fraction of the solid ordered domains as measured from the AFM images on a POPE:POPG 3:1 bilayer on silicon oxide as a function of temperature compared to the analogous measurement by DSC on

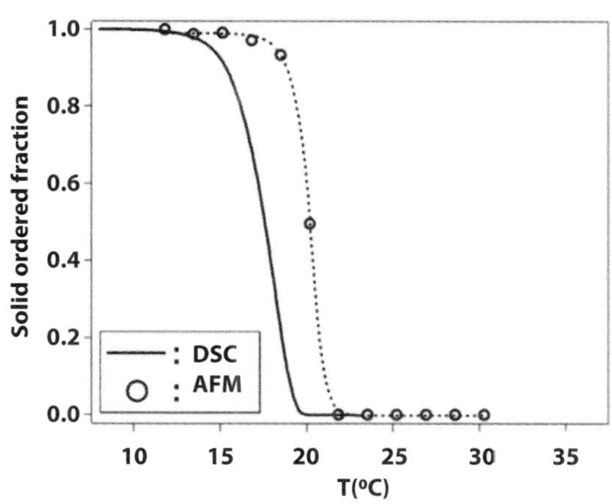

Figure 7.6 The black solid line, representing the inverse transition enthalpy obtained from DSC on SUVs of POPE:POPG 3:1, is compared to the solid ordered fraction (open circles, dashed curve is a guide for the eye) of the SLB of the same lipid composition on silicon oxide. The transition on the silicon oxide support occurs at a slightly higher temperature than the one of the SUVs. (Reprinted with permission from *J. Phys. Chem. B*, Vol. 114, p. 8926, 2010).

liposomes (for DSC data, the integral of the excess specific heat is reported as a function of temperature). In Figure 6 it emerges that the two transitions are separated by only 2°C. These data suggest that SLBs on silicon dioxide present thermodynamical properties similar to liposomes, while preserving the possibility of clearly identifying phase transitions along with the presence of transmembrane proteins by AFM.

7.3.2 Transitions Induced by pH

Especially in the case where charged lipid species are present, the phase state of the lipids is strongly dependent on pH [82, 83]. Indeed, in bilayers composed of charged lipids, the main phase transition results in a variation of the electrostatic free energy due to a difference in the surface charge density following expansion or contraction of the bilayer surface. By varying pH, it is possible to modify the surface charge of the bilayer and to alter the associated electrostatic free energy. From a biological point of view, due to the fact that biological systems are remarkably constant in temperature, a phase transition induced by a local pH variation might have, for the functional properties of the membrane, a more important role than that played by temperature-induced phase transitions. Besides the chemical changes in the relative lipid composition in each domain which occur upon phase transition in a lipid bilayer composed by a mixture of lipids, also the mechanical properties of the lipid bilayer are affected by the phase transition. So, the study by AFM of a phase transition induced by a pH change is very important. Using the same lipid system investigated for the temperature induced main phase transition, POPE:POPG 3:1, it has been shown that it is possible to induce the formation of domains also at constant temperature by changing the pH of the buffer solution [59]. Figure 7a shows two DSC traces obtained on POPE/POPG liposomes at two different pH values. From the plot it is evident that if we start with a SLB at 27°C at pH 7, the bilayer will be in the liquid disordered phase. Keeping constant the temperature and by changing the pH from 7 to around 3, the lipid bilayer is made to enter the phase transition region, with the formation of domains, as observed in the series of images in Figure 7. The formation of solid domains is also coupled to a lipid flip-flop mechanism which shift the transition from one with coupled leaflets to one in which the two leaflets behave differently [59]. The domain marked by the number 1 in Figure 7c is characterized by a flip-flop transition which limits the formation of a solid domain to only one leaflet in Figure 7d. It is to be noted that the obtained transition is completely reversible. In fact, if the pH in again changed to higher values, the solid domains disappear as confirmed by Figures 7.7(e-f).

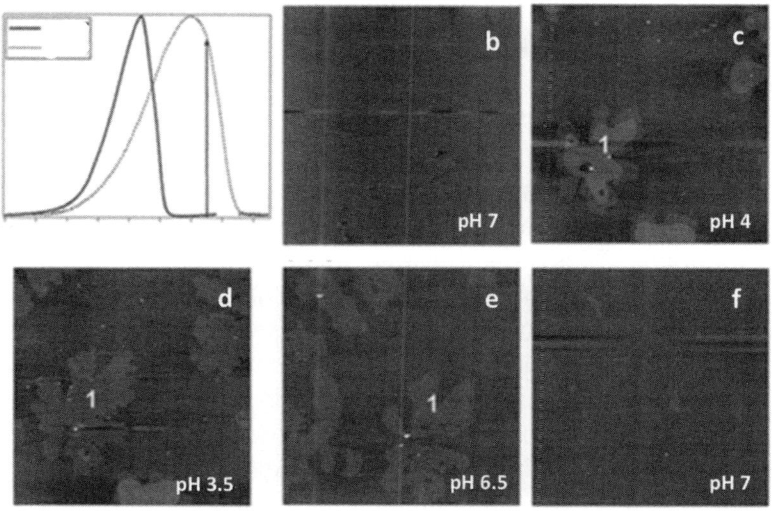

Figure 7.7 pH-Induced Domain Formation. (*a*) Traces of the heat capacity of POPE:POPG 3:1 SUVs at pH values of 7 and 3. The curves have been corrected for the effect of the mica support which leads to an increase of the melting temperature by about 5°C. (*b*) The AFM image (image size: 10 μm x 10 μm) at a pH of 7 does not show any lateral heterogeneity. The membrane was prepared at high temperature to ensure a coupling of both leaflets. (*c*) Starting the buffer exchange, at some point (after reaching a pH of about 4) domains started to form. The different height levels in domain 1 indicate that the transition was not fully coupled. (*d*) Following the evolution of domain 1, it becomes clear that a flip-flop mechanism was present. The initial solid ordered domain changed to an intermediate domain. (*e-f*) After finally reaching a pH of 3, the buffer was again exchanged for a buffer at pH 7. The intermediate domains vanished demonstrating the reversibility of the transition. (Reprinted with permission from *J. Phys. Chem. B*, Vol. 114, p. 8926, 2010).

7.4 Atomic Force Spectroscopy (AFS) of Supported Lipid Bilayers

Recently, the possibility that the mechanical properties of biological membranes can exert a strong influence on the behavior of integral membrane proteins has been increasingly accepted [13]. This scenario points to a role of the physical properties of the bilayer, along with its chemical composition at molecular scale, in affecting membrane functions [84–87]. The investigation of the mechanical properties of lipid bilayers can be performed in analogy to that of a continuous soft material that is, neglecting molecular details. Concepts developed in the field of liquid crystals can so be applied to the investigation of bilayer properties. Quantitative investigation of lipid bilayer material properties started with the pioneering work by Helfrich

[88]. Even if it is not to be excluded that long-range mechanical properties of lipid bilayers, such as those related to undulations, might have an effect on protein activity, it is of extreme relevance to measure membrane mechanical properties at the nanoscale. Indeed, that is the typical scale of membrane heterogeneity believed to be relevant to conformational transitions involved in the functional activities of membrane proteins.

Considering the techniques that can provide valuable information on the mechanical properties of lipid bilayers, the micropipette aspiration technique is one of the main experimental tools [89]. Such a technique enables the retrieval of quantitative information on the elastic and shear moduli of lipid bilayers assembled as Giant Unilamellar Vesicles (GUV) [90]. However, this technique can provide information only on the macroscopic behavior of the bilayer, while it cannot disclose any nanoscopic clue.

A technique that has emerged as a powerful tool in this context is Atomic Force Spectroscopy (AFS) [91]. Being operated with the probe of an AFM, it allows simultaneous lipid bilayer topological characterization and analysis of the local mechanical properties [16]. Particularly, the tip of an AFM is pressed on a supported lipid bilayer until indentation takes place. A discontinuity in the force curve appears when the tip instantaneously punches through the bilayer. Figure 8 shows the schematic principle of the technique along with a typical force curve measured on a lipid bilayer. The force required to punch through a lipid bilayer is in the nN range [91] and depends on environmental conditions such as ionic strength [92], temperature [93] and pH [94] and speed of the indenting tip. It has been suggested by some authors that the indentation force can be considered as a fingerprint of the specific lipid bilayer state [95].

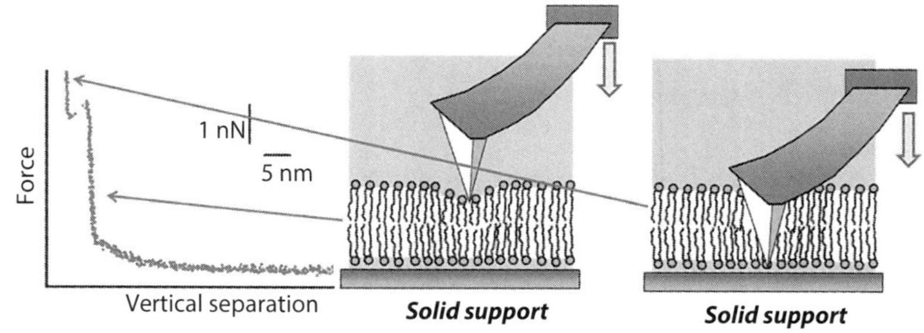

Figure 7.8 Scheme of a force curve performed on a supported lipid bilayer. The cantilever is moving at constant speed towards the bilayer. After a first long range interaction section which might include electrostatic and hydration forces, a contact region between the tip and the bilayer follows. This region can be considered of elastic deformation. After a threshold force is reached, the tip jumps in contact with the substrate by a process of plastic deformation (rupture) of the membrane.

From a theoretical standpoint different approaches have been proposed to interpret the AFM punch-through experiments on lipid bilayers [96, 97]. They are based either on the continuum nucleation theory [98] or on a model in which the molecular details of the interaction among phospholipid molecules and between the bilayer and the solid support are taken into account [98]. The first approach is similar to that developed for interpreting the formation of hydrophilic pores in pure lipid bilayers by the application of transbilayer small or large voltages (electroporation) [99] acting as a sort of electrocompression mechanism. Fluctuations in the bilayer and the associated presence of defects can spontaneously lead to the formation of small holes. Indeed, it has been shown that a mechanical stress, e.g. the application of a lateral tension ramp, can produce pores in a bilayer [100]. These concepts, transferred to the AFS case, suggest that the energetic cost for hole formation depends on its diameter, line tension, and bilayer spreading pressure, which accounts for the tendency of a lipid bilayer to fill the gap between tip and substrate. After reaching a critical value, the pore radius increases spontaneously. In AFS punch-through experiments the energy barrier associated with this critical radius has to be overcome. The force applied by pressing the tip on the bilayer, and the associated deformation, increase the total energy of the system and correspondingly decrease the energy barrier. So, the barrier can be stochastically crossed and, as a result, the tip punches through the bilayer.

Punch through measurements performed on POPE and POPE:POPG bilayers as a function of temperature, in a range spanning the main phase transition of the bilayers, have shown an anomalous behavior of the indentation force in the transition region. In particular, the most probable force required for piercing the bilayer, as measured on fluid domains, appears to drop remarkably, deviating from the monotonic increase that it shows outside that region [101] (Figure 7.9). Such a decrease has been attributed to the enhanced thermal fluctuations, the first step of pore formation process. Also the lateral compressibility is enhanced in the phase transition region. This means that the lateral interactions between the phospholipids have a dominant role in affecting tip indentation results. These data will be correlated to the modified functional activity of ion channels in the phase transition region of the bilayer in the last section of this work.

The bilayer penetration process might also depend on the internal structure of the lipid bilayer such as the degree of coupling between the two leaflets. In this context, it is interesting to relate the features of tip penetration process to the internal organization of the bilayer.

In a recent experiment, it has been shown that the features of the punch through event depend both on the way the investigated lipid bilayer

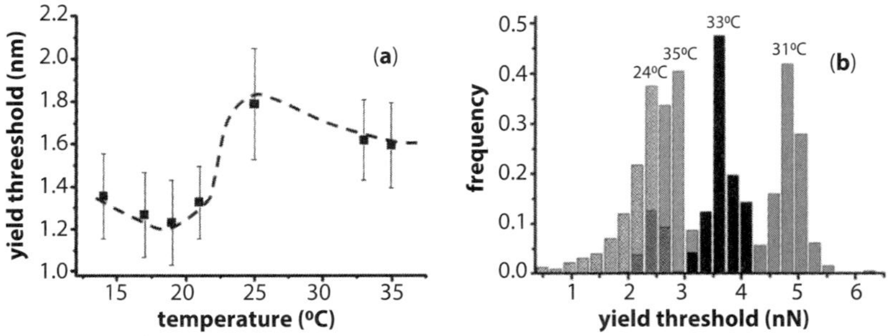

Figure 7.9 (*a*) yield threshold *vs* temperature. The results of an AFS experiment on a POPE SLB at different temperatures are reported. The force spectroscopy experiment was performed in distilled water. The most probable yield thresholds are given (open circles) together with the distributions widths. The dashed curve is a guide for the eye. (*b*) Distributions of the force required to punch-through the bilayer at different temperatures. (Reprinted with permission from *Soft Matter*, Vol. 7, p. 7054, 2011).

has been assembled on the solid support and on the tip speed [102]. Preparation conditions are known to affect the coupling between the two leaflets (see section 7.2.2). Particularly, that experiment has shown two distinct punch-through events in case of an uncoupled bilayer, whereas only one was observed in case of a sample prepared at higher T and higher ionic strength. Interestingly, the total distance travelled by the tip during the penetration events matches in both cases the overall bilayer thickness (about 5 nm).

Finally, since the punch-through reaction is controlled by the application of a force, it might depend on the rate of force application. On the basis of this consideration, it is interesting to study the stress-strain relation of lipid bilayers at different indenting tip speeds. This analysis could be relevant to the elucidation of how a lipid bilayer responds to forces applied at different rates. At the same time, due to the analogy between punch-through experiments performed at different tip speeds and the more general area of Dynamic Force Spectroscopy (DFS), the theories developed to interpret DFS experimental data can be tentatively exploited to reconstruct the energy landscape for the punch-through process [103, 104] (vide infra).

7.4.1 Mechanical Moduli Studied by AFS

Apart from the stability of the lipid bilayers that is testified by the force they can sustain before collapsing, by exploiting AFS it is also possible to obtain an estimate of the elastic constants of these samples. To extract the elastic constants of a lipid bilayer it is important to consider the contact region between the tip and the supported lipid bilayer. The most popular analytical model used to analyze AFS indentation experiments is the Hertz model [105] (or some of its adaptations like the Sneddon's one [106]). However, it has been shown that the above model is not appropriate in case of supported lipid bilayers [107]. The difficulty stems from the small thickness of the lipid bilayer. Indeed, the presence of a rigid substrate supporting the bilayer can strongly influence the value of the Young modulus obtained by interpreting the stress-strain relation on the basis of the Hertz theory, especially in the case of high deformation ratio. Another more plausible model to figure out a value for the spring constant of a lipid bilayer related to the area stretching modulus is that developed by Das *et al.* [108]. This model accounts for the presence of the two leaflets making up the bilayer, although it neglects the strong asymmetry in the physical properties of the two leaflets as due to the presence of the substrate. This model is based on the fact that a spherical indenting AFS tip forces the lipid on a curved surface. As a consequence, the area stretching modulus of the lipid bilayer is involved. By calculating the Gibbs free energy cost for the deformation and its derivative with respect to tip movement, the authors obtained the following analytical expression for the applied force and the bilayer deformation (Equation 7.1):

$$F = \frac{\pi \kappa_A R}{4}\left(\frac{2z_0}{2d - z_0}\right)^2$$

where κ_A is the area stretching modulus, R is the diameter of the apical region of the AFM tip, $2d$ is the thickness of the bilayer and z_0 is the indentation. Figure 7.10 reports a force curve resulting from an average over more than 200 individual force curves obtained on a supported POPE bilayer at 27°C in 50 mM KCl. By fitting the previous equation to the experimental data in the contact region of the force curve (overlaid trace in figure 10) and assuming a tip apical radius ranging between 5 nm and 10 nm it is possible to extract a value for κ_A in the range 0.12 N/m ÷ 0.06 N/m. At 34°C the value for κ_A is between 0.16 N/m and 0.08 N/m whereas at 38°C it is between 0.14 N/m and 0.07 N/m. This analysis can also show that AFS is able to detect the softening effect of the phase transition in the contact region besides the decreased mechanical stability.

As shown in Figure 7.10, the fit of the above model to the experimental data is better for high force values than for small ones. This fact is likely due to the effect of the compression modulus, more evident in the initial part of the force curve than in the higher force range, which is neglected in the model by Das *et al.* One cannot exclude the presence also of electrostatic

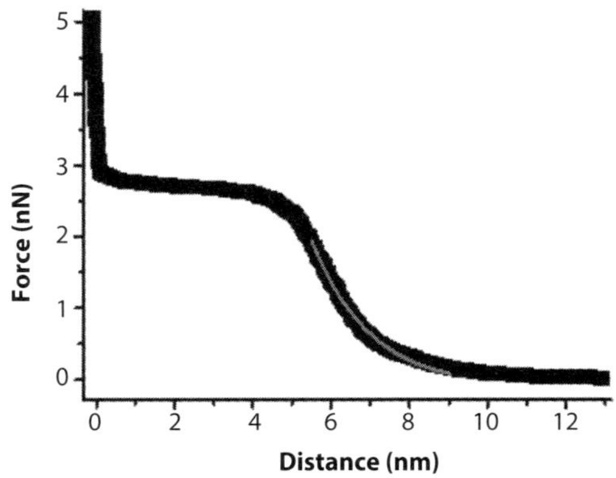

Figure 7.10 Force curve (resulting from an average over more than 200 individual curves) on a supported POPE bilayer at 27°C in 50 mM KCl. The curve overlaid is the fit performed with the model of Das et al. to the experimental data.

forces due to the fact that measurements were performed in a 50 mM KCl solution that assures a Debye length of about 3–4 nm. However, such a possibility has less severe consequences in view of focusing mainly on comparison of the mechanical properties of the bilayer at different temperatures rather than on obtaining an accurate value for the area stretching modulus. Other models have been developed for AFS indentation experiments. Fraxedas *et al.* [109] developed a model in which the lateral interactions among the constituting elements of the sample play a dominant role and they applied it also to indentation experiments on lipid bilayers [92]. In another work, Leonenko *et al.* [107] demonstrated that the Hertz model is not suitable to analyze the deformation of the lipid bilayer induced by the AFM tip. In fact, nanoscale structural rearrangements are involved in the bilayer compression and steric contribution to the repulsive force are expected [110] giving rise to an exponential dependence of the force on the distance. In a recent paper, Picas *et al.* [111] extracted from the force curves the Young modulus of the lipid bilayer and then, invoking the thin shell theory calculated the area stretching modulus and the bending stiffness of the bilayer.

7.4.2 Energy Landscape of Lipid Bilayer Breakthrough and Comparison with Lipid Pore Formation

The punch-through process resembles a mechano-chemical reaction whose energy landscape is characterized by two equilibrium states (the tip on top of the bilayer, and the tip in contact with the underlying solid substrate) separated by an energy barrier. In this case, one can assume that the effect of an applied force is equivalent to a reduction of the activation energy of the reaction [103]. As a consequence, the rate of escape from the initial equilibrium state will increase with the applied force. If it were possible to relate the decrease of energy barrier to the applied force, one could obtain the dependence of the process rate on the applied force. These considerations were first presented in a pioneering series of articles by Butt *et al.* who discussed both a continuum nucleation model based on the analogy with a pore formation process in lipid bilayers and a molecular model [96, 97, 112]. These two models, fitted to experimental data, can provide an estimate of the thermodynamic parameters in the punch-through event.

Experimental data provide both the distribution of the rupture forces at a given tip speed and the dependence of the most probable force upon different tip speeds. In analogy with DFS, the rate of escape from a potential well can be obtained with the Bell-Evans phenomenological model [104, 113], by quantifying the decrease of the activation energy with force as the product of the force and the distance of the transition state from the initial equilibrium state along the pulling coordinate. In this model, if the force increases linearly in time according to the relation $F = kvt$, where k is the elastic constant of the cantilever, v is the tip speed and t is the time (kv being the loading rate), a plot of the most probable force required to cross the barrier as a function of log v results in a linear trend [104]. However, for some systems it has been found experimentally that the relation between the most probable force and speed does not increase logarithmically as expected [114, 115]. The reason for this behavior is unclear. In some cases the existence of two barriers has been evoked in order to explain experimental results in which two different linear trends have been obtained in the most-probable-rupture-force *vs* log v plot [116]. In other cases, the nonlinear behavior has been interpreted as an evidence of the inadequacy of the Bell-Evans model to describe the complete dynamic range of the reaction mechanism [114]. To overcome these difficulties, more sophisticated models, such as the Dudko-Hummer-Szabo (DHS) one, have been introduced of which the Bell-Evans' model represents a particular case valid just in a limited range of force increase rates [117–119].

Without considering any analytical detail of the energy landscape, a monotonic increasing value of the jump-through force *vs* log(tip speed) is to be expected. This consideration is based on the assumption of a random walk of the system on the potential energy surface describing the rising over the activation energy barrier and on the time allowed to the system for the landscape exploration. Measurement performed at 38°C on a POPE bilayer on mica shows a linear increase of the most probable jump-through

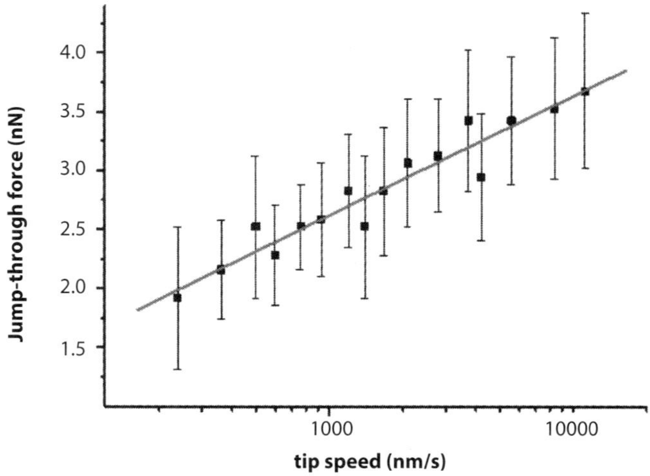

Figure 7.11 Punch-through force values on a POPE bilayer in 50 mM KCl at 25°C as a function of the tip speed. The error bars represent the standard deviations of the measurements. (Reprinted with permission from *Soft Matter,* Vol. 7, p. 7054, 2011).

force with the log(tip speed) (Figure 7.11). The observed trend suggests that the process under investigation could be described by a model where the energy barrier for the tip penetration decreases linearly with compression force. The phenomenological Bell-Evans model, which is based on the fact that the reaction rate constant depends exponentially on the applied force times the distance between the initial state and the transition state, foresees a linear dependence for the above relation. Obviously, the most critical experimental aspect in this kind of measurements is the limited tip speed range that can be explored to significantly distinguish between different analytical dependences.

Interestingly, if one changes temperature, say to 34°C, the relation between the most probable jump-through force and the logarithm of the tip speed is not linear over the whole explored range [102]. As already mentioned, in the context of the Bell-Evans model, the presence of a change of slope in the plot representing rupture force *vs* log(tip speed) is interpreted as due to the existence of more than one activation barrier. In this framework, pore formation in unsupported lipid bilayers has been described as a two-step process in which the first barrier is connected to the spontaneous generation of a pore by fluctuations (related to defects) and the second one accounts for the eventual evolution to an unstable pore [120, 121]. This scenario has been used to describe failure experiments on lipid bilayers under the application of a tension ramp in which the presence of two different linear regimes in the F-*vs*-log(tip speed) plot was found [100]. Indeed, at high loading rates, the limiting factor is supposed to be the formation of defects in the bilayer (the inner barrier), whereas, at low rates, the limiting factor is supposed to be the crossing of the critical pore diameter during the life-time of the defect (the outer barrier). Considering the results obtained at 34°C, it can be stated that the experimental results at issue, obtained by the application of a stress perpendicular to the bilayer plane, are qualitatively similar to those obtained by Evans *et al.* [100]. These experiments seem to suggest that, by changing the mechanical properties of the bilayer due to a variation in temperature, the effect is that of simulating an enlargement of the effective tip speed range. The relevance of the mechanical properties of a lipid bilayer in determining the jump-through force *vs* log(tip speed) relation is further confirmed by measurements at 27°C on the same POPE bilayer [102]. At 27°C, where the bilayer is in the phase transition region and undergoes a softening behavior, the above relation is again linear.

Moreover, considering also the presence of viscous effects, it seems that a deviation from a linear relation between the most probable jump-through force and the log(tip speed) occurs when the viscous effects are more prominent. The Bell model is valid only for weak pulling conditions. It is possible that in the presence of a strongly damped system, where viscoelastic effects are relevant, the Bell-Evans model is no longer valid [114].

7.5 Lipid/Protein Interactions

7.5.1 Protein Partitioning in Membrane Domains

The AFM technique can be used also in the study of lipid-protein interactions. The research field of lipid-protein interactions includes the relations between transmembrane proteins and lipids, the interactions of a lipid bilayer with surface bound proteins, lipoproteins and peptides. Here, we will mainly concentrate on lipoproteins and transmembrane proteins highlighting the information retrieved by AFM on these systems in comparison with other techniques on different model systems. The focus will be on the distribution of membrane proteins relative to phase separation in the bilayer. In the next section we will present a test case in which the redistribution of membrane proteins upon the formation of domains can have effects on the functional behavior of the protein.

As described earlier in section 7.2.3, one strategy to assemble an SLB with reconstituted membrane proteins is that of following the vesicle fusion approach starting from proteoliposomes. In this case, attention has to be paid to the presence of large extramembraneous portions of the membrane proteins that might pin the protein to the substrate. The ideal situation considers transmembrane proteins with a large extramembraneous portion only from one side of the lipid bilayer which, in the assembling of the SLB, end up with this portion facing the AFM tip. These conditions would assure both a retained lateral mobility of the proteins and the possibility of being easily traceable by AFM imaging. In this case, the distribution of proteins along with the phase behavior of the lipid bilayer can be studied by AFM in a physiologic-like environment without the need of a labeling step.

Since the introduction of the idea that domains could appear and could have a functional role in biological membranes, it became evident that understanding the distribution of peptides and proteins in lipid bilayers would have been relevant for elucidating signaling pathways. Among the membrane proteins studied by AFM in relation to their partitioning according to phase changes in the hosting lipid bilayer, GlycosylPhosphatidylInositol (GPI)-anchored proteins (GPI-APs) occupy an important role. GPI-APs are a class of proteins that are anchored to the membrane by means of a posttranslational lipid modification [122]. The lipid modification allows these proteins to be somehow related to the membrane trafficking mechanisms and to domain formation especially in the outer leaflet of the biological membrane. One of the GPI-anchored proteins, which has been studied by AFM, is Placental Alkaline Phosphatase (PLAP) [123, 124]. The aim of these studies was to establish whether PLAP was targeted to raft domains in lipid bilayers. The obtained results constitute a paradigmatic example useful to understand the relationship between different bilayer model systems. Indeed, in AFM experiments, the proteins were mainly found associated with the most ordered domains of the lipid bilayer composed by synthetic mixture of SM and DOPC, according to the fact that the PLAP protein is considered as an example of Detergent Resistance Membrane (DRM) associated protein. However, experiments performed with a similar lipid composition and on the same protein but exploiting GUV, gave the striking result of the proteins mainly concentrated in the liquid disordered regions instead of the ordered ones [125]. The reason of the different behaviors observed could be attributed both to alterations of the thermodynamics of the membrane when detergent is inserted and the temperature is changed to identify DRM areas and to the possible interaction between the proteins and the substrate in SLBs. Additional AFM studied concentrated on other lipidated proteins such as the GPI-anchored intestinal alkaline phosphatase (BIAP) [126] and N-Ras proteins [127]. In both cases, AFM experiments demonstrated that the protein distribution in the lipid bilayer is affected by the presence of different phases in the bilayer. The main driving force for the membrane protein partitioning is typically the hydrophobic matching condition between the hydrophobic surface of the protein and the hydrophobic thickness of the lipid bilayer.

An interesting paradigmatic case is represented by the behavior of the K^+ channel KcsA [128]. KcsA proteins reconstituted in proteoliposomes undergo a vectorial incorporation in the outside-out configuration [129]. This situation enables the formation of SLBs in which the cytoplasmic portion of the channels faces the AFM tip [69]. Figure 7.12 shows a sequence of AFM images in which the white bumps on the lipid bilayers can be associated to KcsA molecules. It is clear from the sequence that, starting from the liquid disordered phase of the bilayer, in which the proteins are uniformly distributed, upon a decrease in temperature and the consequent formation of a solid ordered domain, the proteins are mainly excluded from the solid domain. Accordingly, the proteins will feel both a different lipid environment (the lipid bilayer is composed by a mixture of POPE

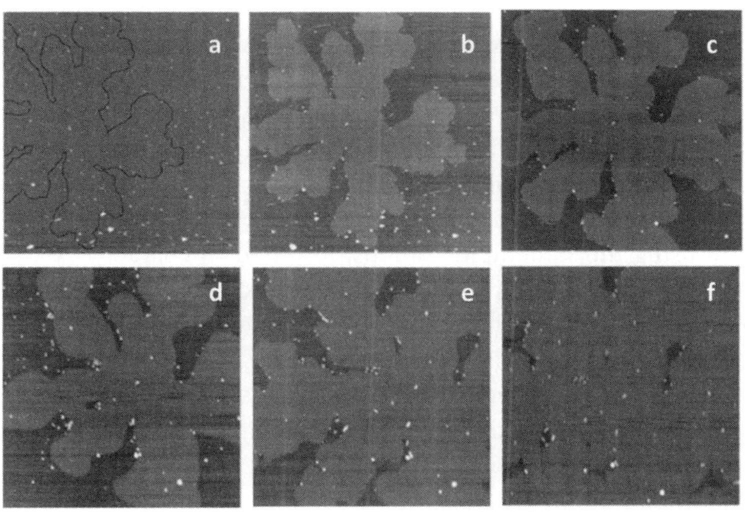

Figure 7.12 Reconstituted KcsA molecules (300 μg/mg) at different temperatures. In *a*): all the bilayer at 28°C is in the liquid disordered phase. Many KcsA molecules are found in the bilayer. In *b*): the temperature was decreased to 26.5°C while imaging the same area. A solid domain developed and the proteins were mainly excluded from the solid phase. The solid domain area has been overlaid on the image in figure a) in order to show the protein exclusion phenomenon. At *c-f*): Images at decreasing temperature (c: 26.5°C, d: 26.0°C, e: 25.0°C, f: 23.0°C) showing the growth of the solid domain and the induced clustering of the proteins confined to the remaining liquid areas. (Reprinted with permission from *J. Mol. Recognit.*, Vol. 24, p. 387, 2011).

and POPG in the proportion 3:1) with respect to the case of the lipid bilayer in the liquid disordered phase and an environment with specific mechanical properties (see above). The variation of the lipid composition around the protein is due to the fact that the liquid disordered domains that coexist with the solid ordered ones are enriched in the lipids with a lower main phase transition temperature, in the case POPG. The variation in the chemical composition of the bilayer should always be considered when alterations in the functional behavior of membrane proteins are observed. By further decreasing temperature, the solid domain area increases and the proteins are concentrated in the decreasing liquid fraction of the bilayer and are eventually induced to form clusters (Figure 12f). Many proteins are also localized in the boundary region between the two domains boundaries. This behavior can be explained by the concept of interfacial adsorption [17, 130].

7.5.2 Functional Relevance of Partitioning

To date, state of the art of the technique is such that AFM cannot determine functional properties of the proteins incorporated in the SLB, but other techniques can be exploited to study these properties. For example, it is possible to study the partitioning of transmembrane ion channels relatively to domain formation in lipid bilayers by AFM and then reconstitute the same protein in Black Lipid Membrane using the same lipid composition to study the functionality of the channels. In the specific case of KcsA, the functionality of the channel has been explored as a function of temperature using the same lipid composition used to characterize the protein partitioning [131], by exploiting the Black Lipid Membrane (BLM) setup [131]. By performing a functional test of KcsA in a planar lipid bilayer (voltage clamp measurements on BLM) it has been demonstrated that, when the lipid bilayer is in the phase transition region, all the functional parameters of the KcsA channel, at the single molecule level, change. In particular, the functional parameters follow the trend of the excess heat capacity (DSC trace) of the lipid bilayer as a function of the temperature. In the case of a multicomponent lipid bilayer, phase transition will produce a redistribution of the relative lipid composition in each different domain, as previously discussed. This phenomenon points to the possible role of the specific lipid composition on membrane protein activity. However, even if this role is definitely relevant, in the specific case at issue, it has been demonstrated that the effect of the enrichment of the domains hosting the proteins in the lipids that favor an increased activity of the channel, cannot explain the magnitude of the observed enhanced activity [131]. Given that the heat capacity of a system is strictly related to the fluctuations of its parameters, the scenario of the mechanical properties of the lipid bilayers influencing the behavior of membrane proteins must be considered (see above). In particular, at the single molecule level, it has been observed that single molecule conductance shows a biphasic behavior when the temperature is decreased with the inversion point near to the onset of the lipid phase transition (Figure 7.13). Moreover, the biphasic behavior is related to the lipid composition of the bilayer. Indeed, Figure 7.13 shows that the onset of the trend inversion of the single channel conductance as a function of temperature depends on the lipid composition accordingly to the phase transition temperature shift of the different compositions. Moreover, if the phase transition is not present in the observed temperature range (i.e. because a lipid with a different melting temperature has been used), no deviation from a linear trend of the conductance is found. According to what has been observed by AFM, the functional behavior in this case can be related to partitioning of the proteins in the different lipid phases or at the domain boundaries. The overall picture appears as follows: 1) upon the beginning of a phase transition, KcsA proteins preferentially distribute

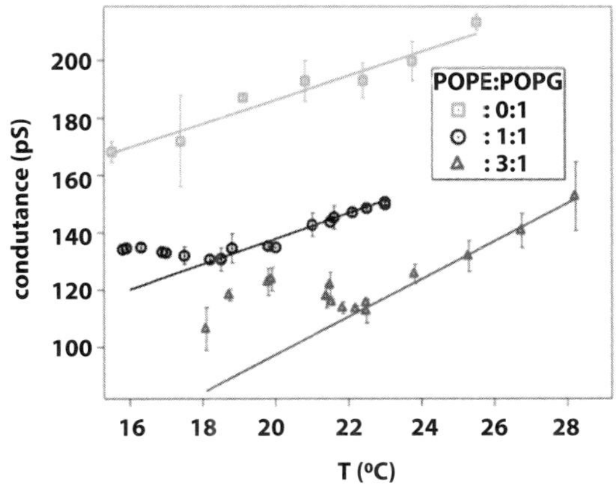

Figure 7.13 Temperature-dependent ion conductance of Kcsa in blms of different POPE:POPG compositions at a holding potential of 75 mV. The solid lines are the respective linear fits in the high temperature region. A non-monotonic behavior is observed for the two POPE:POPG mixtures (○: 3:1; Δ: 1:1), but a pure linear trend is found for POPG (□). The trend inversion happened at a lower temperature when KcsA was reconstituted in POPE:POPG 1:1 than in 3:1, accordingly to the shift of the phase transition region for the two lipid compositions. (Reprinted with permission from *Biophys. J.*, Vol. 99, p. 3675, 2010).

in the still-liquid-disordered domains; 2) these domains undergo a strong increase in fluctuations and in their compressibility at the nanometer scale (mechanical properties); 3) proteins undergo variations of their activity when the lipid bilayer is in the phase transition region and the strength of these variations scales with the excess heat capacity profile of the bilayer as a function of temperature.

7.6 Conclusions

We showed how AFM and AFS experiments on supported lipid bilayers provide useful information about the chemical-physical properties of SLBs and about their mechanical properties at the nanoscale. The high lateral and vertical resolution of AFM and the possibility to work under near physiological conditions make AFM a technique of choice to study SLBs without a staining treatment of the bilayer. The chemical-physical properties of SLBs can be studied as a function of different preparation strategies, different environments such as ionic strength conditions, different pH and temperature. Moreover, the partitioning of membrane-associated proteins relative to heterogeneities in the lateral organization of the bilayer can be studied as a function of lipid composition, phase state of the bilayer and temperature. Moreover, new advancements in the elucidation of lipid/protein interactions exploiting AFM are surely connected to technical improvements. Among these, the increase in time resolution of AFM imaging is a central issue. At the same time, progress in coupling AFM with other techniques such as fluorescence and Raman spectroscopy will increase our knowledge of lipid/protein interaction. As to the possibility of investigating the mechanical properties of lipid bilayers, it is also of great relevance to study how different amphiphilic drugs that partition in the lipid bilayer are able to affect bilayer's mechanical properties. This could be relevant to connect the effect of these molecules on membrane proteins through a mediating role of the lipid bilayer physical properties.

References

1. O.G. Mouritsen, *Life—as a matter of fat. The emerging science of lipidomics*, Heidelberg, Springer, 2005.
2. T. Heimburg, *Thermal biophysics of membranes*, Berlin, Wiley-VCH, 2007.
3. S.J. Singer, and G.L. Nicolson, The fluid mosaic model of the structure of cell membranes, *Science*, Vol. 175, p. 720, 197.2

4. K. Simons, and E. Ikonen, Functional rafts in cell membranes, *Nature*, Vol. 387, p. 569, 1997.
5. D.M. Engelman, Membranes are more mosaic than fluid, *Nature*, Vol. 438: p. 578, 2005.
6. D. Lingwood, and K. Simons, Lipid rafts as a membrane organizing principle, *Science*, Vol. 327, p. 46, 2010.
7. L.J. Pike, Rafts defined: a report on the Keystone symposium on lipid rafts and cell function, *J. Lipid. Res.*, Vol. 47, p. 1597, 2006.
8. S.W. Hell, and J. Wichmann, Breaking the diffraction resolution limit by stimulated emission: stimulated-emission-depletion fluorescence microscopy, *Opt. Lett.* Vol. 19, p. 780, 1994.
9. K.I. Willig, S.O. Rizzoli, V. Westphal, R. Jahn, and S.W. Hell, STED microscopy reveals that synaptotagmin remains clustered after synaptic vesicle exocytosis, *Nature*, Vol. 440, p. 935, 2006.
10. A. Kusumi, T.K. Fujiwara, R. Chadda, M. Xie, T.A. Tsunoyama, Z. Kalay, R.S. Kasai, K.G. Suzuki, Dynamic organizing principles of the plasma membrane that regulate signal transduction: commemorating the fortieth anniversary of Singer and Nicolson's fluid-mosaic model, *Annu. Rev. Cell Dev. Biol.*, Vol. 28, p. 215, 2012.
11. A.R. Honerkamp-Smith, S.L. Veatch, and S.L. Keller, An introduction to critical points for biophysicists; observations of compositional heterogeneity in lipid membranes, *Biochim. Biophys. Acta.*, Vol. 1788, p. 53, 2009.
12. O.G. Mouritsen, Model answers to lipid membrane questions, *Cold Spring Harb Perspect Biol.*, Vol. 3, p. a004622, 2011.
13. R. Phillips, T. Ursell, P. Wiggins, and P. Sens, Emerging roles for lipids in shaping membrane-protein function, *Nature*, Vol. 459, p. 379, 2009.
14. A. Alessandrini, G. Viero, M. Dalla Serra, G. Prévost, and P. Facci, γ-Hemolysin oligomeric structure and effect of its formation on supported lipid bilayers: An AFM Investigation, *Biochim. Biophys. Acta*, Vol. 1828, p. 405, 2013.
15. K. Jacobson, O.G. Mouritsen, and G.W. Anderson, Lipid rafts: at a crossroad between cell biology and physics, *Nature Cell Biol.*, Vol. 9, p. 7, 2007.
16. A. Alessandrini, and P. Facci, AFM:a versatile tool in biophysics. *Meas. Sci & Technol.*, Vol. 6, p. R65, 2005.
17. A. Alessandrini, and P. Facci, Unraveling lipid/protein interaction in model lipid bilayers by Atomic Force Microscopy, *J. Mol. Recognit.*, Vol. 24, p. 387, 2011.
18. K. El Kirat, S. Morandat, and Y.F. Dufrêne. Nanoscale analysis of supported lipid bilayers using atomic force microscopy, *Biochim. Biophys. Acta*, Vol. 1798, p. 750, 2010.
19. E.I. Goksu, J.M. Vanegas, C.D. Blanchette, W.C. Lin, and M.L. Longo, AFM for structure and dynamics of biomembranes, *Biochim. Biophys. Acta*, Vol. 1788, p. 254, 2009.
20. L. Picas, P.E. Milhiet, and J. Hernández-Borrell, Atomic force microscopy: A versatile tool to probe the physical and chemical properties of supported membranes at the nanoscale, *Chem. Phys. Lipids*, Vol. 165, p. 845, 2012.

21. H.M. McConnell, T.H. Watts, R.M. Weis, and A.A. Brian, Supported planar membranes in studies of cell-cell recognition. *Biochim. Biophys. Acta*, Vol. 864, p. 95, 1986.
22. E.T. Castellana, and P.S.Cremer, Solid supported lipid bilayers: From biophysical studies to sensor design, *Surf. Sci. Rep.* Vol. 61: p. 429, 2006.
23. Y.F. Dufrêne, W.R. Barger, J.B.D. Green, and G.U. Lee, Nanometer-scale surface properties of mixed phospholipid monolayers and bilayers, *Langmuir*, Vol. 13, p. 4779, 1997.
24. A.A. Brian, and H.M. McConnell, Allogeneic stimulation of cytotoxic T cells by supported planar membranes. *Proc. Natl. Acad. Sci. U.S.A.*, Vol. 81, p. 6159–6163, 1984.
25. J.M. Crane, V. Kiessling, and L.K. Tamm, Measuring lipid asymmetry in planar supported bilayers by fluorescence interference contrast microscopy, *Langmuir*, Vol. 21, p. 1377, 2005.
26. A.A. Brian, and H.M. McConnell, Allogeneic stimulation of cytotoxic T cells by supported planar membranes, *Proc. Natl. Acad. Sci. U.S.A.*, Vol. 81, p. 6159, 1984.
27. E. Reimhult, M. Zäch, F. Höök, and B. Kasemo, A multitechnique study of liposome adsorption on Au and lipid bilayer formation on SiO_2, *Langmuir*, Vol. 22, p. 3313, 2006.
28. B.W. Koenig, S. Kruger, W.J. Orts, C.F. Majkrzak, N.F. Berk, J.V. Silverton, and K. Gawrisch, Neutron reflectivity and atomic force microscopy studies of a lipid bilayer in water adsorbed to the surface of a silicon single crystal, *Langmuir*, Vol. 12, p. 1343, 1996
29. P.E. Milhiet, F. Gubellini, A. Berquand, P. Dosset, J.L. Rigaud, C. Le Grimellec, and D. Lévy, High-resolution AFM of membrane proteins directly incorporated at high density in planar lipid bilayer, *Biophys. J.*, Vol. 91, p. 3268, 2006.
30. D.J. Muller, Out and in: simplifying membrane protein studies by AFM, *Biophys. J.*, Vol. 91, p. 3133, 2006.
31. D. Levy, and P.E. Milhiet, Imaging of transmembrane proteins directly incorporated within supported lipid bilayers using atomic force microscopy, *Methods Mol. Biol.*, Vol. 950, p. 343, 2013.
32. H.P. Wacklin, Composition and Asymmetry in Supported Membranes Formed by Vesicle Fusion, *Langmuir*, Vol. 27, p. 7698, 2011.
33. M. Tanaka, and E. Sackmann, Polymer-supported membranes as models of the cell surface, *Nature*, Vol. 437, p. 656, 2005.
34. L.J. Jeuken, S.D. Connell, P.J. Henderson, R.B. Gennis, S.D. Evans, and R.J. Bushby, Redox enzymes in tethered membranes, *J. Am. Chem. Soc.*, Vol. 128, p. 1711, 2006.
35. M.G. Friedrich, V.U. Kirste, J. Zhu, R.B. Gennis, W. Knoll, and R.L. Naumann, Activity of membrane proteins immobilized on surfaces as a function of packing density, *J. Phys. Chem. B*, Vol. 112, p. 3193, 2008.
36. I. Mey, M. Stephan, E.K. Schmitt, M.M. Müller, M. Ben Amar, C. Steinem, and A. Janshoff, Local membrane mechanics of pore-spanning bilayers, *J. Am. Chem. Soc.*, Vol. 131, p. 7031, 2009.

37. H.M. Seeger, G. Marino, A. Alessandrini, and P. Facci, Effect of physical parameters on the main phase transition of supported lipid bilayers, *Biophys. J.*, Vol. 97, p. 1067, 2009.
38. M. Deserno, Mesoscopic membrane physics: concepts, simulations, and selected applications, *Macromol. Rapid Commun.*, Vol. 30, p. 752, 2009.
39. C. Xing, and R. Faller, Interactions of lipid bilayers with supports: a coarse-grained molecular simulation study, *J. Phys. Chem. B*, Vol. 112, p. 7086, 2008.
40. C. Xing, and R. Faller, Density imbalances and free energy of lipid transfer in supported lipid bilayers, *J. Chem. Phys.*, Vol. 131, p. 175104, 2009.
41. A.P. Shreve, M.C. Howland, A.R. Sapuri-Butti, T.W. Allen and A.N. Parikh, *Langmuir*, Vol. 24, p. 13250, 2008.
42. L.A. Bagatolli, J.H. Ipsen, A.C. Simonsen, and O.G. Mouritsen, An outlook on organization of lipids in membranes: Searching for a realistic connection with the organization of biological membranes, *Prog. Lipid Res.*, Vol. 49, p. 378, 2010.
43. J. Israelachvili, *Intermolecular and Surface Forces 2nd Ed, London*, Academic Press, 1991.
44. H.-J Butt, M. Jaschke, and W. Ducker, Measuring surface forces in aqueous solution with the atomic force microscope, *Bioelectrochem. Bioenerg.*, Vol. 38, p. 191, 1995.
45. M.D. Collins, Interleaflet coupling mechanisms in bilayers of lipids and cholesterol, *Biophys. J.*, Vol. 94, p. L32, 2008.
46. T.Y. Wang, and J.R. Silvius, Cholesterol does not induce segregation of liquid-ordered domains in bilayers modeling the inner leaflet of the plasma membrane, *Biophys. J.*, Vol. 81, p. 2762, 2001.
47. M.D. Collins, and S.L. Keller, Tuning lipid mixtures to induce or suppress domain formation across leaflets of unsupported asymmetric bilayers, *Proc. Natl. Acad. Sci. U.S.A.*, Vol. 105, p. 124, 2008.
48. V. Kiessling, J.M. Crane, and L.K. Tamm, Transbilayer effects of raft-like lipid domains in asymmetric planar bilayers measured by single molecule tracking, *Biophys. J.*, Vol. 91, p. 3313, 2006.
49. R. Merkel, E. Sackmann, and E. Evans, Molecular friction and epitactic coupling between monolayers in supported bilayers, *J. Phys. France*, Vol. 50, p. 1535, 1989.
50. E. Sackmann, Supported membranes: scientific and practical applications, *Science*, Vol. 271, p. 43, 1996.
51. D. Marguet, P.F. Lenne, H. Rigneault, and H.T. He, Dynamics in the plasma membrane: how to combine fluidity and order, *EMBO J.*, Vol. 25, p. 3446, 2006.
52. H.T. He, and D. Marguet, Detecting nanodomains in living cell membrane by fluorescence correlation spectroscopy, *Annu. Rev. Phys. Chem.*, Vol. 62, p. 417, 2011.
53. M.J. Skaug, R. Faller, and M.L. Longo, Correlating anomalous diffusion with lipid bilayer membrane structure using single molecule tracking and atomic force microscopy, *J. Chem. Phys.*, Vol. 134, p. 215101, 2011.

54. K.J. Seu, A.P. Pandey, F. Haque, E.A. Proctor, A.E. Ribbe, J.S. Hovis, Biophys. J. Vol. 92, p. 2445, 2007.
55. D. Keller, N.B. Larsen, I.M. Moller, and O.G. Mouritsen, Decoupled phase transitions and grain-boundary melting in supported phospholipid bilayers, *Phys. Rev. Lett.*, Vol. 94, p. 25701, 2005.
56. M. Przybylo, J. Sýkora, J. Humpolíĉková, A. Benda, A. Zan, and M. Hof, Lipid diffusion in giant unilamellar vesicles is more than 2 times faster than in supported phospholipid bilayers under identical conditions. *Langmuir*, Vol. 22, p. 9096, 2006.
57. T. Dertinger, I. von der Hocht, A. Benda, M. Hof, and J. Enderlein, *Langmuir*, Vol. 22, p. 9339, 2006.
58. L. Guo, J.Y. Har, J. Sankaran, Y.M. Hong, B. Kannan, and T. Wohland, Molecular diffusion measurement in lipid bilayers over wide concentration ranges: a comparative study, *ChemPhysChem*, Vol. 9, p. 721, 2008.
59. H.M. Seeger, A. Di Cerbo, A. Alessandrini, and P. Facci, Supported lipid bilayers on mica and silicon oxide: comparison of the main phase transition behavior, *J. Phys. Chem. B.*, Vol. 114, p. 8926, 2010.
60. M. Hetzer, S. Heinz, S. Grage, and T.M. Bayerl, Asymmetric Molecular Friction in Supported Phospholipid Bilayers Revealed by NMR Measurements of Lipid Diffusion, *Langmuir*, Vol. 14, p. 982, 1998.
61. L. Zhang, and S. Granick, Lipid diffusion in outer and inner leaflets of planar supported bilayers, *J. Chem. Phys.*, Vol. 123, p. 211104, 2005.
62. K.J. Seu, A.P. Pandey, F. Haque, E.A. Proctor, A.E. Ribbe, and J.S. Hovis, Effect of Surface Treatment on Diffusion and Domain Formation in Supported Lipid Bilayers, *Biophys. J.*, Vol. 92, p. 2445, 2007.
63. C.L. Armstrong, M.D. Kaye, M. Zamponi, E. Mamontov, M. Tyagi, T. Jenkins, and M.C. Rheinstädter, Diffusion in single supported lipid bilayers studied by quasi-elastic neutron scattering, *Soft Matter*, Vol. 6, p. 5864, 2010.
64. J. Salafsky, J.T. Groves, and S.G. Boxer, Architecture and function of membrane proteins in planar supported bilayers: A study with photosynthetic reaction centers, *Biochemistry*, Vol. 35, p. 14773, 1996.
65. P.B. Contino, C.A. Hasselbacher, J.B. Ross, Y. Nemerson, Use of an oriented transmembrane protein to probe the assembly of a supported phospholipid bilayer, *Biophys J.*, Vol. 67, p. 1113, 1994.
66. E. Reimhult, B. Kasemo, F. Höök, Rupture pathway of phosphatidylcholine liposomes on silicon dioxide, *Int. J. Mol. Sci.*, Vol. 10, p. 1683, 2009.
67. T.H. Liu, S.H. Hsu, Y.T. Huang, S.M. Lin, T.W. Huang, T.H. Chuang, S.K. Fan, C.C. Fu, F.G. Tseng, and R.L. Pan, The proximity between C-termini of dimeric vacuolar H+-pyrophosphatase determined using atomic force microscopy and a gold nanoparticle technique, *FEBS J.*, Vol. 276, p. 4381, 2009.
68. N. Vuong, J.E. Baenziger, and L.J. Johnston, Preparation of reconstituted acetylcholine receptor membranes suitable for AFM imaging of lipid-protein interactions, *Chem. Phys. Lipids*, Vol. 163, p. 117, 2010.

69. H.M. Seeger, C.A. Bortolotti, A. Alessandrini, and P. Facci, Phase-transition-induced protein redistribution in lipid bilayers, *J. Phys. Chem. B,* Vol. 113, p. 16654, 2009.
70. D.J. Müller, A. Engel, U. Matthey, T. Meier, P. Dimroth, K. Suda, Observing membrane protein diffusion at subnanometer resolution, *J. Mol. Biol.*, Vol. 327, p. 925, 2003.
71. T. Hughes, B. Strongin, F.P. Gao, V. Vijayvergiya, D.D. Busath, and R.C. Davis, AFM visualization of mobile influenza A M2 molecules in planar bilayers, *Biophys J.*, Vol. 87, p. 311, 2004.
72. I. Casuso, N. Kodera, C. Le Grimellec, T. Ando, and S. Scheuring,. Contact-mode high-resolution high-speed atomic force microscopy movies of the purple membrane, *Biophys. J.*, Vol. 97, p. 1354, 2009.
73. S. Chiantia, J. Ries, N. Kahya, and P. Schwille, Combined AFM and two-focus SFCS study of raft-exhibiting model membranes, *ChemPhysChem.* Vol. 7, p. 2409, 2006.
74. T. Ando, T. Uchihashi, N. Kodera, D. Yamamoto, M. Taniguchi, A. Miyagi, and H. Yamashita, High-speed atomic force microscopy for observing dynamic biomolecular processes, *J. Mol. Recognit.*, Vol. 20, p. 448, 2007.
75. I. Casuso, J. Khao, M. Chami, P. Paul-Gilloteaux, M. Husain, J.P. Duneau, H. Stahlberg, J.N. Sturgis, and S. Scheuring, Characterization of the motion of membrane proteins using high-speed atomic force microscopy, Nat Nanotechnol., Vol. 7, p. 525, 2012.
76. R. Blunck, H. McGuire, H.C. Hyde, F. Bezanilla, Fluorescence detection of the movement of single KcsA subunits reveals cooperativity, *Proc. Natl. Acad. Sci. USA*, Vol. 105, p. 20263, 2008.
77. R.P. Gonçalves, G. Agnus, P. Sens, C. Houssin, B. Bartenlian, and S. Scheuring, Two-chamber AFM: probing membrane proteins separating two aqueous compartments, *Nat. Methods*, Vol. 12, p. 1007, 2006.
78. B. Stottrup, S. Veatch, and S. Keller, Nonequilibrium behavior in supported lipid membranes containing cholesterol, *Biophys. J.*, Vol. 86, p. 2942, 2004.
79. L.A. Bagatolli, and E. Gratton, A correlation between lipid domain shape and binary phospholipid mixture composition in free standing bilayers: A two-photon fluorescence microscopy study, *Biophys. J.*, Vol. 79, p. 434, 2000.
80. F. Tokumasu, A.J. Jin, G.W. Feigenson, and J.A. Dvorak, Atomic force microscopy of nanometric liposome adsorption and nanoscopic membrane domain formation, *Ultramicroscopy*, Vol. 97, p. 217, 2003.
81. A., Charrier, and F. Thibaudau, Main phase transitions in supported lipid single-bilayer, *Biophys. J.*, Vol. 89, p. 1094, 2005.
82. H. Träuble, and H. Eibl, Electrostatic effects on lipid phase transitions: membrane structure and ionic environment, *Proc. Nat. Acad. Sci. USA,* Vol. 71, p. 214, 1974.
83. F. Jähnig, Electrostatic free energy and shift of the phase transition for charged lipid membranes, *Biophys. Chem.*, Vol. 4, p. 309, 1976.

84. D. Marsh, Protein modulation of lipids, and vice-versa, in membranes, *Biochim. Biophys. Acta,* Vol. 1778, p. 1545, 2008.
85. T.J. McIntosh, and S.A. Simon, Roles of bilayer material properties in function and distribution of membrane proteins, *Ann. Rev. Biophys. Biomolec. Struct.*, Vol. 35, p. 177, 2006.
86. A.G. Lee. How lipids affect the activities of integral membrane proteins, *Biochim. Biophys. Acta,* Vol. 1666, p. 62, 2004.
87. A. Alessandrini, and P. Facci, Changes in single K^+ channel behavior induced by a lipid phase transition, *Commun. Integr. Biol.*, Vol. 4, p. 346, 2011.
88. W. Helfrich, Elastic properties of lipid bilayers: theory and possible experiments, *Z. Naturforsch. C*, Vol. C 28, p. 693, 1973.
89. R. M. Hochmuth, Micropipette aspiration of living cells, *J. Biomech.*, Vol. 33, p. 15, 2000.
90. J.R. Henriksen and J.H. Ipsen, Measurement of membrane elasticity by micro-pipette aspiration, *Eur. Phys. J. E*, Vol. 14, p. 149, 2004.
91. Y.F. Dufrene, T. Boland, J. W. Schneider, W. R. Barger and G.U. Lee, Characterization of the physical properties of model biomembranes at the nanometer scale with the atomic force microscope, *Farad. Disc.*, Vol. 111, p. 79, 1998.
92. S. Garcia-Manyes, G. Oncins and F. Sanz, Effect of ion-binding and chemical phospholipid structure on the nanomechanics of lipid bilayers studied by force spectroscopy, *Biophys. J.*, Vol. 89, p. 1812, 2005.
93. S. Garcia-Manyes, G. Oncins and F. Sanz, Effect of temperature on the nanomechanics of lipid bilayers studied by force spectroscopy, *Biophys. J.*, Vol. 89, p. 4261, 2005.
94. S. Garcia-Manyes, G. Oncins and F. Sanz, Effect of pH and ionic strength on phospholipid nanomechanics and on deposition process onto hydrophilic surfaces measured by AFM, *Electrochim. Acta*, Vol. 51, p. 5029, 2006.
95. S. Garcia-Manyes, L. Redondo-Morata, G. Oncins, and F. Sanz, Nanomechanics of lipid bilayers: heads or tails?, *J. Am. Chem. Soc.*, Vol. 132, p. 12874, 2010.
96. H.J. Butt and V. Franz, Rupture of molecular thin films observed in atomic force microscopy. I. Theory, *Phys. Rev. E*, Vol. 66, p. 031601, 2002.
97. S. Loi, G. Sun, V. Franz and H.J. Butt, Rupture of molecular thin films observed in atomic force microscopy. II. Experiment, *Phys. Rev. E*, Vol. 66, p. 031602, 2002.
98. B.N.J. Persson and E. Tosatti, Layering transition in confined molecular thin films: Nucleation and growth, *Phys. Rev. B*, Vol. 50, p. 5590, 1994.
99. U. Zimmermann, G. Pilwat, A. Pequeux and R. Gilles, Electro-mechanical properties of human erythrocyte membranes: the pressure-dependence of potassium permeability, *J. Membr. Biol.*, Vol. 54, p. 103, 1980.
100. E. Evans, V. Heinrich, F. Ludwig and W. Rawicz, Dynamic tension spectroscopy and strength of biomembranes, *Biophys. J.*, Vol. 85, p. 2342, 2003.

101. A. Alessandrini, H.M. Seeger, A. Di Cerbo, T. Caramaschi, and P. Facci, What do we really measure in AFM punch-through experiments on supported lipid bilayers?, *Soft Matter*, Vol. 7, p. 7054, 2011.
102. A. Alessandrini, H.M. Seeger, T. Caramaschi, and P. Facci, Dynamic force spectroscopy on supported lipid bilayers: effect of temperature and sample preparation, *Biophys J.*, Vol. 103, p. 38, 2012.
103. E. Evans, and K. Ritchie, Dynamic strength of molecular adhesion bonds, *Biophys. J.*, Vol. 2, p. 1541, 1997.
104. E. Evans, Probing the relation between force—lifetime—and chemistry in single molecular bonds, *Annu. Rev. Biophys. Biomol. Struct.*, Vol. 30, p. 105, 2001.
105. H. Hertz, Über die Berührung fester elastischer Körper, J. Reine Angew. Mathematik, Vol. 92, p. 156, 1882.
106. I.N. Sneddon, The relation between load and penetration in the axisymmetric boussinesq problem for a punch of arbitrary profile, *Int. J. Eng. Sci.* Vol. 3, p. 47, 1965.
107. Z.V. Leonenko, E. Finot, H. Ma, T.E. Dahms, and D.T. Cramb, Investigation of temperature-induced phase transitions in DOPC and DPPC phospholipid bilayers using temperature-controlled scanning force microscopy, *Biophys J.*, Vol. 86, p. 3783, 2004.
108. C. Das, K.H. Sheikh, P.D. Olmsted, and S.D. Connell, Nanoscale mechanical probing of supported lipid bilayers with atomic force microscopy, *Phys Rev. E*, Vol. 82, p. 041920, 2010.
109. J. Fraxedas, S. Garcia-Manyes, P. Gorostiza, F. Sanz, Nanoindentation: Toward the sensing of atomic interactions, *Proc. Natl. Acad. Sci. U S A*, Vol.99, p. 5228, 2002.
110. B. Cappella, and G. Dietler Force-distance curves by atomic force microscopy, *Surf. Sci. Rep.*, Vol. 34, p. 1, 1999.
111. L. Picas, F. Rico, and S. Scheuring, Direct measurement of the mechanical properties of lipid phases in supported bilayers, *Biophys J.*, Vol. 102, p. L01, 2012.
112. V. Franz, S. Loi, and H.J. Butt, 2002. Tip penetration through lipid bilayers in atomic force microscopy, *Colloids Surf. B Biointerfaces*, Vol. 23, p. 191, 2002.
113. G.I. Bell, 1978. Models for the specific adhesion of cells to cells, *Science*, Vol. 200, p. 618, 1978.
114. O.K. Dudko, A.E. Filippov, J. Klafter, and M. Urbakh, Beyond the conventional description of dynamic force spectroscopy of adhesion bonds, *Proc. Natl. Acad. Sci. USA*, Vol. 100, p. 378, 2003.
115. R.W. Friddle, A. Noy, and J.J. De Yoreo, Interpreting the widespread nonlinear force spectra of intermolecular bonds, Proc. Natl. Acad. Sci. U S A, Vol. 109, p. 13573, 2012; and references therein.
116. R. P. Merkel, P. Nassoy, and E. Evans, Energy landscapes of receptor-ligand bonds explored with dynamic force spectroscopy, *Nature*, Vol. 397, p. 50, 1999.

117. O.K. Dudko, G. Hummer, and A. Szabo, Intrinsic rates and activation free energies from single-molecule pulling experiments, *Phys. Rev. Lett.*, Vol. 96, p. 108101, 2006.
118. G. Hummer, and A. Szabo, Kinetics from nonequilibrium single-molecule pulling experiments, *Biophys. J.* Vol. 85, p. 5, 2003.
119. D. Bartolo, I. Derenyi, and A. Ajdari, Dynamic response of adhesion complexes: beyond the single-path picture, *Phys. Rev. E.* Vol. 65, p. 051910, 2002.
120. R.V. Glaser, S.L. Leikin, and A.I. Sokirko, Reversible electrical breakdown of lipid bilayers: formation and evolution of pores, *Biochim. Biophys. Acta*, Vol. 940, p. 275, 1998.
121. G. Fuertes, D. Gimenez, and J. Salgado, A lipocentric view of peptide-induced pores. *Eur. Biophys. J.* Vol. 40, p. 399, 2011.
122. S. Mayor, and H. Riezman, Sorting GPI-anchored proteins, *Nat. Rev. Mol. Cell Biol.*, Vol. 2, p. 110, 2004.
123. D.E. Saslowsky, J. Lawrence, X Ren, D.A. Brown, R.M. Henderson, and J.M. Edwardson, Placental alkaline phosphatase is efficiently targeted to rafts in supported lipid bilayers, *J. Biol. Chem.*, Vol. 277, p. 26966, 2002.
124. M.C. Giocondi, B. Seantier, P. Dosset, P.E. Milhiet, and C. Le Grimellec, Characterizing the interactions between GPI-anchored alkaline phosphatases and membrane domains by AFM. *Pflugers Arch.*, Vol. 456, p. 179, 2008.
125. N. Kahya, D.A Brown, and P. Schwille, Raft partitioning and dynamic behavior of human placental alkaline phosphatase in giant unilamellar vesicles, *Biochemistry*, Vol. 44, p. 7479, 2005.
126. M.C. Giocondi, F. Besson, P. Dosset, P.E. Milhiet, and C. Le Grimellec, Remodeling of ordered membrane domains by GPI-anchored intestinal alkaline phosphatase, *Langmuir*, Vol. 23, p. 9358, 2007.
127. K. Weise, G. Triola, S. Janosch, H Waldmann, and R. Winter, Visualizing association of lipidated signaling proteins in heterogeneous membranes-Partitioning into subdomains, lipid sorting, interfacial adsorption, and protein association, *Biochim. Biophys. Acta* , Vol. 1798, p. 1409, 2010.
128. D.A. Doyle, J.M. Cabral, R.A. Pfuetzner, A.L. Kuo, J.M. Gulbis, S.L. Cohen, B.T. Chait, and R. MacKinnon, The structure of the potassium channel: molecular basis of K^+ conduction and selectivity, *Science* Vol. 280, p. 69, 1998.
129. L.G. Cuello, J.G. Romero, D.M. Cortes, and E. Perozo, pH-dependent gating in the Streptomyces lividans K+ channel, *Biochemistry*, Vol. 37, p. 3229, 1998.
130. F. Dumas, M.M. Sperotto, M-C. Lebrun, J-F. Tocanne, and O.G. Mouritsen, Molecular sorting of lipids by bacteriorhodopsin in dilauroylphosphatidylcholine/distearoylphosphatidylcholine lipid bilayers, *Biophys. J.*, Vol. 73, p.1940, 1997.
131. H.M. Seeger, L. Aldrovandi, A. Alessandrini, and P. Facci, Changes in single K^+ channel behavior induced by a lipid phase transition, *Biophys. J.*, Vol. 99, p. 3675, 2010.

Part 3

DIRECTED MOLECULAR SEPARATION

8

Self-Assembled Nanoporous Membranes for Controlled Drug Release and Bioseparation

Dominique Scalarone*, Pierangiola Bracco and Francesco Trotta

Department of Chemistry and NIS Centre of Excellence, University of Torino, Torino, Italy

Abstract
This chapter highlights recent activity in the area of nanoporous membranes templated by block copolymers. Asymmetric composite membranes and stand-alone membranes, membranes templated by the cylindrical morphology and bicontinuous nanoporous structures are presented. Fabrication procedures of nanoporous membranes based on self-assembling block copolymers are described, as well as their transport properties and size-selectivity. Since membranes derived from block copolymers have accessible pore size approaching the size scale of biological components, focus is placed especially on applications for controlled release of drugs and effective separation of biomolecules.

Keywords: Nanoporous membranes, self-assembly, block copolymers, bioseparation, drug delivery

8.1 Introduction

Polymer membranes having high and controlled porosity, narrow pore size distribution and tunable surface properties are ideal materials for applications in separation processes, microfluidics and for the fabrication of miniaturized devices and sensors for controlled transport and release of molecular species. Recently significant advances have been achieved

**Corresponding author*: dominique.scalarone@unito.it

Annarosa Gugliuzza (ed.) Smart Membranes and Sensors, (229–254) 2014 © Scrivener Publishing LLC

in membrane science, but some limitations still prevent the development of novel high performance membranes. These limitations are intrinsic to the materials and technologies used for membrane production. Among solvent-based techniques, phase inversion is largely used for the production of microfiltration and ultrafiltration membranes [1–3]. Despite showing high flux and good retention of some biomolecules, the applicability of microfiltration membranes is restricted to large molecules (i.e. >100 nm) due to their relatively large pore dimension. On the other hand, ultrafiltration membranes having smaller pores than microfiltration ones, have broad size distributions of pores that limit their efficacy in separation processes. Track-etched nanoporous polymer membranes having nearly monodisperse pores and sharp molecular weight cut-off are now available, but their pore density ($<10^9$ pores per cm^2) is not very high and some pores could merge during the etching step resulting in a much lower flux than that of conventional ultrafiltration membranes [4, 5]. Also lithographic techniques, that are usually used for the processing of inorganic materials, have been adapted to pattern polymer and to fabricate nanoporous materials with uniform pore shape and size, but again with low porosity (10^6 to 10^{10} pores per cm^2) [6].

An interesting strategy to obtain membranes with improved morphological features in the nano dimension, namely uniform pore size and high pore density ($>10^{11}$ pores per cm^2), is based on the use of self-assembling block copolymers. The capability of block copolymers to spontaneously arrange into well-defined ordered structures with nanoscopic size has proven to be particularly advantageous for templating nanoporous membranes. In the last decade several types of block copolymer-based membranes for filtration and selective transport of target molecules have been prepared and successfully tested at laboratory level, but there are still some problems that must be solved in order to proceed with their industrial scale up. This chapter wants to give an overview of the state of the art on nanoporous membranes derived from self-assembling block copolymers, briefly describing theoretical concepts concerning self-assembly, the experimental procedures used to prepare such membranes, their main strengths and weaknesses.

The review focuses especially on recent researches on membranes in the form of thin films to be used in biofiltration processes, sensors and in implantable long-term controlled drug release systems, while does not consider micellar assemblies based on amphiphilic block copolymers mimicking biomembranes, such as polymersomes, that have been extensively reviewed elsewhere [7, 8]

8.2 General Aspects of Block Copolymer Self-Assembly

The self-assembling process is a bottom-up chemical approach that consists in the spontaneous and reversible organization of molecules, or parts of them, in ordered aggregates through non-covalent interactions and without external intervention. The self-assembly of macromolecular compounds is used by nature to form a number of biological structures with different levels of hierarchical organisation and varying from proteins and viruses to the mineral structures in bones and shells [9].

In the past decade the discovery and development of new radical living polymerization techniques [10], such as atom transfer radical polymerization (ATRP), have allowed the synthesis of self-assembling block copolymers with well-defined microstructure and narrow molecular weight distribution. The intrinsic dimension and the self-assembling capability of these copolymers make them ideal tools for the fabrication of a variety of periodic structures in the nanoscale range.

The driving force for self-assembling of block copolymers is the thermodynamic incompatibility of the different blocks in the polymer chains, that drives them to spontaneously segregate in spatial domains whose extension is controlled by the chemical interactions existing between the blocks. At the interface between two different blocks, the system tends to reduce the contact surface in order to minimize the energy content. Different blocks are then subjected to a strong reciprocal repulsion so that energetically unfavorable contacts are avoided. This results in the spontaneous formation of well-defined structures (i.e., spheres, cylinders, lamellae, gyroids) with peculiar periodicity.

As shown in Figure 8.1, the self-assembling behaviour of block copolymers mainly depends on three experimental parameters: the degree of polymerization (N), the composition (f), and the Flory-Huggins interaction parameter (χ) [11]. Moreover, shape and extension of the organized nanodomains can be easily tuned by varying the stoichiometric ratio of the monomers and the chain length of each polymer block [12], by preparing blends with homopolymers of the same nature of the minority block, and exploiting supramolecular interactions between one of the blocks and small organic molecules [13, 14]. In the last two cases molecules able to specifically interact with one of the blocks are expected to preferentially localize in the corresponding domains, possibly influencing their size, shape and orientation.

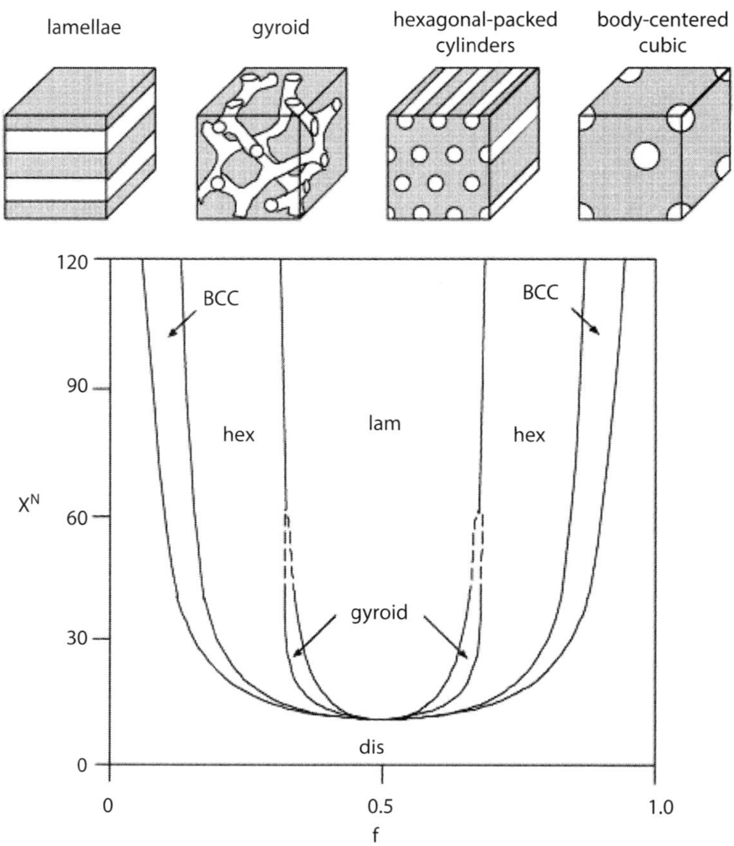

Figure 8.1 Schematic representations of the equilibrium morphologies of diblock copolymer melts and theoretical phase diagram, illustrating the regions of stability of disordered (*dis*), lamellar (*lam*), gyroid (*gyr*), hexagonal (*hex*) and body-centered cubic (*BCC*) phases (Reprinted from [11] with permission from Elsevier).

The main limit to the use of block copolymers in nanotechnology lies in the poor control of the order that the nanophases exhibit in the whole mass of the material. Nanodomains accidentally develop in space, but to optimize the physical properties of such materials, as requested by their potential applications in nanotechnology, their orientation and ordering should be controlled. In general, orientation and lateral order can be optimized by thermal treatments at low temperature, by treatments under solvent vapours or by application of unidirectional external forces [15–20]. In effect, the large number of thermodynamic and kinetic factors affecting nanophase segregation makes this task very difficult.

8.3 Block Copolymer-Based Membranes

The main advantages of block copolymer-templated membranes over other types of polymer membranes are their narrow pore size distribution, high void fraction and smooth surfaces, resulting in superior selectivity, high fluxes and fouling resistance. Pore size can be easily varied to optimize size selectivity, and additional separation mechanisms can be introduced by tuning the membrane surface properties through functionalization of the pore walls.

The most interesting morphologies addressed to the preparation of nanoporous membranes are bicontinuous and cylindrical morphologies. In general, membranes are obtained from self-assembled block copolymer materials by removing the minor component, while the major component forms a continuous matrix. Bicontinuous morphologies give membranes with nanosized homogeneous porosity, pore percolation, but low periodicity, while cylindrical morphologies allow the fabrication of membranes with highly ordered channels spanning the whole membrane thickness [21].

At present the main critical factors limiting the production and applicability on a large scale of such membranes are the achievement of proper pore alignment and the poor mechanical integrity of the membrane. Structural order over large extension strongly depends on a number of molecular interactions, including hydrophilic and hydrophobic effects, hydrogen bonding, Coulomb interactions and van der Waals forces, and the capability to control ordering and orientation of block copolymer materials is especially complicated when they are in the form of thin films. The copolymer confinement between the substrate and the air interface results in interfacial forces that can induce changes in the periodicity. Also the tendency of polymer domains to orientate along a preferential direction depends on interfacial and surface energies, as well as on the film thickness [22, 23].

Most preparative protocols for nanoporous membranes derived from block copolymers are restricted to thin films that are not enough mechanically robust for separation processes and require additional and highly resistant support membranes. To improve mechanical properties, the matrix of the nanoporous material is preferentially a cross-linked polymer. For instance, PS nanoporous membranes obtained from PS-*b*-PMMA block copolymers and subsequently cross-linked by UV irradiation exhibit better adhesion to supporting membranes and improved mechanical and solvent resistance than non-irradiated membranes [24].

Furthermore, the fabrication procedures that have been proposed till now often consists of too many steps that are difficult to be automated and prevent the large-scale membrane production.

Strengths and weaknesses of nanoporous self-assembled membranes will be further discussed in section 8.4, describing the experimental procedures that have been developed so far to fabricate block copolymer-derived membranes.

8.4 Fabrication of Nanoporous Membranes Derived from Block Copolymers

The most common strategy to prepare nanoporous membranes from self-assembling block copolymers involves 1) the deposition of block copolymer solutions on proper supports, 2) the annealing of the resulting thin films in order to optimize phase segregation and orientation of polymer domains, and 3) the removal of the minority block to obtain a porous structure.

8.4.1 Structure of Nanoporous Membranes: Composite and Stand-alone Membranes

As already said, block copolymer-derived membranes are generally too thin (i.e. tens to few hundreds of nanometers in thickness) for applications in separation processes and they must be adequately supported. Highly selective separations and high flux have been obtained by membranes with an asymmetric film geometry consisting in a thin nanoporous layer, prepared from a block copolymer template, and a support membrane. Both polymeric [25–26] and silicon microporous supports [27] have been proposed for the fabrication of composite membranes.

Yang *et al.* fabricated composite membranes for virus filtration by a multi-step procedure that includes two etching treatments and the transfer of the active layer onto a polymer support [26]. At first a 80 nm thick nanostructured film is obtained by spin-coating a PS-*b*-PMMA and PMMA mixture on a modified silicon wafer. The film is floated onto the surface of a buffered HF solution able to dissolve silicon oxide, and then is transferred onto a microfiltration polysulfone membrane providing enough mechanical integrity for the subsequent fabrication steps, as depicted in Figure 8.2. The reduced thickness of the selective top layer determines a low resistance against mass transfer, thus membranes prepared according to this procedure show both high areal density of pores and high flux.

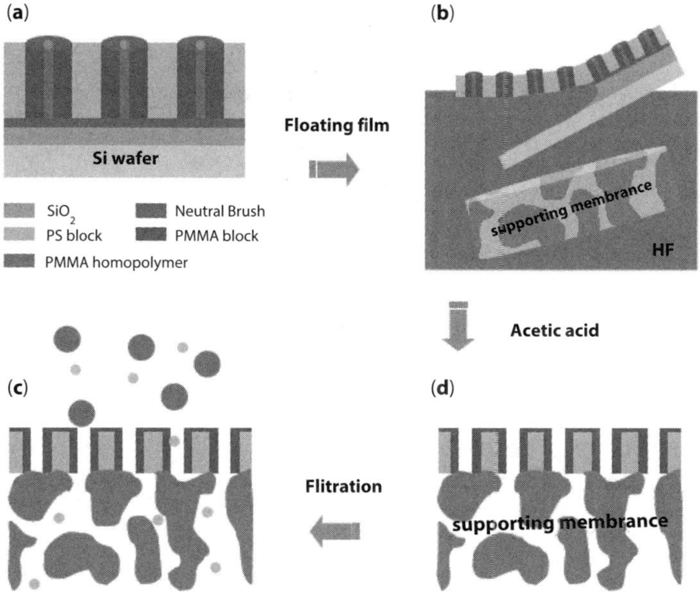

Figure 8.2 Schematic representation of the procedure for the fabrication of composite membranes consisting in a nanoporous top layer and a microporous polymer membrane (reprinted from [26] with permission from Wiley-VCH).

Despite the high selectivity demonstrated in the filtration at low pressure of human rhinovirus type 14, which is one of the major responsible for common cold, wider applicability of these membranes was limited by poor dimensional stability at high filtering pressures. Moreover the PS nanoporous layer did not show good adhesion to the microfiltration polysulfone and was poorly resistant to organic solvents.

Phillip *et al.* were able to prepare crack-free, robust and highly selective water filtration membranes consisting in a nanoporous top layer derived from PS-*b*-PLA templates and a microporous poly(ether sulfone) support [28]. These composite membranes were prepared by drawing the block copolymer solution across the pre-wetted support membrane. The solvent used for the block copolymer solution was water-immiscible to assure exclusion of the casting solution from the water-filled pores of the support. Fast evaporation of the solvent yielded an ordered array of perpendicularly oriented PLA cylinders that were then transformed into the desired porous structure by etching with a dilute alkaline solution. Figure 8.3 compares the surface morphology of the microporous poly(ether sulfone) membrane with that of the copolymer-templated selective layer.

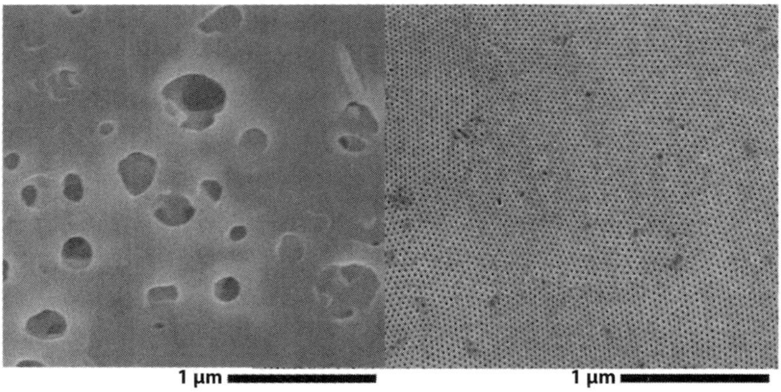

Figure 8.3 SEM micrographs showing the different porosity of the supporting microporous membrane (on the left) and of the block copolymer-templated selective layer (on the right) of a composite polymer membrane (reprinted from [28] with permission from the American Chemical Society).

Despite this fabrication procedure requires several steps, it appears less complicated and time consuming than that described in Figure 8.2. Anyway, the overall performances of these membranes were compromised by relatively low fluxes due to the fact that not all the pores spanned the entire selective layer and a large number of nanopores were blocked by the poly(ether sulfone) support, which has a considerably lower void fraction that the selective layer.

In alternative to microporous polymeric membranes, microfabricated silicon supports have been also proposed for the fabrication of composite membranes. Silicon grids having 20 μm x 20 μm pores, and prepared via photolithographic and reactive ion etching patterning of a silicon wafer, were used to support nanoporous polymer films derived from PS-*b*-PI-*b*-PLA triblock copolymers [27] (Figure 8.4). Transport studies demonstrated lower flow resistance in comparison to microporous poly(sulfone) supporting membranes. In particular, diffusion experiments of large molecules demonstrated that size-selective transport through the membrane could be activated with a simple chemical signal, that is very interesting for sensing applications and controlled drug delivery.

In another interesting fabrication procedure the self-assembly of block copolymers was combined with conventional non-solvent-induced phase separation. In phase inversion processes a concentrated polymer solution is immersed in a non-solvent leading to phase-separation of a polymer-rich phase, that forms the matrix, and a polymer-poor phase, that generates pores. If the polymer is a block copolymer the denser surface layer can self-assemble into an ordered nanoporous structure on top of a mechanically

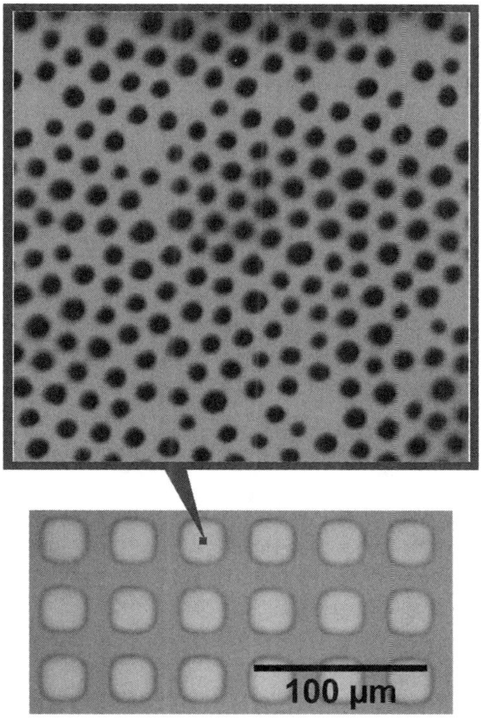

Figure 8.4 Tapping-mode AFM height image (top) of the nanoporous block copolymer film deposited onto a microporous silicon support (bottom). Nanopores of the block copolymer-derived top layer have mean pore sizes of 43 nm ± 11% RSD. The silicon support have 20 μm squared pores spaced 20 μm apart (adapted from [27] with permission from the American Chemical Society).

robust sponge-like layer. Following an integrated phase inversion procedure, double layer phase-inversion membranes derived from PS-*b*-P4VP copolymers and having a top layer with well-ordered cylindrical pores were obtained [29]. Upscaling of this membrane formation process appears simple in comparison to other fabrication techniques, though at present costs related to the raw polymer materials are still too high for the convenient production of separating membranes.

In alternative to composite membranes, stand-alone membranes based on PE or on thermosetting polymers have been proposed. These are based on block copolymers giving bicontinuous morphologies. For instance, highly resistant bicontinuous nanoporous polyethylene membranes were fabricated from PE-*b*-PS precursors [30, 31], or by compression molding of PLA-*b*-PE-*b*-PLA triblock copolymers, followed by etching of the PLA block [32] (Figure 8.5). Robust thermoset membranes templated

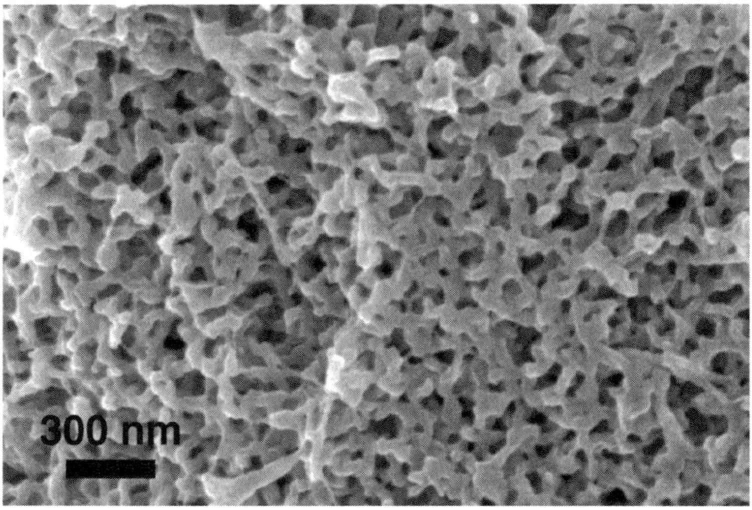

Figure 8.5 SEM image of a freeze-fractured surface from a PE membrane with a disordered bicontinuous porous structure derived from etched PLA-*b*-PE-*b*-PLA samples (reprinted from [32] with permission from the American Chemical Society).

by block copolymers and with percolating pores were synthetized by reaction of a thermosetting monomer, DCPD, with a doubly reactive block copolymer [33]. Others were obtained by cross-linking 1,2-PB-*b*-PDMS films with a gyroid nanostructure at 140°C with dicumyl peroxide and quantitavely removing the PDMS phase by chemical etching with fluorine compounds [34, 35].

All types of stand-alone membranes are tough, exhibit excellent chemical and thermal stability, high void fraction, narrow pore size distribution, their preparation do not encounter a pore alignment step, but the tortuosity of the bicontinuous structure leads to lower permeability compared to normally aligned cylindrical pores.

8.4.2 Controlling Ordering and Orientation in Block Copolymer-Derived Membranes

In principle, membranes templated by the cylindrical morphology and with cylinder orientation normal to the film surface should be superior in terms of selectivity, high fluxes and also fouling resistance than bicontinuous porous networks. On the other hand, to properly align cylindrical domains, the fabrication procedure of membranes with normally oriented cylinders requires more steps than those involved in the fabrication of bicontinuous structures.

Different techniques have been used to optimize the alignment and to control the orientation in block copolymer films, such as application of electric fields [16, 36], chemically patterned substrates [37, 38], shear forces [39, 40], controlled interfacial interactions [41, 42] thermal and solvent annealing [43–46].

Controlled evaporation of solvents has been often applied for the fabrication of ordered nanoporous membranes. Both the rate of solvent evaporation and solvent annealing have been invoked for interpreting strong effects of ordering and orientation on the nanostructured morphologies of block copolymers [47]. In particular, solvent evaporation forms an ordering front that propagates through the entire film and improves the long range lateral order of copolymer films [48–50]. The solvent imparts mobility to the polymer chains thus enabling a rapid removal of defects and orienting cylindrical or lamellar domains perpendicular to the substrate surface, in opposition to the general tendency of the domains to align parallel to the surface because of preferential wetting with one of the blocks. The effects of solvent vapor treatments appear to be time-dependent, showing high initial rates in the ordering behavior and differences in the nanostructures with treatment times [51, 52].

Another approach to orient nanodomains normally to the film surface is to coat the substrate with a random copolymer in order to produce a neutral surface [18, 41]. By applying this strategy PMMA nanocylinders standing perpendicular to the substrate and surrounded by a PS continuous phase were obtained from PS-*b*-PMMA solutions. Subsequently PMMA was removed by UV degradation leaving hollow channels crossing the PS matrix (Figure 8.6).

Supramolecular approaches to block copolymer self-assembly have also been proposed. These are generally based on the complexation by hydrogen bonding of one of the blocks with small molecules bearing hydroxyl and/or carboxylic groups [13], on attractive hydrogen bonding interactions between different block copolymers [53] or on preferential molecular interactions with homopolymers miscible with one of the blocks [54]. Addition of a third component capable of strongly interact with one of the blocks is able to affect the final morphology not only because of the volume fraction increase in one component, but especially for the modification of surface and interface interactions. This approach appears to provide a simple and powerful technique for fine tuning of the block copolymer morphologies, and has been successfully applied in bulk and in thin films.

Nanoporous membranes derived from the supramolecular assembly of PS-*b*-P4VP and 2-(4'-hydroxybenzenazo) benzoic acid (HABA) were

(a)

(b)

Figure 8.6 FESEM image of the top view (*A*) and of the cross-sectional view (*B*) of a thin PS-*b*-PMMA-derived membrane with ordered perpendicularly oriented channels crossing the entire film thickness (reprinted from [41] with permission from Wiley-VCH).

fabricated by Sidorenko *et al.* [55]. By hydrogen bonding HABA to the 4VP blocks they were able to prepare highly ordered thin films consisting of cylindrical P4VP-HABA nanodomains surrounded by a PS matrix. Subsequent HABA washing with a selective solvent resulted in nanostructured membranes with hexagonally packed hollow cylinders of 8 nm in diameter and crossing the membrane from top to bottom (Figure 8.7).

Similarly the supramolecular assembling of PS-*b*-P4VP and 1,5-dihydroxynaphthalene (DHN) was used to prepare PS films with pores of approximately 20 nm and penetrating deeply the film [56].

Phillip *et al.* [28] used controlled solvent evaporation to align in a single step PS-*b*-PLA with perpendicularly oriented PLA cylinders. Fast drying of the block copolymer film yielded a hexagonally packed perpendicular cylindrical structure, while low evaporation rates resulted in a parallel cylindrical structure at the membrane surface. This suggests that the

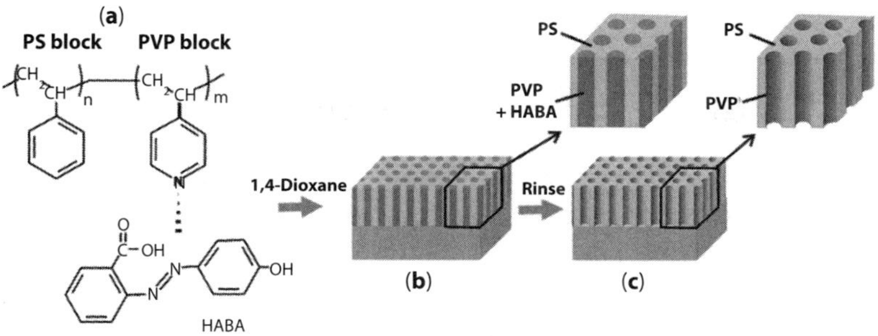

Figure 8.7 Scheme of the membrane fabrication approach based on the supramolecular assembly of PS-*b*-P4VP and HABA: *a*) formation of hydrogen bonds between HABA and pyridine rings of P4VP; *b*) film deposition from 1,4-dioxane and solvent annealing; and *c*) methanol rinsing (adapted from [55] with permission from the American Chemical Society).

perpendicular orientation is a kinetically trapped non-equilibrium structure and that cylinders grow through a nucleation and growth mechanism that begins at the free surface when the solvent evaporates.

Shear forces were also used to align block copolymers. When block copolymers are subjected to shear deformation, they store a shear stress that the material will dissipate through molecular interactions. For samples of cylindrical morphology the alignment of cylinders along the direction of the shear relieves the stress stored during the deformation. Ordering of cylindrical domains was obtained through reciprocating shear using a rheometer in the shear-sandwich configuration, or through shear flow by extrusion through a capillary rheometer or by channel die processing [57]. Monoliths obtained by pressing powdery block copolymers in a mold and forcing the molten material through a channel die were used to prepare thick membranes with hexagonally packed cylindrical pores aligned in the direction of the flow [58].

Ikkala *et al.* combined the supramolecular approach and shear orientation to prepare functional membranes with hollow cylindrical pores derived from PS-*b*-P4VP/pentadecyl phenol (PDP) complexes. In particular, high orientation of the cylinders in the tangential direction was obtained by imposing an oscillatory shear flow at 125°C [59].

8.4.3 Pore Generation in Nanostructured Polymer Films

Block copolymer films are transformed in nanoporous membranes by elimination of a minor component through methods that do not compromise the integrity of the matrix material. This may be accomplished by selective degradation of the minority block domains or by selective dissolution of minor components embedded in cylindrical domains.

UV degradation [36], ozonolysis [60], plasma etching [61] and chemical etching [62, 63] have been applied for the fabrication of several block copolymer nanostructured materials and templates.

For instance PMMA domains in PS-*b*-PMMA films degrade fast under UV irradiation while PS crosslinks [41], PLA is etched by alkaline solutions while in the same conditions PS is stable.

In section 8.4.2 we have already reported several examples of hierarchical nanostructured materials containing small molecules which selectively interact with the pyridine nitrogen of PS-*b*-P4VP and we already anticipated that in such materials pore generation occurs via selective dissolution of the additional small molecules [55, 56]. The major advantage of the supramolecular approach is that the low molecular weight molecule can be easily removed by solvent rinsing leaving a nanoporous material with the same morphology as the parent material. The same strategy was also successfully applied using homopolymer instead of small molecules. The addition of a homopolymer to a block copolymer has been proposed as an easy way to modulate the size of cylindrical nanodomains in block copolymer films without perturbating their order and orientation. Jeong *et al.* [64] were able to modulate the diameter of cylindrical PMMA nanodomains by incorporating PMMA into PS-*b*-PMMA. In a second step ordered nanopores spanning the film thickness were obtained by selective removal of the homopolymer by acetic acid rinsing.

8.5 Tunability of Surface Properties

The possibility to establish and control specific interactions between analytes and the pore walls of nanoporous membranes is crucial for many applications. In the case of membranes for controlled transport and separation processes hydrophylicity of the pore surface is an important requirement because it can improve fouling resistance.

To successfully modify the surface chemistry of the pore walls the selected functionalization procedure must not change the mechanical and morphological properties of the membrane. This means that the reaction conditions used to insert or transform functional groups must be selective and solvents must have a limited impact on the integrity of the polymer matrix. Any slight plasticization effect can trigger the collapse of the nanoporous network compromising the membrane performances. In addition, the functional groups to be inserted must be compatible with the membrane processing and with the etching steps performed to generate the nanoporous structure. For instance, annealing at relatively high temperatures, chemical etching, UV irradiation and other treatments to remove the minority block can degrade functional groups inserted in previous steps of the fabrication procedure.

New functionalities can be inserted on the pore walls during the etching treatments or by post-etching reactions. One of the first strategy to be exploited was based on the introduction of a specific functionality at the covalent junction between the blocks. This functional group becomes exposed when the minority block is removed, as in the case of PLA and PEO, whose selective chemical etching from PS matrices generates hydroxyl groups at the pore walls. The limit of this strategy is that the areal density of the new functionalites is normally too low to change significantly the hydrophilicity of the pore surface [65, 66].

In this context the best performances are those exhibited by membranes with nanopores coated by hydrophilic polymers. PS monoliths with PEO coated pores were obtained from PS-*b*-PEO/PS-*b*-PLA blends after selective degradation of PLA domains. The advantage of this strategy is that the function group density in the pore can be easily tuned by controlling the length of the PEO block and the overall composition of the blend [67, 68].

Another strategy entails selective degradation of PLA from triblock copolymers forming core-shell cylinders, thus generating nanoporous materials with internal hydrophilic coatings. According to this strategy nanoporous PS monoliths with PDMA coated nanochannels were prepared using PLA-*b*-PDMA-*b*-PS triblock copolymer precursors. This approach appears especially effective for the tailoring of the surface chemistry of the pores. Under acidic conditions pendant amide groups were transformed into carboxylic acids generating an internal polyacrylic acid coating, that was then modified with a variety of functional groups, such as pyridine, chiral hydroxyl and alkene [69].

PS nanoporous matrices having alkene groups into the pore interior have been prepared from PS-*b*-PI-*b*-PLA triblock copolymers forming a core-shell cylindrical structure with PLA cores and PI coronas. The exposed alkene groups of the PI shell were then epoxidized providing convenient sites for further chemical modifications, such as the introduction of alcohols and amines through addition reactions [70].

Another interesting strategy to change the surface chemistry of originally hydrophobic membranes has been recently proposed by Ndoni *et al.* who were able to hydrophilize the nanopores of 1,2-PB membranes via UV-photooxidation [71] or thiol-ene photochemistry [72].

8.6 Application of Block Copolymer-Derived Membranes to Bioseparation and Controlled Drug Release

Nanoporous membranes based on self-assembling block copolymers have proven to be effective in preventing passage of large molecules (i.e. proteins, viruses, antibodies), while they allow passage of small molecules. As the range of pore sizes accessible by self-assembling block copolymers approaches the size scale of biomolecules, they are of particular interest for biofiltration and drug delivery applications.

The possibility to employ nanoporous films in the separation of biomacromolecules was assessed by measuring cyclic voltammetry of ferritin molecules adsorbed onto gold electrodes coated with a porous film derived from PS-*b*-PMMA and having vertically aligned cylindrical pores. PEG functionalization of the nanopore surface was required for the penetration of ferritin through the membrane and size-selectivity of the PEG-modified nanoporus membranes was controlled by changing the molecular weight of the diblock copolymer precursor [73].

Nanoporous membranes derived from a mixture of PS-*b*-PMMA and PMMA homopolymer proved excellent selectivity also for the separation of human rhinovirus type 14 (HRV14). These membranes, having crossing channels with diameters of approximately 15 nm, allow proteins, such as bovine serum albumin (BSA), to pass through the pores, while HRV14 varions are blocked (Figure 8.8). The tunability of the pore size makes these

Figure 8.8 Scanning electron microscopy (*a*) and scanning force microscopy (*b*) images of HRV14 virions that have been blocked by a nanoporous membrane derived from PS-*b*-PMMA (reprinted from [26] with permission from Wiley-VCH).

systems very interesting for filtration, purification and concentration of various types of viruses, and especially for the elimination of viral contamination from biotherapeutic proteins [24, 26].

Transport properties of block copolymer-derived membranes have been systematically investigated by Phillip [58]. He studied the flow of liquid water through nanoporous membranes derived from PLA-*b*-PDMA-*b*-PS obtaining results that are consistent with theoretical predictions based on the Hagen-Poiseuille law. Data indicate that block copolymer membranes with nearly monodisperse pores give sharp molecular weight cut-offs, comparable to those observed in track-etched membranes, while permeance is over 100 times faster, even though the block copolymer membrane has smaller pores than track-etched membranes. The potentialities of these membranes for ultrafiltration of liquid solutions were evaluated considering the case of BSA. The separation factor and the hydraulic permeability were calculated by assuming a pore diameter and the estimates for such membranes had both faster flow and more abrupt increase in separator factor than data for commercially available membranes.

In a subsequent work [28] the water permeability of a composite membrane having a selective nanoporous layer with cylindrical morphology was determined experimentally and resulted to be 1,15 $Lm^{-2}h^{-1}bar^{-1}$, that is approximately 400 times lower than the value predicted by Equation 8.1:

$$v = \frac{\varepsilon}{\tau}\langle v \rangle = \frac{\varepsilon}{\tau}\frac{\varepsilon a^2 \Delta p}{32\eta I}$$

where v is the fluid velocity at the membrane surface, $\langle v \rangle$ is the Hagen-Poiseuille velocity in a single pore, τ is the tortuosity (and is set to 1 for perfectly perpendicular pores), ε the void fraction, d is the pore diameter, Δp the pressure drop across a membrane of thickness l and η the liquid viscosity.

The reduction of flux in comparison to the model was ascribed to a reduction of the effective void fraction due to the presence of nanopores that are blocked by the support membrane and to the presence of pores that do not span the entire membrane thickness or that are not perfectly oriented. Size-selectivity of PS-*b*-PLA-templated membranes with a pore diameter of 24 nm was proved by measuring the rejection of PEO samples. Rejection increases with the sample molecular weight and is complete for 100 kDa PEO samples having a calculated hydrodynamic radius of 10.6–12.1 nm.

Phillip also studied transport and separation properties of nanoporous membranes with a bicontinuous structure [74]. Selectivity of polycyclopentadiene-based membranes, with nanopores templated by a poly(norbornenylethyl-s-styrene)-polylactide block copolymer, was demonstrated through transport studies with polyethylene oxides and dextrans. The polycyclopentadiene (PDCPD) membrane was found to reject PEO samples of different molecular weight as expected, approaching the 100% rejection for the 35 kDa sample. Moreover, ultrafiltration experiments carried out using solutions of mixed dextrans showed sharper molecular weight cut off for PDCPD membranes than for commercial phase inversion membranes (Figure 8.9). Diffusion of gases and water flow measurements confirmed pore percolation and the presence of pores spanning the entire membrane thickness. Unfortunatly, flux measured through a 100 μm thick membrane was between 1×10^{-6} and 3×10^{-6} $m^3/(m^2s)$, that is from one to two order of magnitude lower than those through membranes made

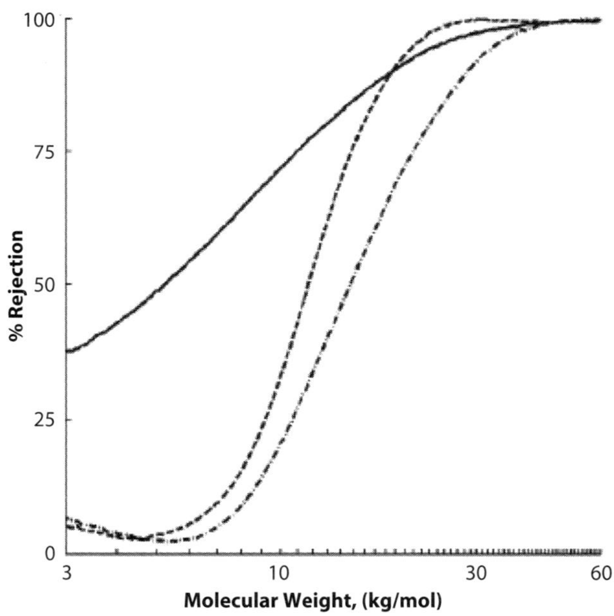

Figure 8.9 Rejection curves of a commercial membrane made by phase inversion (*solid line*) and of two PDCPD membranes, one with a 32 kDa etchable PLA block (*dashed line*) and the other with a 65 kDa block (*dash-dotted line*). Membranes based on self-assembled block copolymers show a rejection sharper than those of the commercial ultrafiltration membrane. Moreover, the data for a membrane with a 32 kD PLA block show a cut-off at lower molecular weight than for a 65 kD PLA block, suggesting that pore size can be modified by tailoring the molecular weight of the etchable block (reprinted from [74] with permission from the American Chemical Society).

by phase inversion. Such values of flux could be considerably increased in thinner membranes as flux is normally inversely proportional to the membrane thickness. The predicted flux through a 0.5 μm thick membrane at a pressure drop of 200 kPa is 2.8×10^{-3} $m^3/(m^2s)$, that is over ten times faster than commercially available membranes, but such thin membranes need to be supported in order to ensure mechanical integrity.

Li *et al.* have demonstrated that the permeation rate of glucose through 1,2-polybutadiene gyroid membranes could be tuned by manipulating their surface morphology or the surface chemistry of the nanopores [34]. In particular, the glucose diffusion flux was observed to decrease from nonskin, to one-sided skin to two-sided skin membranes. The selectivity of nonskin gyroid nanoporous membranes was studied through diffusion tests using a selection of antibiotics, proteins and other biomolecules. For ciprofloxacin, vancomycin, cytochrome C and myoglobin, selectivity resulted higher than expected from the pore size, meaning that the permeation of solutes across the membrane does not only depend on the size of solutes and pores, but also on physicochemical parameters influencing specific solute-solute and solute-membrane interactions.

Depending on the membranes structure (i.e. nonskin, one-sided skin, two-sided skin) sensors and hemodialysis have been suggested as possible applications, but at present only their performances in ultrafiltration processes have been investigated in details. In particular, the effect of membrane fouling has been extensively examined in terms of flux and rejection properties. Adsorption of nearly monodisperse PEG samples from water solutions was observed to severely reduce flux through the hydrophobic 1,2-PB membranes under both static and dynamic conditions. The hydrophilization of the membranes reduced fouling. However, the best antifouling performances were obtained changing the filtration solution by adding high concentration of ethanol [75], thus proving that rejection properties also depend on the solvent chemistry. The differences in rejection profiles and molecular weight cut-offs have been explained by invoking the intervention of selectivity mechanisms other than the intrinsic size properties of the nanopores, such as chemical affinity between membrane and solutes, and the reduction in the effective pore size due to functionalization of the pore wall and to binding with water molecules. Of course, the improvement in mechanical properties and the capability to control the separation mechanisms is a necessary requirement to propose these systems as tunable membranes for specific applications.

As anticipated, a very promising application field is drug delivery. The feasibility of using nanoporous templates as drug-eluting coatings on implantable devices for the controlled release of drugs has been recently investigated by several research groups. The capability to modulate the size, geometry and depth of the nanopores has been exploited to produce pore-filling nanoporous templates that can effectively deliver the required amount of drug over an extended period.

Nanoporous, tough, flexible and chemically resistant polyethylene membranes prepared by PE-*b*-PS precursors have been proposed for implantable sensor applications and as components of medical devices for glucose separation and controlled transport [31]. Size-selectivity of such membranes was demonstrated by diffusion experiments of bovine serum albumin and glucose. While glucose molecules, being much smaller than any of the pores, effectively diffuse through the PE membrane, BSA was retained, depending on the pore size and holding time. Full retention was observed for membranes having a pore width of 5–10 nm, that is comparable to the size of albumin.

Recently, a block copolymer membrane derived from the one described in Figure 8.2 has been integrated into drug delivery devices providing a long-term controlled delivery of protein drugs without denaturation nor deactivation [25]. The pore size of the membrane was tuned by gold deposition, resulting in smaller pores with increasing the thickness of the golden layer deposited on the membrane surface. Controlled release of protein drugs based on the single-file diffusion [76] was observed in in vitro release tests of BSA and human growth hormone (hGH) by adjusting the pore size to 15 and 6 nm respectively, that is less than twice the hydrodynamic diameter of the protein drugs. Moreover in vivo tests demonstrated that the hGM release from an implant titanium device with a Au-deposited block copolymer membrane having pore size of 6 nm was maintained fairly constant up to 20 days, proving the feasibility for the treatment of various chronic diseases (Figure 8.10).

Another interesting application regard the controlled release of sirolimus. A surprisingly extended duration of sirolimus elution was obtained by using thin films with oriented cylindrical or lamellar nanopores derived from self-assembled PS-*b*-PLLA block copolymers [77]. Following a specifically designed pore-filling process, sirolimus, an immunosuppressive agent with anti-inflammation and anti-proliferation characteristics, was effectively entrapped into the nanochannels and subsequently released according to a well-controlled elution profile. The nanoscale size effect on drug release produced in the sirolimus-loaded nanoarrays results in prolonged and sustained-released profiles, in contrast to steeper release profiles encountered once the pore size of the template is over hundreds nanometers, as for AAO/sirolimus hybrids and PS/sirolimus blends (Figure 8.11)

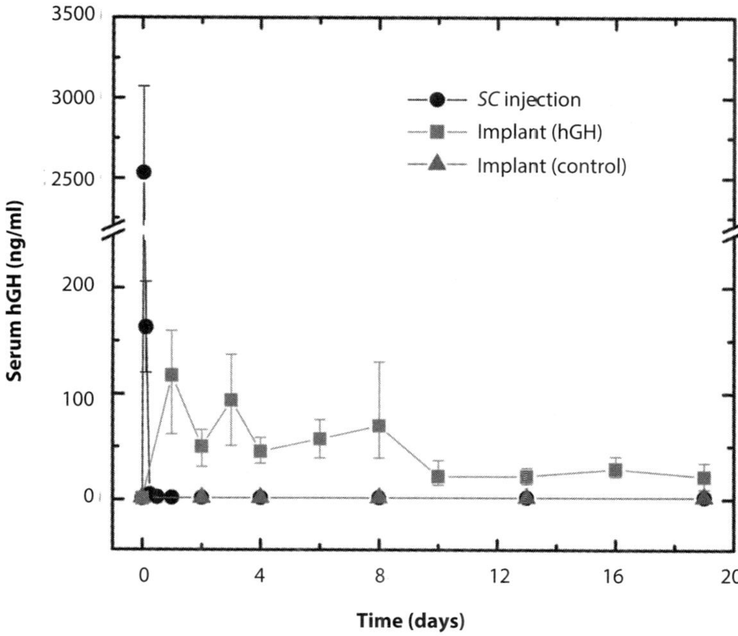

Figure 8.10 Pharmacokinetics of *in vivo* released hGH after a single subcutaneous injection of aqueous hGH (circle) and from a subcutaneous titanium implant device with Au-deposited cylindrical nanochannels having a diameter of 6 nm (rectangle). The drug delivery nanodevice without hGH was used as a control (triangle) (reprinted from [25] with permission from the American Chemical Society).

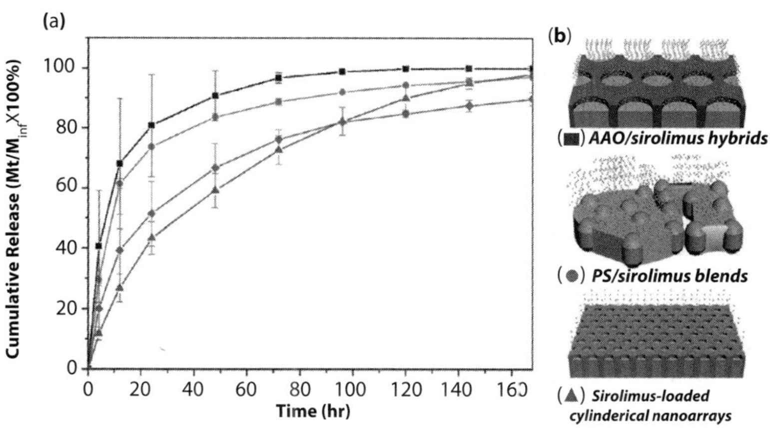

Figure 8.11 (*a*) Cumulative release profiles of sirolimus from sirolimus-loaded cylindrical nanoarrays (triangle), sirolimus-loaded lamellar nanoarrays (diamond), AAO/sirolimus hybrids (square), and PS/sirolimus blends (circle). M_{inf} is the infinite amount of loaded drugs in test templates; M_t is the cumulative amount of drug released at time *t*. (*b*) Schematic illustration of the sirolimus-loaded cylindrical nanoarrays, the AAO/sirolimus hybrids, and the PS/sirolimus blends (reprinted from [77] with permission from the American Chemical Society).

8.7 Conclusion

The self-assembling of block copolymers is emerging as a powerful methodology to fabricate nanoporous membranes with unique properties, namely controlled pore size within the nanoscale range, narrow pore size distribution, high pore density and tunability of surface properties. Nanoporous membranes templated by block copolymers promise to be especially advantageous in separation processes involving biomolecules and in drug delivery applications. The major drawback of most of the preparative protocols proposed till now is the number of steps required to obtain mechanically robust membranes, with uniform pore shape and, if required, pore alignment for large areas. At present this prevents the scaling up of the membranes production, but work is in progress in this research field, promising results have already been published and there is no reason to doubt that further improvements will be achieved in the next future.

References

1. R.W. Baker, *Membrane Technology and Applications*, 2nd ed., John Wiley & Sons Ltd., Chichester, 2004.
2. M. Ulbricht, *Polymer*, Vol. 47, p. 2217, 2006.
3. S.P. Nunes, *Trends in Polymer Science*, Vol. 5, p. 187, 1997.
4. R.L. Fleischer, P.B. Price, and R.M. Walker, *Nuclear Tracks in Solids. Principles and Applications*, University of California Press, Berkeley, 1975.
5. P. Apel, *Radiation Measurements*, Vol. 34, p. 559, 2001.
6. F. Martin, R. Walczak, A. Boiarski, M. Cohen, T. West, C. Cosentino, and M. Ferrari, *Journal of Controlled Release*, Vol. 102, p. 123, 2005.
7. X. Zhang, P. Tanner, A. Graff, C.G. Palivan, and W. Meier, *Journal of Polymer Science Part A: Polymer Chemistry*, Vol. 50, p. 2293, 2012.
8. J. S. Lee, and J. Feijen, *Journal of Controlled Release*, Vol. 161, p. 473, 2012.
9. G.A. OzinK. Hou, B.V. Lotsch, L. Cademartiri, D.P. Puzzoa, F. Scotognella, A. Ghadimi, and J. Thomson, *Materials Today*, Vol. 12, p. 12, 2009.
10. K. Matyjaszewski, ed., *Controlled/Living Radical Polymerization: Progress in ATRP, NMP, and RAFT*, ACS Symposiun Series, Vol. 768, American Chemical Society, Washington D.C., 2000.
11. V. Castelletto, and I.W. Hamley, *Current Opinion in Solid State & Materials Science*, Vol. 2, p. 426, 2004.
12. I.W. Hamley, *The Physics of Block Copolymers*, Oxford University Press, Oxford, 1998.
13. G. ten Brinke, J. Ruokolainen, and O. Ikkala, *Advances in Polymer Science*, Vol. 207, p. 113, 2007.

14. J. Tata, D. Scalarone, M. Lazzari, and O. Chiantore, *European Polymer Journal*, Vol. 45, p. 2520, 2009.
15. S.H. Kim, M.J. Misner, T. Xu, M. Kimura, and T.P. Russell, *Advanced Materials*, Vol. 16, p. 226, 2004.
16. T.L. Morkved, M. Lu, A.M. Urbas, E.E. Ehrichs, H.M. Jaeger, P. Mansky, and T.P. Russell, *Science*, Vol. 273, p. 931, 1996.
17. M. Lazzari, and C. De Rosa, "Methods for the alignment and the large-scale ordering of block copolymer morphologies", in M. Lazzari, G.J. Liu and S. Lecommandoux, eds., *Block Copolymers in Nanoscience*, Wiley-VCH, Weinheim, p. 191–231, 2006.
18. E. Huang, L. Rockford, T.P. Russell, and C.J. Hawker, *Nature*, Vol. 395, p. 757, 1998.
19. R.J. Spontak, and P. Alexandridis, *Current Opinion in Colloid and Interface Science*, Vol. 4, p. 140, 1999.
20. K.W. Guarini, C.T. Black, and S.H.I. Yeuing, *Advanced Materials*, Vol. 14, p. 1290, 2002.
21. E.A. Jackson, and M.A. Hillmyer, *ACS Nano*, Vol. 4, p. 2548, 2010.
22. H.Yokoyama, T.E. Mates, and E.J. Kramer, *Macromolecules*, Vol. 33, p. 1888, 2000.
23. A. Knoll, A. Horvat, K.S. Lyakhova, G. Krausch, G.J.A.Sevink, A.V. Zvelindovsky, and R. Magerle, *Physical Review Letters*, Vol. 89, p. 035501–1, 2002.
24. S.Y. Yang, J. Park, J. Yoon, M. Ree, S.K. Jang, and J.K. Kim, *Advanced Functional Materials*, Vol. 18, p. 1371, 2008.
25. S.Y. Yang, J.A. Yang, E.S. Kim, G. Jeon, E.J. Oh, K.Y. Choi, S.K. Hahn, and J.K. Kim, *ACS Nano*, Vol.4 , p. 3817, 2010.
26. S.Y. Yang, I. Ryu, H.Y. Kim, J.K. Kim, S.K. Jang, and T.P. Russell, *Advanced Materials*, Vol. 18, p. 709, 2006.
27. E.E. Nuxoll, M.A. Hillmyer, R. Wang, C. Leighton, and R.A. Siegel, *ACS Applied Materials & Interfaces*, Vol. 1, p. 888, 2009.
28. W.A. Phillip, B. O'Neill, M. Rodwogin, M.A. Hillmayer, and E.L. Cussler, *ACS Applied Materials & Interfaces*, Vol. 2, p. 847, 2010.
29. K.-V. Peinemann, V. Abetz, P.F.W. Simon, *Nature Materials*, Vol. 6, p. 992, 2007.
30. H. Uehara, T. Yoshida, M. Kakiage, T. Yamanobe, T. Komoto, K. Nomura, K. Nakajima, and M. Matsuda, *Macromolecules*, Vol. 39, p. 3971, 2006.
31. H. Uehara, M. Kakiage, M. Sekiya, D. Sakuma, T. Yamonobe, N. Takano, A. Barraud, E. Meurville, and P. Ryser, *ACS Nano*, Vol. 3, p. 924, 2009.
32. L.M. Pitet, M.A. Amendt, and M.A. Hillmyer, *Journal of the American Chemical Society*, Vol. 132, p. 8230, 2010.
33. L. Chen, W.A. Phillip, E.L. Cussler, and Marc A. Hillmyer, *Journal of the American Chemical Society*, Vol. 129, p. 13786, 2007.
34. L. Li, L. Schulte, L.D. Clausen, K.M. Hansen, G.E. Jonsson, and S. Ndoni, *ACS Nano*, Vol. 5, p. 7754, 2011.
35. L. Schulte, A. Grydgaard, M.R. Jakobsen, P.P. Szewczykowski, F. Guo, M.E. Vigild, R.H. Berg, and S. Ndoni, *Polymer* , Vol. 52, p. 422, 2011.

36. T. Thurn-Albrecht, J. Schotter, G.A. Kastle, N. Emley, T. Shibauchi, L. Krusin-Elbaum, K. Guarini, C.T. Black, M.T. Tuominen, and T.P. Russell, *Science*, Vol. 290, p. 2126, 2000.
37. L. Rockford, S.G.J. Mochrie, and T.P. Russell, *Macromolecules*, Vol. 34, p. 1487, 2001.
38. S.O. Kim, H.H. Solak, M.P. Stoykovich, H.J. Ferrier, J.J. de Pablo, and P.F. Nealey, *Nature*, Vol. 424, p. 411, 2003.
39. M.A. Villar, D.R. Rueda, F. Ania, and E.L. Thomas, *Polymer*, Vol. 43, p. 5139, 2002.
40. R.J. Albalak, and E.L. Thomas, *Journal of Polymer Science Part B: Polymer Physics*, Vol. 32, p. 341, 1994.
41. T. Thurn-Albrecht, R. Steiner, J. DeRouchey, C.M. Stafford, E. Huang, M. Bal, M. Tuominen, C.J. Hawker, and T.P. Russell, *Advanced Materials*, Vol. 12, p. 787, 2000.
42. T. Hashimoto, J. Bodycomb, Y. Funaki, K. Kimishima, *Macromolecules*, Vol. 32, p. 952,1999.
43. H.C. Kim, and T.P. Russell, *Journal of Polymer Science Part B: Polymer Physics*, Vol. 39, p. 663, 2001.
44. G. Krausch, and R. Magerle, *Advanced Materials*, Vol. 14, p. 1579, 2002.
45. J. Peng, Y. Han, W. Knoll, and D.H. Kim, *Macromolecular Rapid Communications*, Vol. 28, p. 1422, 2007.
46. J. Zhao, S. Jiang, X. Ji, L. An, and B. Jiang, *Polymer*, Vol. 46, p. 6513, 2005.
47. G. Kim, and M. Libera, *Macromolecules*, Vol. 31, p. 2569,1998.
48. K. Fukunaga, T. Hashimoto, H. Elbs, and G. Krausch, *Macromolecules*, Vol. 35, p. 4406, 2002.
49. H. Elbs, C. Drummer, V. Abetz, and G. Krausch, *Macromolecules*, Vol. 35, p. 5570, 2002.
50. S.H. Kim, M.J. Misner, T. Xu, M. Kimura, and T.P. Russell, *Advanced Materials*, Vol. 16, p. 226, 2004.
51. P. Černoch, P. Štěpánek, J. Pleštil, M. Šlouf, A. Sidorenko, and M. Stamm, *European Polymer Journal*, Vol. 43, p. 1144, 2007.
52. Y. Xuan, J. Peng, L. Cui, H. Wang, B. Li, and Y. Han, *Macromolecules*, Vol. 37, p. 7301, 2004.
53. S. Jiang, A. Gopfert, and V. Abetz, *Macromolecules*, Vol. 36, p. 6171, 2003.
54. J. Zhao, S. Jiang, X. Ji, L. An, and B. Jiang, *Journal of Polymer Science Part B: Polymer Physics*, Vol. 42, p. 3496, 2004.
55. A. Sidorenko, I. Tokarev, S. Minko, and M. Stamm, *Journal of the American Chemical Society*, Vol. 125, p. 12211, 2003.
56. A. Laforgue, C.G. Bazuin, and R.E. Prud'homme, *Macromolecules*, Vol. 39, p. 6473, 2006.
57. A.S. Zalusky, R. Olayo-Valles, J.H. Wolf, and M.A. Hillmyer, *Journal of the American Chemical Society*, Vol. 124, p. 12761, 2002.
58. W.A. Phillip, J. Rzayev, M.A. Hillmyer, and E.L. Cussler, *Journal of Membrane Science*, Vol. 286, p. 144, 2006.

59. R. Maki-Ontto, K. de Moel, W. de Odorico, J. Ruokolainen, M. Stamm, G. ten Brinke, and O. Ikkala, *Advanced Materials*, Vol. 13, p. 117, 2001.
60. J. Lee, A. Hirao, and S. Nakahama, *Macromolecules*, Vol. 21, p. 274, 1988.
61. M. Park, C. Harrison, P.M. Chaikin, R.A. Register, and D.H. Adamson, *Science*, Vol. 276, p. 1401, 1997.
62. J.M. Leiston-Belanger, T.P. Russell, E. Drockenmuller, and C.J. Hawker, *Macromolecules*, Vol. 38, p. 7676, 2005.
63. S. Ndoni, M.E. Vigild, and R.H. Berg, *Journal of the American Chemical Society*, Vol. 125, p. 13366, 2003.
64. U. Jeong, H.-C. Kim, R.L. Rodriguez, I.Y.Tsai, C.M. Stafford, J.K. Kim, C.J. Hawker, and T.P. Russell, *Advanced Materials*, Vol. 14, p. 274, 2002.
65. J.H. Wolf, and M.A. Hillmyer, *Langmuir*, Vol. 19, p. 6553, 2003.
66. A.S. Zalusky, R. Olayo-Valles, J.H. Wolf, and M.A. Hillmyer, *Journal of the American Chemical Society*, Vol. 124, p. 12761, 2002.
67. H. Mao, and M.A. Hillmyer, *Soft Matter*, Vol. 2, p. 57, 2006.
68. H. Mao, P.L. Arrechea, T.S. Bailey, B.J.S. Johnsonw, and M.A. Hillmyer, *Faraday Discussions*, Vol. 128, p. 149, 2005.
69. J. Rzayev, and M.A. Hillmyer, *Journal of the American Chemical Society*, Vol. 127, p. 13373, 2005.
70. T.S. Bailey, J. Rzayev, and M.A. Hillmyer, *Macromolecules*, Vol. 39, p. 8772, 2006.
71. S. Ndoni, L. Li, L. Schulte, P.P. Szewczykowski, T.W. Hansen, F.X. Guo, R.H. Berg, and M.E. Vigild, *Macromolecules*, Vol. 42, p. 3877, 2009.
72. A. Berthold, K. Sagar, and S. Ndoni, *Macromolecular Rapid Communications*, Vol. 32, p. 1259, 2011.
73. Y. Li, and T. Ito, *Analytical Chemistry*, Vol. 81, p. 851, 2009.
74. W.A. Phillip, M. Amendt, B. O'Neill, L. Chen, M.A. Hillmyer, and E.L. Cussler, *ACS Applied Materials & Interfaces*, Vol. 1, p. 472, 2009.
75. L. Li, P. Szewczykowskia, L.D. Clausenb, K.M. Hansenb, G.E. Jonssonc, and S. Ndoni, *Journal of Membrane Science*, Vol. 384, p. 126, 2011.
76. Q.-H. Wei, C. Bechinger, P. Leiderer, *Science*, Vol. 287, p. 625, 2000.
77. K.-H. Lo, M.-C. Chen, R.-M. Ho, and H.-W. Sung, *ACS Nano*, Vol. 3, p. 2660, 2009.

Abbreviations

AAO	anodized aluminum oxide
ATRP	atom transfer radical polymerization
BSA	bovine serum albumin
DCPD	dicyclopentadiene
DHN	1,5-dihydroxynaphthalene
HABA	2-(4'-hydroxybenzenazo) benzoic acid

HF	hydrofluoric acid
hGH	human growth hormone
HRV14	human rhinovirus type 14
P4VP	poly(4-vinylpyridine)
PB	polybutadiene
PDCPD	polycyclopentadiene
PDMA	polydimethylacrilamide
PDMS	poly(dimethylsiloxane)
PDP	pentadecyl phenol
PE	polyethylene
PEG	polyethylene glycol
PEO	poly(ethylene oxide)
PI	polyisoprene
PLA	polylactide
PLLA	poly(L-lactide)
PMMA	poly(methyl methacrylate)
PS	polystyrene

9

Hybrid Mesoporous Silica for Drug Targeting

Luigi Pasqua[1,*], Rosangela Piluso[1], Ilenia Pelaggi[1] and Catia Morelli[2]

[1]*Department of Environmental and Chemical Engineering, University of Calabria, Italy*
[2]*Department of Pharmacy and Health and Nutritional Sciences, University of Calabria, Italy*

Abstract
The major limitation of traditional chemotherapeutic agents is their poor selectivity for cancer cells and their severe toxicity to normal cells. Therefore, localized drug delivery would, ideally, improve the therapeutic efficacy, minimizing side effects. The properties of mesoporous silica nanoparticles (MSNs) seem to cope with this aim. MCM-41 type MSNs were synthesized using a PEG surfactant-based interfacial synthesis procedure, and successively grafted with folic acid (FOL), a small molecule used as targeting function, being the natural ligand for the folate receptor (FR). In fact, folate-bound nanoparticles (MSN-FOL), loaded with an antineoplastic drug, are exclusively internalized by cancer cells overexpressing FR (FR+ cells) through a highly specific, receptor-mediated, endocytotic process, while not-functionalized MSNs show the unique feature of being not internalized by cells. Moreover, once MSN-FOL reach the cytoplasm of FR+ tumor cells, the antineoplastic drug is released, causing cell death. Thus, such a conceived device could represent a promising tool for targeted chemotherapy on FR+ tumors.

Keywords: Mesoporous silica nanoparticles, target therapy, folate receptor, endocytosis

**Corresponding author*: l.pasqua@unical.it

Annarosa Gugliuzza (ed.) Smart Membranes and Sensors, (255–268) 2014 © Scrivener Publishing LLC

9.1 Introduction

The hybridization of inorganic precursor allows the development of innovative multifunctional materials, combining properties of organic, biological and inorganic components. Since the discovery of MCM-41-type ordered mesoporous silica (MS) by Mobil corporation scientists in 1992, there has been large amount of research concerning the synthesis and applications of MS. Great endeavors have been made in the control of particle size, pore diameter, morphology, structure, surface properties and functionalization of MS to make them useful tools in the fields of biomedicine, catalysis, environmental protection, optics, etc. In particular, for biomedical applications, mesoporous silica nanoparticles (MSNs) have been proposed as carriers in controlled drug/gene release, biosensors, bio-markers, supports for enzymes, etc. Indeed, proteins and enzymes can be adsorbed on unmodified hydroxylated silica surface or functionalized silica walls of mesoporous materials [1–3]. Furthermore, MSNs have been investigated for hosting non-steroidal anti-inflammatory drugs bearing a carboxylic acid through a confinement procedure consisting in either physisorption on the pure silica surface[4, 5]or via chemical anchoring on modified silica surface[6, 7]. Well-ordered dye-functionalized MSNs have been reported to undergo internalization into cells and consequently they might be useful as markers in cell tracking and in drug delivery [8]. Moreover, silica nanoparticles were tested for their ability to transfer DNA both *in vitro* and in animal models. Very low or no cell toxicity was observed, indicating mesoporous silica as potential alternatives for gene transfection [9–11].

The use of MSNs as efficient drug delivery carriers have attracted great attention in the last decade, especially in the oncologic field. In fact, the major limitation of current chemotherapy is the dose-responsive effect, so that cell kill is proportional to drug exposure. A highly aggressive style of dosing is thus necessary to eradicate tumors, although it is hindered by poor selectivity for cancer cells and severe toxicity to normal cells. Therefore, localized drug delivery would, ideally, improve the therapeutic efficacy, minimizing side effects. MSNs seem to cope with this aim, since their structure, morphology, size and surface properties have been found to be easily tunable for drug loading, controlled drug release and delivery purposes. Meanwhile, the *in vivo* behaviour of MSN-based nano drug delivery systems (nano-DDSs), including biocompatibility (cytotoxicity and blood and tissue compatibility) and pharmacokinetics (biodistribution, biodegradation, retention, excretion, blood circulation) is also drawing increasing attention for MSNs clinical application perspectives and has become a current hot topic in MSN biosafety research [12].

In recent years a great effort has been made to develop tumour-selective devices by conjugating anti-cancer drugs to hormones, antibodies and vitamin derivatives. Among them, a low molecular weight vitamin compound, folic acid, shows great promise as a tumour-homing agent. This essential vitamin has a high affinity for the folate receptor (FR), a tumour associated glycosylphosphatidylinositol anchored protein, which can actively internalize bound folates and folate conjugated compounds via receptor-mediated endocytosis. The receptor for folic acid constitutes a useful target for tumor-specific drug delivery [13], primarily because:

- the folic acid receptor is up-regulated in many human cancers, including malignancies of the ovary, colorectal tract, brain, kidney, breast, myeloid cells and lung, enabling the malignant cell to compete successfully for the vitamin when supplies are limited;
- access to the folate receptor in those normal tissues that express it can be severely limited due to its location on the apical (externally-facing) membrane of polarized epithelia;
- folate receptor density appears to increase as the stage/grade of the cancer worsens.

On the basis of these statements, it has been hypothesized that folate conjugation to anti-cancer drugs will improve drug selectivity and decrease negative side effects [14]. To this aim, a mesoporous silica based system, potentially useful as drug targeting device, has been recently developed in our laboratory. Interestingly, these MSNs, synthesized using a PEG surfactant-based interfacial synthesis procedure, do not enter cells unless opportunely functionalized. Indeed, a highly specific, receptor mediated, cellular internalization of folic acid (FOL) grafted MSNs (MSN-FOL) occurs exclusively in FR expressing cells.

9.2 Synthesis and Characterization of Bifunctional Hybrid Mesoporous Silica Nanoparticles Potentially Useful for Drug Targeting

Mesoporous silica nanoparticles used in our study were synthesized by modifying an interfacial synthesis procedure carried out at room temperature (RT) that employs a PEG-type surfactant and introduce decane as the organic phase. The aim was to limit mesoporous silica particle size, an essential parameter in drug targeting applications.

The mesoporous silica was employed in the coupling reaction before the removal of the templating agent so that the ligand could be addressed preferentially on the external surface.

Non functionalized, as synthesized, particles (MSNsurf) were either calcined (MSN) and used as blank in all *in vitro* experiments, or else extracted (MSN-E) for pore structure characterization.

Aminopropyl-functionalized particles (MSN-AP) were prepared by covalent grafting of aminopropyltriethoxysilane (APTES) on MSNsurf, non-functionalized, as synthesized, particles. Folic acid functionalized particles (MSN-FOL) were synthesized by amide bond formation between the amino group of MSN-AP and folic acid. To track the particles route within the cells, also a fluorescent tracer (fluorescein isothiocyanate, FITC) can be employed to obtain MSN-FITC and MSN-FITC-FOL materials.

The following removal of surfactant by solvent extraction in very mild conditions produced a porous matrix provided with a biologically active group and, if desired a fluorescent tracer on the external surface.

Folic acid molecule (Fig. 9.1) was covalently bonded to the mesoporous silica particle through the carboxylic acid groups, while the purinic part is oriented towards the external, so it can interact with the receptors on the cell membrane.

The general scheme that represents the preparation of bifunctional mesoporous silica potentially useful for drug targeting is reported below (Scheme 9.1).

The starting material is functionalized with 3-aminopropyl group, then carbodiimide-type condensing agents (Scheme 9.1), such as react

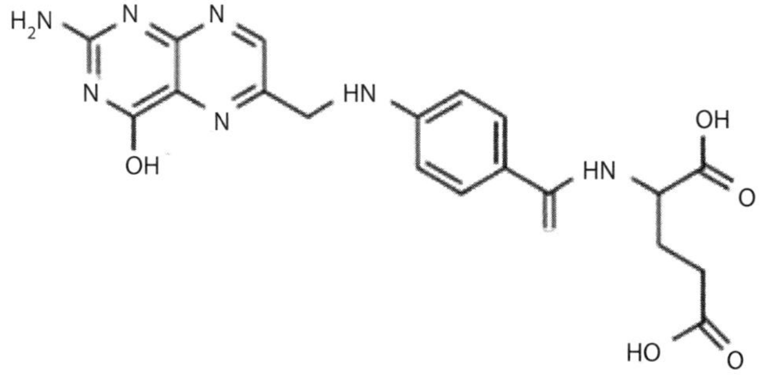

Figure 9.1 Folic Acid chemical structure. (Reproduced with permission from L. Pasqua, F. Testa, R. Aiello, S. Cundari and J.B. Nagy, *Microporous and Mesoporous Materials*, 2007, 103, 1. ©2007, Elsevier [15]).

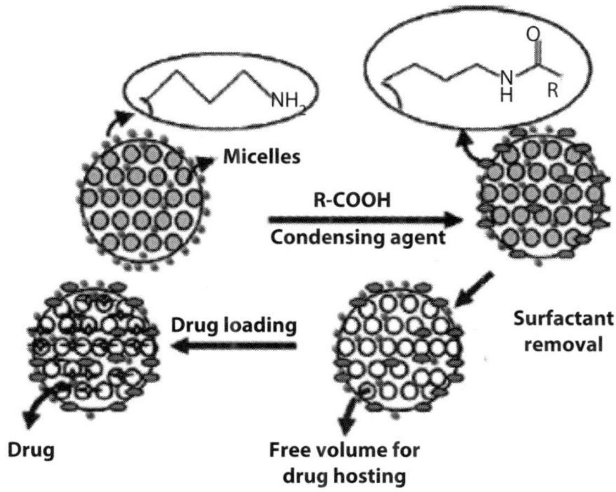

Scheme 9.1 General scheme that represents the preparation of bifunctional mesoporous silica potentially useful for drug targeting. (Reproduced with permission from L. Pasqua, F. Testa, R. Aiello, S. Cundari and J.B. Nagy, *Microporous and Mesoporous Materials*, 2007, 103, 1. ©2007, Elsevier [15]).

with carboxylate compound to form an initial adduct which reacts with the amino derivatized silica surface to produce the amide derivative and a substituted urea as by-product.

The obtained hybrid Mesoporous silica shows the single reflection of XRD pattern typical of MSU-type mesoporous materials. The uniform mesoporous structure of Mesoporous material is quite maintained after coupling reaction and extraction procedure, although a partial degradation of the porous structure of silica, as a consequence of the different chemical treatments, can be noted.

Mesoporous materials with a higher cristallinity could be easily obtained using ionic surfactant as templates but, in this case, solvent extraction should require more drastic conditions with respect to those normally used for materials obtained with neutral surfactant. In addition these materials exhibit high wall thickness and chemical and mechanical stability.

The folic acid-functionalized Mesoporous materials exhibit yellow colour.

Fig. 9.2 shows nitrogen adsorption–desorption isotherms of folic acid-functionalized mesoporous silica before (bottom curve) and after surfactant extraction (top curve). As expected, the pore volume increases considerably after surfactant removal. The sample was activated before the analysis at the temperature of 110°C to preserve the molecular integrity of the folic acid.

After surfactant removal, folic acid-functionalized mesoporous silica (SBET 252 m2/g; pore volume 0.2 cm3/g) presents a Type IV nitrogen isotherm typical of mesoporous materials. The very low BET surface area and pore volume can be explained considering the contributes of the organic

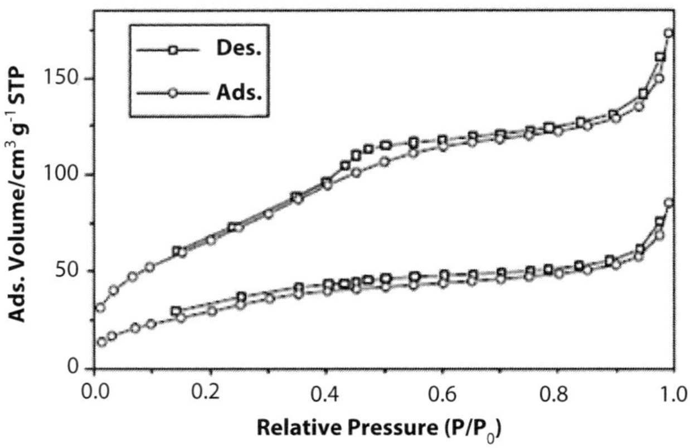

Figure 9.2 Nitrogen adsorption–desorption isotherms of folic acid-functionalized mesoporous silica before (bottom curve) and after surfactant extraction (top curve) {Reproduced with permission from L. Pasqua, F. Testa, R. Aiello, S. Cundari and J.B. Nagy, *Microporous and Mesoporous Materials*, 2007, 103, 1. ©2007, Elsevier [15]}.

fraction on the overall mass of the hybrid materials and the low activation temperature before nitrogen adsorption to prevent degradation of organic functions. The low temperature is probably responsible for a partial water retaining inside the pores. The adsorption and desorption steps generate an hysteresis loop for relative pressures higher than P/P0 = 0.43, which represents the lower hysteresis closure point for nitrogen.

An average pore diameter around 40 Angstrom is calculated according to the BJH desorption model. Even though it is not easy to determine the percentage of folic acid coverage on the external surface of the particles it can be assumed that a large part of the folic acid is external, because the presence of surfactant micelles at least makes difficult its diffusion into the pores. Thermogravimetric analysis shows a decrease of organic/SiO2 mass ratio from 0.8 to 0.3 for surfactant-extracted materials.

Successively, the drug-loading procedure of the matrix completed the preparation of the bifunctional material. The so obtained system can be useful as drug targeting device [15].

9.3 Drug-Loaded Folic-Acid-Grafted MSNs Specifically Target FR Expressing Tumour Cells [16]

A schematic representation of the MSN-FOL is reported in Fig. 9.3, panel a; panels b and c show the Transmission Electron Microscopy (TEM) micrographs of MSN-FOL material that exhibits a porous texture in adherence with materials of the MSU family, while Scanning Electron Microscopy

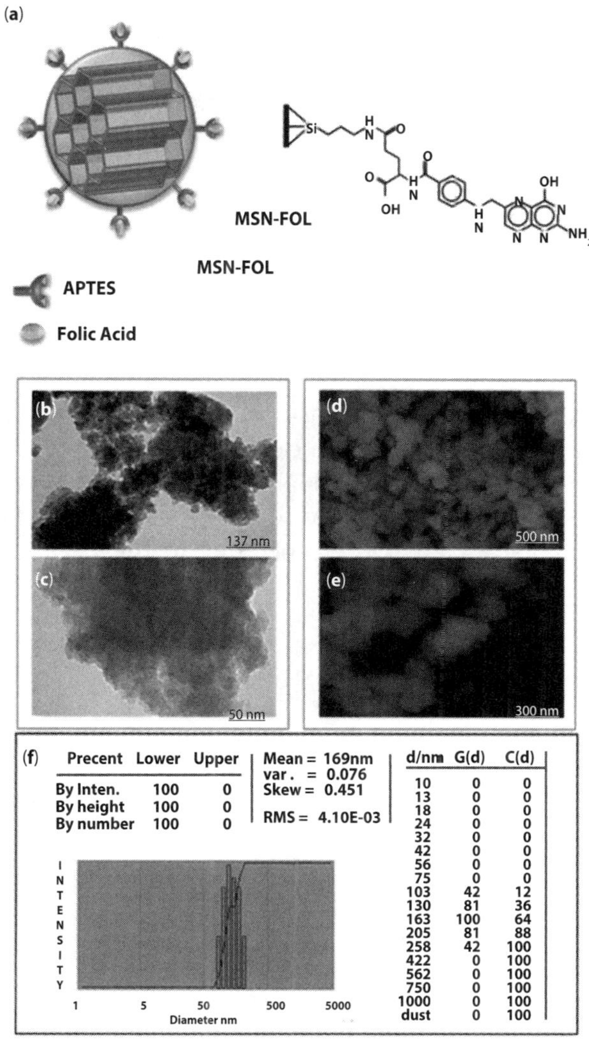

Figure 9.3 Folate conjugated mesoporous silica nanoparticles for drug targeting applications. (*a*) Schematic Representation; (*b*)-(*c*) TEM; and (*d*)-(*e*) SEM Micrographs of MSN-FOL. [In (*f*), 120μg Of MSN-FOL were added to 4ml of pure water (30μg ml^{-1} final concentration) and resuspended by gentle stirring; particle size distribution was then determined through DLS.] (Reproduced from Ref. 16 with permission from The Royal Society of Chemistry).

(SEM) images (panels d and e) show that this synthesis procedure yields nanoscaled particles without a regular morphology appearing as aggregates of up to 500 nm. Indeed, Dynamic Light Scattering (DLS) measurements of all functionalized particles showed that, when resuspended in water, their average size distribution ranges between 100 to 300 nm (Fig. 9.3 f and data not shown), presumably for aggregates disassembly deriving from repulsive hydration forces occurring between silanol groups and hydrogen bonded water molecules.

All synthesized MSNs are clearly non-toxic up to 30 μg/10^6 cells, while higher amounts show cytostatic effects at later time points. Thus, the choice of an adequate range of concentrations is an essential requirement for a proper use of the material.

Fluorescent Assisted Cell Sorting (FACS) analysis conducted on folate receptor positive (FR+) HeLa cervix cancer cells (Fig. 9.4a) incubated with MSN-FITC-FOL, evidenced that MSNs are internalized by cells.

The degree of endocytosis was determined by quantifying living cells that exhibited green fluorescence due to the FITC dye: only the

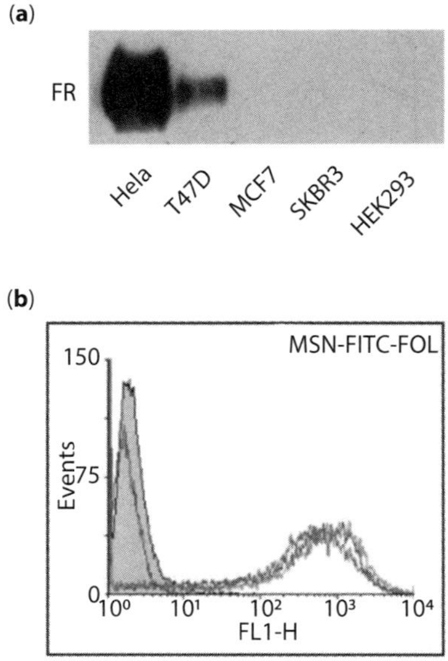

Figure 9.4 (*a*) Protein expression of FR in different cell lines. 50 μg of non-denatured proteins from total lysates were loaded and subjected to WB analysis. (*b*), MSN-FITC-FOL are internalized by HeLa cancer cells. Flow cytometry assay of HeLa cells incubated with MSN-FITC-FOL for 1h, 6h, 24h and 96h. Samples were analyzed by measuring the relative fluorescence intensity of FITC per cell at the FL-1H channel. Data were normalized versus an untreated control. The results are representative of three independent experiments. (Adapted from Ref. 16 with permission from The Royal Society of Chemistry).

fluorescence deriving from inside the cells can be revealed, since that coming from the outer surface is quenched by trypan blue treatment. A consistent positive distribution of particles into the cells was evident already after 1 h of incubation, which persisted up to 24 h, while a prolonged exposure (96 h) led to a complete loss of fluorescence, probably due to MSN-FITC-FOL expulsion into the medium, as discussed later (Fig. 9.4b).

Specific endocytic pathways are activated depending on MSN surface functionalization, as demonstrated by TEM observations on differently functionalized nanoparticles at various time points. Interestingly, MSN never entered HeLa (Fig. 9.5a), nor other cancerous (T47D, MCF-7 and SKBR3) and normal (HEK293) cell lines (Fig. 9.5b), indicating that calcined, non-functionalized, particles are inert and do not interact with the biological systems.

This result corroborates several recent reports, indicating that silanol groups on mesoporous silica nanoparticles tend to establish electrostatic interactions with the lipid head groups of the biological membranes[17], and such interaction has been proposed to be accountable for particles cellular internalization. It can be speculated that MSN do not enter cells most likely because silanol groups condensed upon calcination, thus preventing any possible interaction with membrane surface constituents. On the contrary, extracted particles (MSN-E) enter cells in great amount, confirming that silanol groups are responsible for cellular uptake (data not shown).

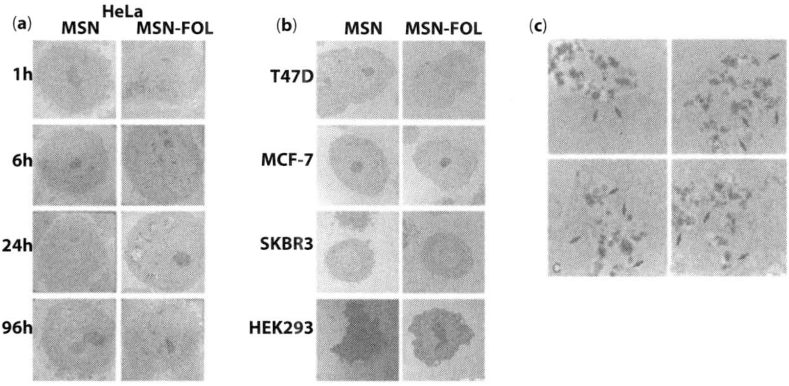

Figure 9.5 Effect of functionalization on MSNs uptake.TEM investigation of MSN and MSN-FOL uptake in *(a)*, FR expressing HeLa cells at indicated time points, and *(b)*, other cells lines at 6h. Images were taken at 5000x magnification. *(c)*, Colloidal-gold immunocytochemistry for FR (black dots indicated by arrows) in MSN-FOL treated HeLa cells. Images were taken at 40.000x magnification. (Adapted from Ref. 16 with permission from The Royal Society of Chemistry).

However, it is worth mentioning that water suspensions of calcined, non-functionalized MSNs, show a particle size distribution similar to MSN-E and to all the other functionalized materials (exemplified by MSN-FOL in Fig.), evidencing the absence of water-induced hydrophobic interactions among calcined particles, which could potentially lead to their assembly into bigger aggregates. This means that differences in behaviour of calcined MSN compared to extracted/functionalized MSNs in biological contexts, cannot be ascribed to larger aggregate formation in aqueous suspensions, but only to the differences in their surface functionalization.

Although it can be questioned whether the cellular uptake could be ascribed to electrostatic interactions between the protonated (AP) groups bound to the particles surface and the negatively charged cell membranes, TEM observations showed that AP groups do not significantly affect the endocytotic process, since only a negligible amount of particles is internalized by the different cell lines (data not shown). However, the saturation of AP sites by the folic acid function is sufficient to completely abrogate the already low reactivity of the AP groups. In fact MSN-FOL, where AP sites are mostly bound to folate function, enter only FR+ HeLa and T47D cells, but cannot enter FR negative (FR-) cells MCF-7, SKBR3 and HEK293 (Fig. 9.5b).

Particle uptake occurs as soon as after 1 h of incubation in all cell lines and the majority of cells appear free of particles within 96 h, clearly underlining MSNs biocompatibility (data not shown). Colloidal-gold labeling for FR unambiguously demonstrates that MSN-FOL enter HeLa cells through FR mediated endocytosis. Fig. 9.5c shows how MSN-FOL A) approach the cell membrane within a region enriched in FR clusters, B) are engulfed through folic acid recognition by FR, C) are internalized by the cell and D) co-localize with FR, most likely in glycosylphosphatidylinositol (GPI) enriched early endosomal compartments (GEECs) as recently proposed [18]. Indeed, a strong FR immunopositive reaction is always evident on the edges of the membranes involved in the overall MSN-FOL endocytic process.

Once clarified how MSNs enter the cells, what is their fate?Are particles are retained in the cells or rather expelled outside? DLS measurements of MSN-FOL dispersed in culture medium collected up to 96 h following 1 h treatment of HeLa cells, did show substantial difference between treated samples and control samples starting from 48 h to 96 h after treatment. MSN-FOL distributions ranged around two main peaks at 25 nm and 250 nm, which were absent in controls (Fig. 9.6).

On the basis of these evidences most of the particles could be exocytosed from the cells starting from 48 h either as unmodified units (as suggested from the peak around 250 nm, corresponding to the average size of particle aggregates as they are added to the cultures: see Fig. 9.3f), or as smaller units, probably due to aggregate disassembly (peak around 25nm). On the other hand, the presence of a tail exceeding 250 nm in the particle size distribution could be attributed to aggregate formation in the culture medium, since they do not appear either in control samples or in water suspensions (see Fig. 9.6 and Fig. 9.3f). Particle expulsion from the cell is in line with the dramatic drop of MSN positive cells observed at 96 h (Fig. 9.4b and Fig. 9.5a).

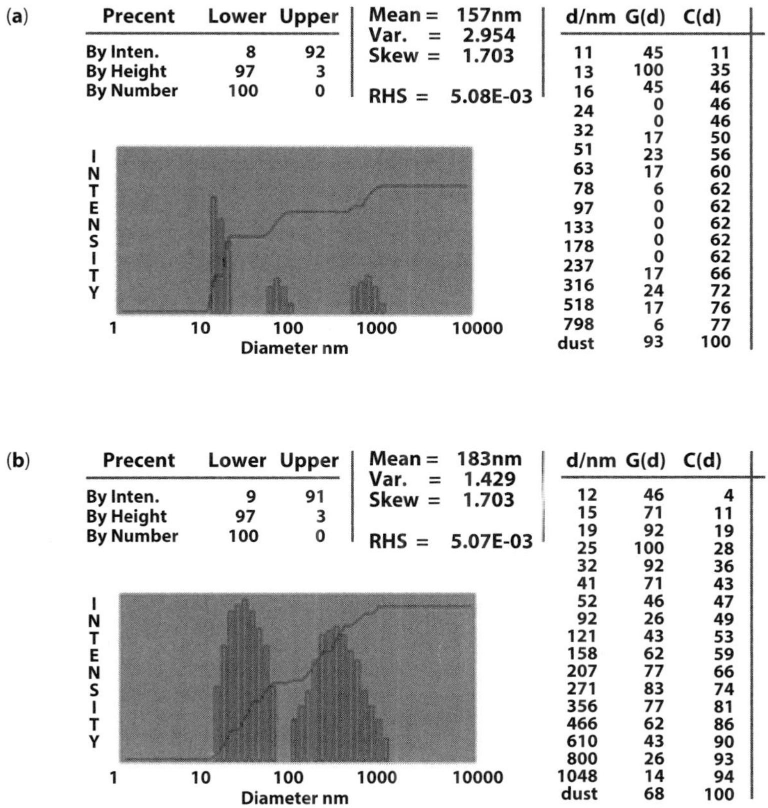

Figure 9.6 MSN-FOL are mostly exocytosed from the cells within 48h. HeLa cells were seeded in 60mm culture dishes, and treated the following day with 30μg/10^6 cells MSN-FOL for 1h or left untreated. The medium was, then, replaced with fresh growing medium and collected after 48h. Size distribution of particles dispersed in (*a*) control sample (medium only) and (*b*) MSN-FOL treated sample was determined by DLS. (Reproduced from Ref. 16 with permission from The Royal Society of Chemistry).

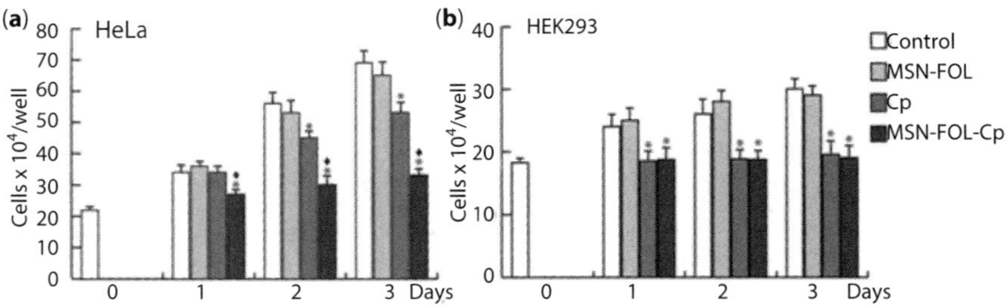

Figure 9.7 Cp loaded MSN-FOL inhibit growth in FR-expressing HeLa cancer cells: (*a*) cancerous FR+ HeLa and (*b*), normal FR- HEK293 cells were separately treated for 1h with MSN-FOL, Cp alone, MSN-FOL-Cp or left untreated. Viability was determined after 1, 2 or 3 days. Values represent the mean± s.d. of four triplicate independent experiments. *$p < 0.01$ vs the respective control; ◆ $p < 0.01$ vs the corresponding free Cp treated samples. (Reproduced from Ref. 16 with permission from The Royal Society of Chemistry).

MSN-FOL could represent a reliable tool for drug targeting purposes, since particles loading with the chemotherapeutic drug cisplatin (MSN-FOL-Cp) is able to induce a strong growth arrest as soon as after one day in FR+HeLa cells, if compared to the less dramatic growth retardation caused by comparable amounts of free Cp, while MSN-FOL does not affect cell proliferation (Fig. 9.7a). On the contrary, normal FR- HEK293 cells, show no difference between the growth inhibition observed in MSN-FOL-Cp and free Cp treated samples, with respect to control and MSN-FOL samples (Fig. 9.7b).Thus, folic acid grafting on the Cp-loaded MSN selectively increases the delivery of the drug to cells that over-express FR, but not to FR- cells. Moreover, it is worth underlining that toxicity of MSN-FOL-Cp can only in part be ascribed to the release of small amounts of drug in the medium occurring during the 1 h exposure of the cells to the particles. To overcome this inconvenience, a great step forward could be the development of an intracellular pH-sensitive drug delivery device, able to release the drug (covalently bound to the vehicle) once reached the more acidic environment of the endosomal cavities.

9.4 Conclusion

As schematically illustrated in Fig. 9.8, PEG-templated mesoporous silica nanoparticles with controlled size, ranging around 200 nm (Fig. 9.3f), as required for drug targeting purposes, represent a great potential for localized drug release. In fact, while the vehicle alone (MSN) is not able to enter cells, unless opportunely functionalized, when folic acid function is

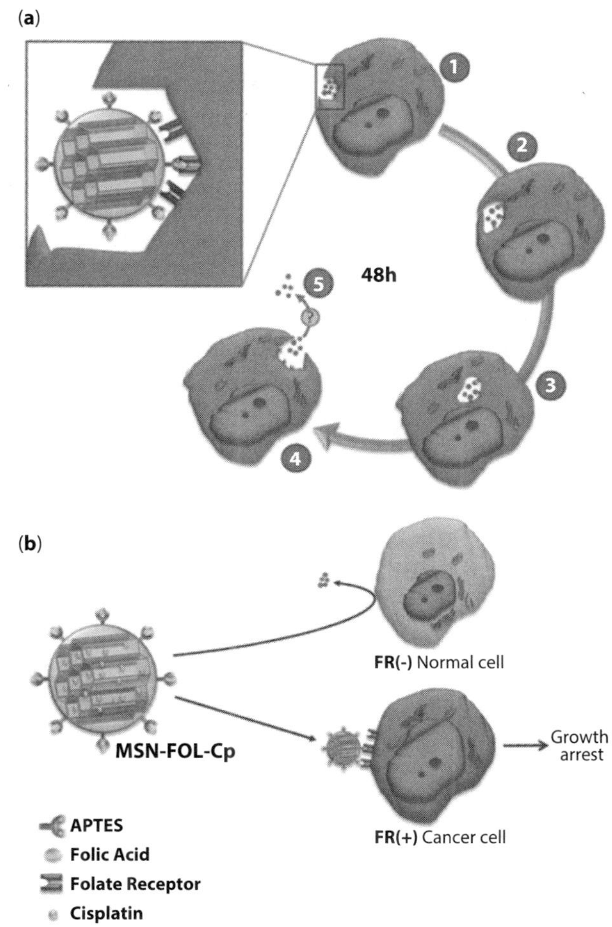

Figure 9.8 Schematic representation of MSN-FOL intracellular fate and mechanism of action. (*a*), MSN-FOL enter the cell via FR mediated endocytosis [1], are sequestered together with FR in endosomal structures [2–3], are probably exocytosed outside the cell within 48 h [4–5]; (*b*), MSN-FOL-Cp do not enter FR-negative cells, while induce growth arrest only in FR-positive cells. (Reproduced from Ref. 16 with permission from The Royal Society of Chemistry).

present on the particles surface (MSN-FOL), they become able to be internalized, in a highly specific way, exclusively by FR over-expressing cancer cells, from where they appear to be mostly exocytosed as soon as after 48 h following particles uptake.

Moreover, the toxic effect of MSN-FOL-Cp is much stronger than free Cp exclusively in FR-expressing cells, while it is comparable in FR negative cells, confirming that the folic acid function is the activator in the cellular uptake process.

References

1. J.F. Diaz and K.J. Balkus, *J Mol Catal B-Enzym*, Vol. *2* (2–3), p. 115–126, 1996.
2. L. Washmon-Kriel, V.L. Jimenez and K.J. Balkus, *J Mol Catal B-Enzym*, Vol. *10* (5), p. 453–469, 2000.
3. J. Deere, E. Magner, J.G. Wall and B.K. Hodnett, *Chem Commun*, Vol. (5), p. 465–466, 2001.
4. M. Vallet-Regi, A. Ramila, R.P. del Real and J. Perez-Pariente, *Chem Mater*, Vol. *13* (2), p. 308–311, 2001.
5. G. Cavallaro, P. Pierro, F.S. Palumbo, F. Testa, L. Pasqua and R. Aiello, *Drug Delivery*, Vol. *11* (1), p. 41–46, 2004.
6. C. Tourne-Peteilh, D. Brunel, S. Begu, B. Chiche, F. Fajula, D.A. Lerner and J.M. Devoisselle, *New J Chem*, Vol. *27* (10), p. 1415–1418, 2003.
7. B. Munoz, A. Ramila, J. Perez-Pariente, I. Diaz and M. Vallet-Regi, *Chem Mater*, Vol. *15* (2), p. 500–503, 2003.
8. Y.S. Lin, C.P. Tsai, H.Y. Huang, C.T. Kuo, Y. Hung, D.M. Huang, Y.C. Chen and C.Y. Mou, *Chem Mater*, Vol. *17* (18), p. 4570–4573, 2005.
9. D. Luo, E. Han, N. Belcheva and W.M. Saltzman, *J Control Release*, Vol. *95* (2), p. 333–341, 2004.
10. M.N.V.R. Kumar, M. Sameti, S.S. Mohapatra, X. Kong, R.F. Lockey, U. Bakowsky, G. Lindenblatt, H. Schmidt and C.M. Lehr, *J Nanosci Nanotechno*, Vol. *4* (7), p. 876–881, 2004.
11. R.A. Gemeinhart, D. Luo and W.M. Saltzman, *Biotechnol Progr*, Vol. *21* (2), p. 532–537, 2005.
12. Q.J. He and J.L. Shi, *J Mater Chem*, Vol. *21* (16), p. 5845–5855, 2011.
13. Y. Lu and P.S. Low, *Adv Drug Deliv Rev*, Vol. *54* (5), p. 675–693, 2002.
14. P.S. Low and A.C. Antony, *Adv Drug Delivery Rev.* , Vol. *56*, p. 1057–1238, 2004.
15. L. Pasqua, F. Testa, R. Aiello, S. Cundari and J.B. Nagy, *Micropor Mesopor Mat*, Vol. *103* (1–3), p. 166–173, 2007.
16. C. Morelli, P. Maris, D. Sisci, E. Perrotta, E. Brunelli, I. Perrotta, M.L. Panno, A. Tagarelli, C. Versace, M.F. Casula, F. Testa, S. Ando, J.B. Nagy and L. Pasqua, *Nanoscale*, Vol. *3* (8), p. 3198–3207, 2011.
17. Y. Roiter, M. Ornatska, A.R. Rammohan, J. Balakrishnan, D.R. Heine and S. Minko, *Nano Letters*, Vol. *8* (3), p. 941–944, 2008.
18. P. Bhagatji, R. Leventis, J. Comeau, M. Refaei and J.R. Silvius, *J Cell Biol*, Vol. *186* (4), p. 615–628, 2009.

10

Molecular Recognition-driven Membrane Processes

Laura Donato, Rosalinda Mazzei, Catia Algieri, Emma Piacentini, Teresa Poerio and Lidietta Giorno*

Institute on Membrane Technology, National Research Council of Italy, ITM-CNR. Rende (CS), Italy

Abstract

High-precision, selective and efficient systems are strongly demanded in all productive sectors, including chemical manufacturing, pharmaceutical, medical treatment, food, energy, diagnostic, analytical tools. Precision separation, conversion, formulation at molecular level may rely on various mechanisms and properties, such as molecular sieving, electrical charge, chemical solubility, molecular recognition. Membrane operation can promote mass transport, separation, conversion, formulation on the basis of the mentioned mechanisms. Biomimic and biohybrid funtionalised membranes are particularly suitable for separations based on molecular recognition.

Membranes can be formed by polymers or monolythes bearing the (biomimic) recognition sites. In alternative, biomimic, biomolecules and recognition sites can be loaded on a previously formed membrane support. In such cases, surface functionalization methods can improve the loading efficiency. Molecularly imprinted membranes; membranes bearing affinity ligands (Antigen-Antibody, avidin/-ligand, DNA-protein, sugar-lectin, RNA-ribosome,...); membranes functionalized with zeolite and nanoparticles; membranes with immobilized enantioselective enzyme for kinetic resolutions, are among examples of selective and efficient operations driven by molecular recognition. This is the mechanism by means of which molecules are able to identify each other using non-covalent - (hydrogen bonding, metal coordination, hydrophobic forces, van der Waals forces, π-π interactions and electrostatic effects) and covalent interactions.

**Corresponding author*: l.giorno@itm.cnr.it, Lidietta.Giorno@cnr.it

Annarosa Gugliuzza (ed.) Smart Membranes and Sensors, (269–300) 2014 © Scrivener Publishing LLC

In this chapter, the various examples including membrane preparation, surface functionalization to introduce recognition sites, membrane performance and application will be discussed.

Initially, molecular recognition based separation promoted by molecularly imprinted membranes will be discussed. Then, membrane processes based on affinity interactions between receptor and target ligand molecule will be presented, highlighting the difference with the previous separation mechanism.

Keywords: Molecular recognition, molecularly imprinted membranes, affinity membranes, zeolites, functional microsphere, surface functionalization

10.1 Molecular Imprinting Technique

The study of interactions between chemical species at molecular and supramolecular level as well as the possibility to produce synthetic materials able to specifically recognize bio-active molecules is one of the most challenging scientific and technological research topic. Among the different strategies used to prepare these artificial systems (thus mimicking biological mechanisms) one of the most promising is the molecular imprinting approach. This technique is based on the introduction of the molecular memory of the substrate to be recognized ("*target molecule*" or "*template*") in a polymeric material during its preparation. It is based on the system used by enzymes for substrate recognition and called the "*lock and key*" model. This objective is achieved by means of the addition of the template to the reaction mixture, which consists of a functional monomer, a cross-linking agent and a solvent. During the polymerization, the template is integrated in the polymer matrix and functional monomers are arranged around the template by chemical interactions. The following extraction of the template (with suitable solvents) from the polymer network gives rise to the formation of specific recognition sites with high substrate selectivity. The recognition is due to the complementarity existing between template and functional monomer in terms of spatial arrangement, size and chemical functionality (Fig. 10.1).

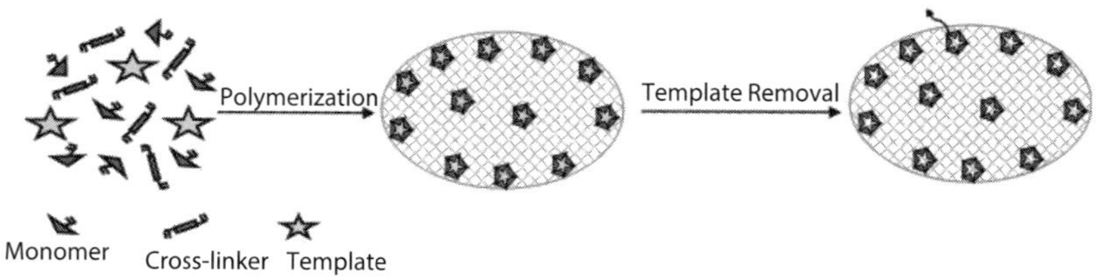

Figure 10.1 Scheme of the imprinting process.

10.1.1 Molecularly Imprinted Membranes (MIMs)

A significant contribution to the studies pursued in the field of molecular recognition is given by the development of molecularly imprinted membranes (MIMs). In fact, the possibility to introduce specific molecular recognition sites into a synthetic membrane opened a new way in the role of the transport or retention of particular substances and shortly it has led to high performance separation membranes able to discriminate between target molecules and other analytes. Two different kind of molecular interactions may be exploited in the preparation of MIMs: covalent and non-covalent binding. Although the first type permits to obtain more homogeneous binding sites (thanks to the possibility to control the stoichiometry of the imprinting materials) the non-covalent binding approach, firstly applied by Mosbach and Ramström [1], is the most widely used road. This is because, dissolving monomers and template in a solvent is easy and a wide range of monomers having functionalities that complement every needed target molecule is available. Additionally, non-covalent binding and interaction constitute the basis of the recognition of biochemical systems in the cell and biological assay systems. In fact, enzymes, antibodies and receptors recognize their target molecule *via* complementary ionic, hydrogen-bonding and hydrophobic interactions like non-covalent molecularly imprinted materials. A combination of covalent and non-covalent imprinting approach can be also promoted. In this case, during the imprinting process, monomer and template interact *via* covalent bond but the subsequent recognition takes place *via* non-covalent interaction.

10.1.2 MIMs Preparation: Methods And Materials

First molecularly imprinted membranes have been prepared by Piletsky *et al.* in 1990 [2] *via* in situ bulk polymerization of acrylate monomers using as template adenosine monophosphate (AMP). Thereafter, a lot of MIMs have been successfully prepared using different materials and techniques. Different methods have been developed to prepare an imprinted membrane. They include:

- the simultaneous formation of imprinted sites and membrane morphology of a self-supported membrane
- the preparation of the membrane from a pre-synthesised imprinted polymer
- the preparation of a composite imprinted membrane by surface imprinting of a pre-existing polymeric membrane

The simultaneous formation of molecular recognition sites and membrane structure is mostly achieved by means of the "*in situ cross-linking polymerization*" and the so-called "*alternative molecular imprinting*". The in-situ cross-linking polymerization is performed using a mixture solution of template, functional monomer, cross-linker and initiator in a suitable solvent. Plasticizer agents are also added to the mixture in order to obtain more flexible membranes.

The alternative molecular imprinting, is deemed to be an extension of a bio-imprinting in which existing recognition sites in an enzyme might be modified through the presence of a target molecule. According to this approach, an imprinted membrane is prepared in the presence of target molecule *via* the phase inversion technique. A polymer *ad hoc* synthesised, which contains functional groups able to interact with the template is used as membrane forming material. The process involves the transformation of the polymer from the liquid phase to the solid state either by the "*dry*" or the "*wet*" method; in alternative, a combination of the two methods can be also used. Membranes prepared *via* the dry method generally exhibit a dense structure due to the progressive increase of the polymer concentration in the forming membrane as a consequence of the solvent evaporation. On the opposite, membranes prepared using the wet method show a porous structure due to the rapid liquid-liquid de-mixing typical of the non-solvent induced phase separation. The phase inversion technique permits to prepare thin and flexible imprinted membranes that can be used in many industrial and research areas. The use of this technique is extended also for the preparation of MIMs from previously synthesised cross-linked imprinted polymers. The process, named "*hybrid molecular imprinting*" is based on the dispersion of a pre-synthesised imprinted polymer powders into a commonly used polymer matrices and membranes are prepared *via* phase inversion. By using this route it is possible to obtain highly selective hybrid materials. In another approach imprinted polymeric nanoparticles can be used as a filter cake between two microfiltration membranes. Composite MIMs may be also prepared by covering the surface of a microfiltration membrane with imprinted polymer particles (i.e. nano-sphere) *via* deposition or cross-flow filtration. Selected examples of "hybrid molecular imprinting" technique for the preparation of particles-loaded membranes will be illustrated in the next paragraphs.

In addition, composite imprinted membranes may be prepared by functionalising the surface of a porous polymeric membrane using an imprinted polymer *via* photo/thermal co-polymerization. In this way, molecular recognition sites are introduced on the surface of the base membrane which is entirely covered by a thin polymer layer and preserves its bulk structure. Nano-fibers and carbon nano-tubes have also potential to be applicable for the molecular recognition.

Many polymers may be used for preparing MIMs having both, flat-sheet and hollow fiber configuration. Some of the mostly used materials are acrylic copolymers [3], polyvinilidene fluoride and polyethersuphone [4], polystyrene, nylon 66 and polysulphone [5], cellulose acetate/sulphonated polysulphone blends [6], etc.

Two different template transport mechanisms may occur in a molecularly imprinted membrane: "*facilitated permeation*" and "retarded *permeation*". In the first case the template passes more quickly through the membrane. Thus, a preferential path for the template molecules is formed *via* binding to and dissociating from neighboured recognition sites in the membrane allowing to facilitate their permeation while the non-specific transport of other solutes is slow.

In the second case the permeation of the template is retarded due to its affinity binding to imprinted sites located on the membrane surface and on transmembrane pores until their saturation occurs. These membranes are regarded as "*adsorber devices*" and their separation efficiency depends on the binding capacity exhibited by the imprinted sites.

This ability of MIMs to exert a selective transport or retention of specific molecules makes them good candidates for the development of highly innovative membrane processes.

10.1.3 Application Of MIMs

Due to their high specificity, MIMs have a great potential for applications in molecular separation and purification of various substances in the environmental, food and pharmaceutical fields. The most prominent areas are sensors and separation technology wherein they are widely used in affinity separations, medical diagnostics, drug delivery, enzymatic catalysis, enantiomeric separation, etc.

Various kind of compounds are used as template molecule including amino acids, proteins, flavonoids, nucleotide bases, hormones, pesticides, etc. Some selected examples which may represent the wide range of MIMs application are listed in Table 10.1.

As it is shown, MIMs are successfully used in enantiomeric separations of drugs, for concentration of specific substances used as food additives, for water purification and much more.

For the first time Kalim *et al.*, developed molecularly imprinted polyvinyl alcohol membranes incorporating imprinted particles with catalytic properties [14]. The membranes were applied as catalysts

Table 10.1 Examples of application of the molecularly imprinted membranes.

Material	Template	Application	Ref.
Poly(4-vinylpyridine)/polypropylene)	S-Naproxen	Enantiomeric separation	7
Poly(acrylontrile-co-acrylic acid)	Naringin	Removal from juices	8
Poly(acrylic acid)/polysulphone	Erytromicin	Recovery from water	9
Poly(ethylene-co-vinyl-alcohol)	Creatinine	Sensing kidney dysfunction	10
Poly(methyl methacrylate-co-methacrylic acid)/ polyurethane	Theophylline	Purification of Drugs	11
Poly(acrylonitrile-co-acrylamide)	Folic acid	Enrichment of food additives	12
Poly(methacrylic acid)/oligourethaneacrylate	Atrazine	Decontamination of water	13

for the dehydrofluorination of 4-fluoro-4-(p-nitrophenyl)-2-butanone. N-benzylisopropylamine was chosen as template. Authors performed catalytic reactions according to the total recycle mode operation using both, membranes containing imprinted particles (MIP) and membranes containing non-imprinted particles (CP). They observed a clear catalytic effect of kMIP/kCP = 1.6.

The application of MIMs for the production of optical isomers is of particular interest due to the possibility to overcome the difficulty to obtain a high yield of an optically pure compound in a single batch operation which is typical of the traditional processes. Donato *et al.* [7] developed the first enantioselective MIMs with specific molecular recognition for the anti-inflammatory drug S-naproxen. Membranes were prepared by photo-copolymerization of polypropylene membranes with the functional monomer 4-vinylpiridine. Permeation studies revealed as the S-naproxen passed at high rate through the imprinted membrane than its opposite isomer (α=1.6).

In a recent published work Garcia *et al.* reported an example of separation by molecular recognition in organic environment [15]. In this paper, the influence of different functional monomers on the recognition capacity of 4,4'-methylendianiline–imprinted membranes in isopropanol was investigated. Flat-sheet membranes were prepared *via* phase inversion technique from acrylonitrile copolymers synthesised using acrylic acid, itaconic acid and methacrylic acid as functional co-monomers. Membranes made of poly(acrylonitrile-co-acrylic acid) showed the highest specific binding (2.6 μmol/g_{memb}.). Furthermore they exhibited a selectivity factor of 1.6 towards the structural analogue 4,4'-ethylendianiline and were completely selective *vs* aniline.

The 4,4'-methylendianiline is a member of the big family of the primary amines. They are used as intermediates reagents in the different steps of synthesis and therefore could be present at low concentrations in solvents due to a possible un-complete reaction. So, It would be very significant to develop and exploit MIMs also in organic media, particularly to remove impurities and health harmful compounds like primary amines. In fact, the purification and recycling of organic solvents may well contribute to save energy and resources used to produce new solvents and last but not least to preserve the environment.

In a simultaneous application of the alternative molecular imprinting and of the electrospray deposition imprinted nano-fiber membranes are also prepared with the aim to obtain systems having high concentration of specific molecular recognition sites and promising for application as sensor chips [16].

Due to the multiplicity of its potential application, the development of imprinted membranes for recognition studies at molecular level has already contributed positively to the research progress this area. In fact, the preparation of membranes with important binding and transport properties can make them competitive compared with the conventional separation devices, which do not have the membrane ability to operate in continuous.

10.2 Affinity Membranes

Surface modifications of membranes has been developed for a number of applications in biotechnology and biomedicine, including protein purification, isolation of biomolecules from complex biologically derived fluids. These types of separations are already being performed and they will become increasingly important over the next 10 years.

Recently many bioaffinity materials were proposed for use in clinical practice as well as for diagnostics. The usual procedure to obtain such materials comprises the covalent binding of specific ligand (antigens, antibodies, non-immunological) to porous inert carriers [17]. The high specificity of affinity membrane is due to the strong interaction between the ligand and the molecule to be separated

In literature there are various routes for ligand immobilization on supports, the principal strategies are based on chemical grafting or molecular recognition [18] on porous supports. In molecular recognition technology organic ligand are created that selectively bind targeted molecules in solution. The binding is based on charge, size, and shape and is specific for the target molecule.

Approaches aiming at creating bio-compatible micro-environments consist in modifying the surface of polymeric membranes by attaching functional groups like sugars, polypeptides and then to adsorb the ligand.

Another way considered as of bio-mimetic inspiration and which was shown to be efficient for enzyme attachment, it consists in using the very strong and specific interaction of the small protein avidin for the biotin. The tetrameric structure of avidin permits itself to interact with four different molecules of biotin at the same time. Various proteins and enzyme could be easily biotinylated, and this mode of enzyme grafting has already been used for electrodes production as well as for membranes made up of conducting fibres.

The main difference between imprinted and affinity membranes is that in the first case the target molecules (template) used to create the recognition sites is removed from the polymeric matrix before testing the membrane performance. In the affinity membrane processes the biomolecules used to create the affinity sites remain heterogenized to the membrane. On the other hand, in both cases the interaction between membrane and the investigated analyte is based on the specific recognition mechanism.

10.2.1 Preparation Of Affinity Membranes

Similarly to the preparation of affinity stationary phases for packed columns, affinity membranes are generally obtained via three principal methods: preparation of the basic membrane, activation of the basic membrane, coupling of affinity ligands to the activated membrane [19]. Ideal membrane should fulfil the following conditions: proper pore structure and mechanical strength for use at high-flow rates and low back pressure in rapid processing; availability of reactive groups (such as -OH, -NH_2, -SH, -COOH) for the further coupling of spacer arms or ligands; chemical and physical stability under harsh conditions of high temperature or chemical sterilization; a hydrophilic surface for higher recovery of protein activity [20]. There are many commercially available materials to choose from, and they include organic, polymeric, inorganic, and composite materials.

Table 10.2 Methods used for membrane activation.

Activation substance	Reference
Cyanogen bromide activation	21
Epoxide activation	22
Periodate oxidation	23
Triazine activation	24
Carbodiimide coupling	25
Carbonylation	26
Succinimide esters activation	27
Tosyl chloride activation	28
Glutaraldeide activation	

Once the base membranes are prepared, they must be activated to acquire reactive groups for the coupling of ligands. Some methods used for membrane activation are reported in Table 10.2.

After the membranes are activated, various ligands can be coupled to them. Occasionally, when the immobilized ligand is a small molecule, such as a receptor substrate or a chemical antigen, steric hindrance will occur between the immobilization support and the substrate to be isolated. This phenomenon will cause a reduction or complete lack of specific binding. To overcome this problem it is necessary to bind a spacer molecule to the support prior to attaching the ligand [28].

The selection of any ligand should meet the following conditions: the ligand must specifically and reversibly bind the substances to be isolated and must contain group that can be chemically modified to obtain attachment to the support; the chemical modification must not impair or damage the specific binding activity of the ligands. The materials used as affinity ligands can be divided into two categories: general ligands (such as dyes, amino acid, Protein A and G, coenzyme, etc.), specific ligand (such as enzymes and substrates, antibodies and antigens) [20].

Immobilized biomolecule such as enzymes, antibodies or nucleic acids are often used for the development of biosensors to detect analytes by converting a biological response into an electrical or optical signal but also for protein separation. Functionalized membranes are optimal support to immobilize biomolecules which can be performed by physical adsorption, including electrostatic and hydrophobic interaction, covalent bonding and specific interactions such as biotin-avidin, antibody-antigen interaction and DNA hybridization [29]. Basing on the interaction membrane surface / biomolecules, different surface properties can be created, some of that are reported in Table 10.3

The attachment o biomolecule via biochemical interaction can be substantially dived in: molecular recognition between biotin-avidin, molecular recognition by protein A/G and antibody, sequence specific DNA hybridization and more recently by the system aptamer-receptor [29]. The bound between membrane and biomolecule by chemical attachment can cause the formation of non-specific interaction and sometimes in order to accommodate the bound with the functionalized surface it is possible to observe denaturation of the same molecule. In order to solve this problem strategies focusing on site-specific immobilization were developed. The specific interaction on molecular recognition of avidin-biotin provides a facile approach for the immobilization of biomolecule on support.

Table 10.3 Surface property after membrane functionalization for biomolecule immobilization.

Surface property	**Interaction with biomolecules**
Positive charge ($R-(CH_2)n-NH_3^+$)	• Charge-charge interaction • Chemical bonding with carboxyl group (R-COOH) of biomolecules • Chemical bonding with amine group ($R-NH_2$) of biomolecules glutaraldeyde mediated
Negative charge ($R-(CH_2)n-COO^-$)	• Charge-charge interaction • Chemical bonding with amine group ($R-NH_2$) of biomolecules
Hydrophobic	• Hydrophobic interaction
Aldehyde NSH	• Chemical bonding with amine group ($R-NH_2$) of biomolecules
Maleimide	• Chemical bonding with sulfhydryl group (R-SH) of biomolecules
Epoxy	• Chemical bonding with hydroxyl (R-OH) amine and sulfhydryl groups o biomolecules
Biotin	• Specific interaction with avidin and streptavidin

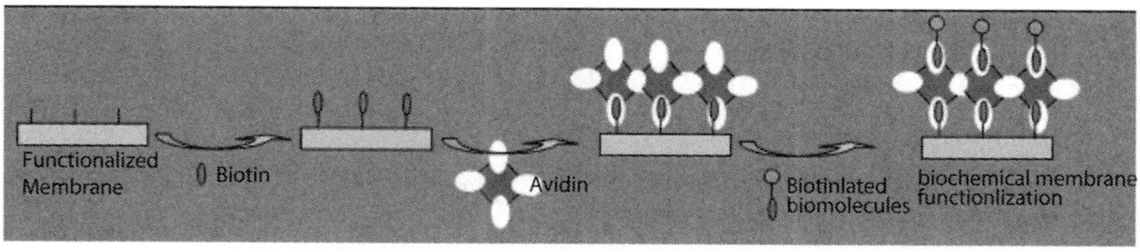

Figure 10.2 Mechanism Of Biomolecule Immobilization Mediated By Biotin-Avidin System Through Molecular Recognition.

The avidin based system is ideal for well-controlled immobilization of biomolecules due to the specific and strong interaction between avidin and biotin. Avidin has four binding site for biotin (Fig. 10.2) and is used to anchor biotinylated biomolecules such as protein and DNA acting as a biocompatible linker between biotin and biotinylated biomolecules.

Avidin–biotin interaction has been used by many researchers to immobilize enzymes on the membrane in a site-specific manner [30–32].

An interesting application was carried out by Dattaa and coworkers [32] who developed a recognition based separation technique to separate Tat protein, which has been proposed as the specific target for AIDS vaccine. The mentioned protein was separated and purified from a complex mixture of proteins, known as bacterial lysate (BL) using avidin–biotin interaction in 4-stack microfiltration membranes system.

Another strategy, mainly used in biosensors development is based on the molecular mechanism that involve immobilized antibody-binding proteins, such as protein A or G that recognize the Fc fraction of antibody, orienting the Fb portions towards the mixture to be recognized leading to highly improved antigen detection [33].

Sequence specific DNA hybridization can be also used to attach specific antibody on membrane surface that can recognize in specific way antigens in complex solutions. This kind of approach is mainly used in biosensor membrane systems.

10.2.2 Affinity Membranes For Chiral Separation

Affinity ultrafiltration uses a large stereospecific binding agent, they can be used to separate enantiomeric mixtures. Several investigators have demonstrated the feasibility of this process: Romero and Zidney [34] studied the separation of D and L-tryptophan by affinity ultrafiltration using bovin serum albumin (BSA) as stereoselective binding agent, they were able obtain 50-fold purification of L-triptophan, but the product yield was less than 50%. Poncet *et al.* [35] also examined the use of affinity ultrafiltration for the separation of trytophan enantiomers, obtaining 91% optical purity and 89% recovery of the D-tryptophan. Romero and Zidney [36] examined the performance of multi-stage diafiltration process for chiral separation of D- and L-tryptophan. The two stage system gave purification factor of more than 20 % at greater than 90% yield. Hadik *et al.* [37] studied D- and L-lactic acid resolution by chiral membrane obtained by drying a supported liquid membrane containing the chiral selector, on the membrane solid surface. The obtained membrane had high selectivity with low mass transfer rates.

More recently Higouchi *et al.* [38] used affinity ultrafiltration membranes with immobilized DNA for the separation of racemic tryptophan, phenylglycine and phenylalanine.

The interaction between the ligand and ligate is expected to have high specificity when it is based on molecular recognition. One of the most specific molecular recognition process is the antibody-ligate reaction [39]. Urmenyi *et al.* investigated the immobilization of antibody onto the surface of poly(ethylene vinyl alcohol membrane. The antibody used was against 17-β-estradiol, an estrogenic hormone and they studied the application of the functionalized membranes as affinity binders for estrogenic hormones. The choice of the antibody against 17-β-estradiol was based on: a possible applications in pharmaceutical industry or drinking water production or as an analytical tool for quantification of estrogenic compounds in wastewater streams.

10.2.3 Affinity Membranes For Protein Separation

Another important application based on molecular recognition is protein purification.

Isolation of protein or a group of proteins from body fluids for therapeutic uses requires a high level of purification and attendant validation. There are numerous papers on the purification of gamma globulin fraction, h-serum-albumin, and various proteins purifications methods. An interesting application of affinity membranes was reported by Hou and Zaniewski [40] who purified crude urokinase from human urine through a multistep process to prepare this plasminogen-activating enzyme for therapeutic use. The membranes used in this work were cation exchange matrices prepared on cellulose fibers. Weiner *et al.* [41] described the separation of IgG with a novel protein An affinity matrix prepared from two-dimensional protein crystals. Castilho *et al.* [42] compared 10 affinity membranes for the purification of immunoglobulins, prepared by coupling various affinity ligands to different microfiltration membranes.

Kubota *et al.* studied the recovery of serum proteins using cellulosic affinity membranes with the immobilization of Cu^{2+} [43] and the modification of the membrane with tannic acid [44]. Albumin and γ-globulin are the major components of serum proteins, and the play an important roles in human body. It is well know that the loss of albumin in renal disease causes serious problems.

10.3 Cyclodextrins As Molecular Recognition Elements

Molecular recognition may be also achieved by cyclodextrins (CDs) which are cyclic compounds consisting of sugar units joined to form a cone shape. Traditional cyclodextrins are constituted by 6–8 number of monomer units and are constitued of a hydrophilic outer surface and a hydrophobic inner cavity. The ring structure and the hydrophobic inner cavity permit them to house guest molecules and to form inclusion complexes whit organic compounds which contains hydrophobic groups capable to fit their cavity [45, 46]. In Fig. 10.3 is shown the chemical structure of the most simple natural CD named α-cyclodextrin. It is constituted of six α-1,4-linked-D-glucopyranose units. Other common CDs are β- and γ-, which are formed of seven, and eight D-sugar units, respectively.

Due to their properties cyclodextrins have been found application in food and pharmaceutical industries. They are used as such, crosslinked to

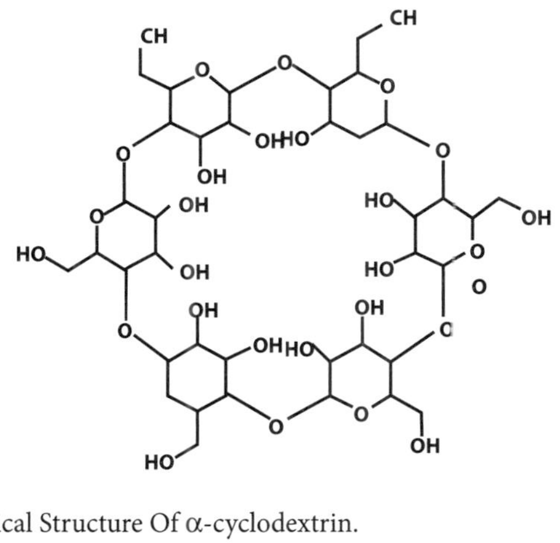

Figure 10.3 Chemical Structure Of α-cyclodextrin.

form polymers, entrapped in polymeric membranes and/or molecularly imprinted for the recognition of target molecules at nano-scale.

Some authors reported the application of cyclodextrins as ligands in the purification of amylase enzymes [45, 47]. Sulfated and phosphated β-cyclodextrins have been also employed as chiral selectors for capillary electrophoresis [48]. In this case, anodic detection was used for the sulfated β-CDs and cathodic detection was used for the phosphated β-CDs. The degrees of substitution of the CD, pH of the conditions electrolyte as well as the concentration of the functionalized β-cyclodextrin, each had an important effect on the success of the enantiomeric separation of the investigated drugs.

Interestingly, chitosan/*β*-cyclodextrin and carboxymethylchitosan/*β*-cyclodextrin microspheres loaded with theophylline have been developed as pulmonary sustained drug delivery carriers. Studies on the characteristics and ciliotoxicity of these systems evidenced an high drug loading and encapsulation efficiency and suggested that them should be effective as for sustained pulmonary drug carriers [49].

Due to their excellent capacity to form complexes with a wide variety of substances the CDs have been also used as coating or covalently linked layer of silica gel and entrapped in membranes for improving the separation efficiency of chromatographic and driven membrane processes. Fontananova *et al.*, successfully assessed the recognition properties of O-octyloxycarbonyl-β-CD derivative immobilized in polyetherehterketone (PEEK-WC)-based membrane towards the flavonoid naringin. Membranes have been prepared in flat sheet configuration according to the phase inversion method. The entrapment of β-CD in membrane optimised the interaction with the flavonoid and allowed to easily regenerate the system. The maximum amount of retained naringin in water was 3.74 $mol/g_{membrane}$ [46]. β-CDs have been also used as functional monomers to interact with naringin molecules *via* hydrogen bonding in molecularly imprinted polymers synthesised using the flavonoid as target molecule. In a recent study [50], Ma *et al.*, developed an imprinted β-CD polymer capable to bound 50.13 $\mu mol/g_{polymer}$ of naringin.

The strategy of molecular imprinting was also employed to synthesised polymers containing vinyl monomers of CDs exhibiting molecular recognitions towards different antibiotics and oligoppeptides used as target molecules [51].

The great potentiality of membrane technology and the molecular recognition properties of cyclodestrins are promising for the development of molecularly imprinted membranes containing CDs as synthetic specific recognition elements able to mimic biological systems.

10.4 Zeolite Membranes as Molecular Recognition Devices: Preparation and Characterization

Zeolite molecular sieves present various features: crystalline nature, high surface area and the capacity to adsorb free radicals. Besides, changing the Si/Al ratio during the synthesis their adsorption properties can be modified. Taking into account these characteristics, considerable attention was focused on the production of zeolite membranes capable to recognize, in a continuous way, molecules on the basis of different size and shape and adsorption properties [52]. Zeolite membranes are also interesting as membrane reactors due to their thermal and chemical resistance [53]. Self-standing zeolite membranes are very fragile and so the film is grown on inorganic (e. g. ceramic and metal) and organic (e. g. plastic and wood) supports to confer mechanical resistance to the membranes. The application of these membranes at industrial level is hindered by the poor synthesis reproducibility and for their high costs (influenced for about 70% by the support price). Another big problem is the presence of intercrystalline defects into the zeolitic layer that has negative effect on the separation performance of the membranes. Considering the difficulties referred to above, the scale-down of the zeolite membrane systems seems to be a good way to solve the problems mentioned before. In fact, in these last years, different papers illustrated the applications of zeolite layers in small and micro-scale applications (as for example microreactors and sensors).

The common methods used for the preparation of the membranes are: the one-step and the secondary growth. However, the second one is more reproducible because separating the nucleation from the crystal growth permits the optimization of each step independently. This method presents two steps: seeding and growth. In the first, zeolite nano-particles are deposited on the support surface. Subsequently, the hydrothermal treatment favors the growth of the crystals present on the support. Very critical step is the seeding, in fact, it influences the quality of the membranes. Many seeding procedures (some of these are listed in Table 10.4) improve the quality of the membranes with respect to those prepared with the one step method. Among these, however, those more controllable provide the filtration of a zeolite water slurry through the support. Considering the two filtration modes (dead-end and cross-flow) a more thin zeolite layer on the support is formed with the second mode because the zeolite slurry is pumped tangentially along the surface of the support. In addition, when tubular membranes are prepared, it is necessary to combine the filtration with the support rotation and tilting to achieve a uniform zeolite layer on the whole surface area available [57].

Table 10.4 Seeding procedure reported in various papers.

Seeding procedure	References
Rubbing	54
Dip-coating	55
Crossflow filtration	56
Dead-end	57

The morphological quality of the zeolitic membranes is evaluated by using different techniques. Top-view, cross-section and crystals size are observed by scanning electron microscopy (SEM). Energy dispersive Xray (EDX) is used to determine the bulk Si/Al and the penetration of siliceous species in the porous support. The X-ray diffraction analysis is used to identify the zeolite topology present on the support. The three-dimensional network of the crystalline grain boundaries is observed with the fluorescence confocal optical microscopy. This technique detects also the internal defects of the zeolite layer that are not observed with the SEM analysis.

10.4.1 Zeolite Membranes In Pharmaceutical Field

Enzymes as catalysts are usually used in a batch reactor in free form. However, these bio-catalysts are sensitive to inactivation by pH and temperature extremes, organic solvents and mechanical stress. To extend the use of these biocatalysts for practical applications, the technology of enzymes immobilization on suitable supports was developed. Molecular sieves are very attractive supports for enzyme immobilization because they offer interesting properties such as: high surface area, hydrophobic or hydrophilic behavior, electrostatic interaction and also mechanical and chemical resistance. The possibility to immobilize small enzymes within the mesopore structure was demonstrated [58]. However, enzymes with larger dimensions are only adsorbed on the surface of the mesoporous materials. Zeolites were also used for the adsorption of amino acids, proteins and cells [59, 60]. Experimental results indicated that different types of interactions are established between the biomolecules and these inorganic particles. The adsorption of the molecules on the zeolite surface depends on their characteristics as zeolite structure and composition, crystal size etc. Recently, Donato *et al.* [61] demonstrated the possibility to produce L-DOPA by covalent immobilization of tyrosinase on the surface of tubular zeolite membranes. In this approach the zeolite membrane exhibits a coupled actions: as a support for the enzyme and as scavanger to block the radicals produced during the reaction. The experimental results showed that the covalent cross-linking of tyrosinase was an efficient method for the enzyme immobilization. In fact, the specific activity of the tyrosinase immobilized on zeolite membrane surface is 2.76 times higher than that obtained with the free enzyme. This result can be attributed to the scavenger action of the zeolite versus the radicals that have a negative effect on the enzyme activity.

The problems (reproducibility during the synthesis and high costs) present in the zeolite membranes field can be overcome dispersing the zeolites in a host polymer (mixed matrix membranes (MMMs)). These systems exhibit higher selectivity and permeability than the pure polymers because combine their processability and flexibility with the typical characteristics of the inorganic materials (zeolites, carbon molecular sieves, silica and carbon nanotubes). In the preparation of these membranes, a key factor is represented by the affinity between the two phases. In particular, the MMMs do not present defects at the polymer-zeolite interface when rubbery polymers are used due to the high mobility of the polymeric chains. On the contrary, the membranes present defects using glassy polymers as a consequence of the rigidity of their chains. New NaAgY-PVDF membranes, prepared by phase inversion method, exhibited an increase of the water membrane permeability (zeolite increases surface hydrophilicity) and of the fouling resistance (Ag^+ reduces propagation rate of the bacteria) [62]. In fact, it is well known that silver is a quite inert and safe metal with a broad spectrum of antimicrobial properties. Besides, zeolites are good inorganic reservoirs for these ions regulating their release. These inorganic particles give also the possibility to change the surface properties of the polymer. The use of the zeolites as membranes for antimicrobial applications is reported in Table 10.5. Lately, for the first time, Algieri *et al.* [63] demonstrated the possibility to use the mixed matrix membranes for the

Table 10.5 Zeolites used for antimicrobial application.

Zeolite	Membrane Type	References
AgNa-Y	MMMs	64
AgNa-A	MMMs	65
AgNa-Y	ZMs*	66

ZMs*: Zeolite membanes

controlled release of drugs. Experimental results showed as adding the NaX zeolite in the PDMS-based membranes a more linear and slower release of the ibuprofen was obtained.

10.4.2 Zeolite: Materials For Sensors

Sensor is a device that converts chemical or biochemical information into an analytical signal. Very important sensor features are: selectivity, sensitivity, speed of response and stability. Zeolites and zeolite membranes for the characteristics referred to before are good candidates as enzyme-based biosensors and gas sensors. Biosensors based on immobilized enzymes give the possibility to solve numerous problems such as: loss of enzyme, reduced enzyme stability and sensitivity. Considering the intrinsic characteristics of the zeolites, the enzymes immobilization on their surface permits to obtain microdevices with high-sensitivity and long life properties. For example, a new approach for the determination of the urea was developed preparing electrodes coated with zeolite films [67]. Experimental results showed that an increase of the Si/Al ratio of the zeolite enhanced the response of the electrode. This is due to the increase of the hydrophobicity that improves the adsorption of the enzyme on the zeolite surface. In different papers, present in the open literature, it was also demonstrated that the presence of zeolite in gas sensors determines an increase of their selectivity versus different gas species. Semiconductor gas sensors coated with zeolite (MFI and LTA) films were prepared and used to sense different species (methane, propane and ethanol) [68]. The response of these sensors was compared with a sensor without the zeolite coating. Experimental data indicated as the sensors with the zeolite reduced the response versus the paraffine but at the same time increased the selectivity to the alcohol. On the contrary, the sensors without the zeolite layer did not distinguish between the different molecules. Rauch and Liu [69] prepared a zeolite NaA film to cover the surface of the sensor for oxygen recognition. The experimental results showed that the sensor without zeolite answered strongly to both oxygen and carbon dioxide. On the contrary, the sensor covered with the zeolite film has the capability to discern between the two gases (O_2 passes through the zeolite film more easily than the CO_2) even if the response to the oxygen is decreased. Other applications of the zeolites as biosensors and gas sensors are reported in Table 10.6.

These examples show very promising micro-scale applications of the zeolite membranes where the intrinsic properties of the zeolites are exploited. A big advantage of the scale-down is the possibility to prepare defect free zeolite layers reducing the membrane area. Besides, the production of

Table 10.6 Zeolites used as sensors.

Zeolite	Membrane Type	Applications	References
NaA BEA	ZMs*	Gas Sensor	63
MFI LTA	ZMs*	Gas Sensors	64
Clinoptiolite	MMMs	Biosensors	65

ZMs*: Zeolite membranes

specific zeolite films on sensor substrates generates new possibilities for increasing the molecular recognition of many molecules for a wide range of practical applications.

10.5 Functionalized Particles-loaded Membranes For Selective Separation Based On Molecular Recognition

Polymeric matrices containing functionalized micro-nanospheres or micro-nanocapsules represent an interesting and innovative approach in molecular recognition-driven membrane separation (Fig. 10.4).

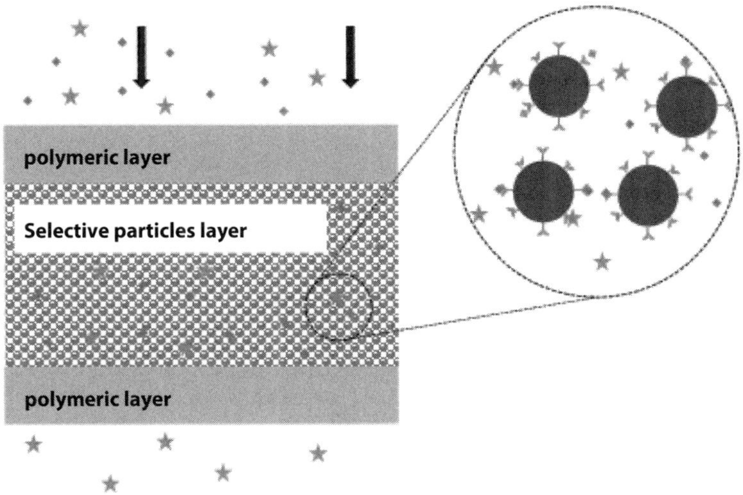

Figure 10.4 Schematic representation of particles-loaded membranes. target molecules are selectively captured by particles having specific recognition properties

The composite material shows the same affinity for the target molecule as does the free micro-nanospheres or micro-nanocapsules. On the other hand, the incorporation of particles within a membrane matrix is a versatile and effective way to simplify the handling of micro- and nanoparticles (e.g. no centrifugation) while keeping their functionality which depends on their large surface area. Particles loaded membranes, depending upon the particles selectivity properties, have a wide variety of applications. Applications include separation technology, sensors and controlled release.

Different strategy have been developed to produce particles-loaded membranes:

- A suspension of dissolved polymer and a particles dispersion are cast as a flat film or spun into a fiber and then solidified by phase inversion and particles are entrapped inside the matrix during coagulation;
- A particles suspension is deposited on the membrane surface and after solvent evaporation the membrane–particles system was lightly washed with distilled water to remove not perfectly attached particles;
- A layer of charged particles is deposited onto the membrane surface oppositively charged and a multilayer composite membrane is obtained by electrostatic self-assembly (ESA) technique;
- A particles suspension and an opposite charged polyelectrolyte solution are alternatively filtered through the membrane and a multilayer composite membrane is obtained by a dynamic layer-by-layer technique.

Adsorbent like ion exchange particles or carbon could be dispersed within the membrane matrix to achieve a specific molecular recognition. Imprinted particles or particles having stimulus-responsive properties (pH, temperature, biochemical stimulus) could be also used. Such particles could be immobilized in solid or hollow fibers porous matrix. The different geometries allow organizing the particles into a 3D superstructure with tailored properties with potential advances in intensified and integrated separation and conversion processes.

Molecularly imprinted polymer microspheres have been used in the preparation of composite membranes with specific recognition abilities. Lehmann *et al.* [73] introduced a new approach to produce composite membrane. Firstly, a chiral amino acid derivative imprinted nanoparticles were synthesized by mini-emulsion polymerization. Then, a water suspension of nanoparticles was pressed through a polyamide membrane, thereby depositing the nano-particles on the membrane surface. Finally, the nanoparticles layer were covered with another porous polyamide membrane. Although the multilayer composite membranes containing imprinted nanoparticles demonstrated high selective rebinding because of the presence of a dense particle layer, the flow rate is limited and improvement was required. For large-scale applications a promising technique is the immobilization of molecularly imprinted polymeric microspheres in membranes made of electrospun nanofibers. These membranes have a very large porosity required for an efficient access of the filtrate. The substrate can easily access to the functional molecularly imprinted polymers surface because the smaller nanofibers not cover the entire surface of the larger microspheres. The use of electrospinning technique to encapsulate premade molecularly imprinted nanoparticles into composite polymer nanofibers is reported in several papers. Molecularly imprinted nanoparticles were prepared with theophylline and 17α-estradiol as template compounds using a precipitation polymerization method of methacrylic acid (MAA) and trimethylolpropane trimethacrylate (TRIM) [74]. The nanoparticles were encapsulated in Poly(ethylene terephthalate) (PET) nanofibers by electrospinning. The composite nanofibers maintained a favorable molecular recognition capability. Molecularly imprinted microspheres have been synthesized by precipitation copolymerization of methacrylic acid (MAA) and divinylbenzene (DVB) for the molecular recognition of (–)-cinchonidine, an alkaloid having antimalaria activity [75]. The microspheres were immobilized within a polyacrylonitrile (PAN) nanofiber membrane by electrospinning. Immobilized and free microspheres demonstrated the same target affinity toward (–)-cinchonidine. Composite membrane was obtained through the deposition of theophylline imprinted poly(methyl methacrylic acid-co-acylic acid) (PMMACAA) nanoparticles in the bulk or on the surface of PMMACAA porous support membranes fabricated by the phase inversion process [76]. The binding capacity of the composite membrane was 6 times that of the membrane with non-imprinted nanoparticles. Poly(acrylic acid-co-methylmethacrylate) membranes superficially modified by the deposition of cholesterol-imprinted methacrylic acid nanospheres have been produced to develop a device for extracorporeal affinity adsorption for atherosclerosis treatment [77].

Composite membranes, obtained by self-assembled multilayer of chitosan microspheres and PAA (poly(acrylic acid)) onto charged surface of polyacrylonitrile (PAN) membranes, have been used for wastewater treatment [78]. The layer-by-layer self-assembled deposition of chitosan microspheres was proposed as a new approach to functionalize membranes with high adsorption capability for metal ions such as copper ions. The binding ability for copper ions of the multilayer composite membranes is mainly due to the amine groups of chitosan which serve as coordination sites for the sequestration of copper ion and it is influenced by the pH. In particular, multilayer composite membranes with chitosan microspheres/PAA showed relatively higher adsorption capability compared to multilayer composite membranes with chitosan/PAA. The incorporation of ion exchange particles into a porous polymeric membrane is demonstrated to be suitable to isolate biomolecules [79, 80]. Different types of Lewatit ion-exchange resins were incorporated into an ethylene–vinyl alcohol copolymer porous structure as a particulate material to prepare heterogeneous mixed matrix adsorber membranes to separate two similarly sized proteins, bovine serum albumin (BSA) and bovine hemoglobin (Hb). A glucose-sensitive polymeric composite membrane was developed by utilizing pH-sensitive poly nanoparticles and immobilized enzyme [81]. Membranes were cast from a mixture of glucose oxidase (GOD), catalase, and poly(N-isopropylacrylamide-co-methacrylic acid) (poly(NIPAm/MAA)) nanoparticles dispersed in a solution of a hydrophobic polymer. The glucose sensitivity was introduced by GOD and catalase while the insulin release is a result of shrinking or swelling of the embedded pH-sensitive nanoparticles. Insulin release is controlled by glucose concentration. A polyethersulfone (PES)/polyvinylpyrrolidone (PVP) polymer blend was used for the preparation of the macro-porous membrane matrix and activated carbon is used as adsorptive particle for blood purification [82]. Dual-layer mixed matrix membranes consisting of a particle-free membrane layer and a mixed-matrix membrane layer, containing activated carbon, demonstrated diffusion and adsorption capacity of uremic retention solutes in one step.

The selective adsorption of target molecules on particles is of great interest as this adsorption enhances the possibility of removing these with low pressure filtration such as microfiltration and ultrafiltration. The main principle of this process is to increase the size of molecules to remove by forming a complex with polymer [83] or surfactant [84]. The method was used for contaminants removal such as metal or endocrine disrupters or pharmaceuticals. The polymers widely used for metal removal include weakly basic, cationic chitosan, poly(ethylenimine) and poly(diallyl dimethylammonium chloride) [85]. Cationic or anionic surfactants in form of micelles are used for the removal of inorganic pollutants. The main concern is to find a suitable polymer or emulsifier to achieve effective complex formation with a target molecule. Particles with high electrical potential on their surface have been used to bound anionic or cationic pollutants, depending upon their charge characteristic.

10.6 Biphasic Enzyme Membrane Systems with Enantioselective Recognition Properties ror Kinetic Resolution

Many interesting biocatalytic reactions involve organic compounds that are poorly soluble in water and they are carried out in biphasic systems. A biphasic bioreactors consists of a dispersed organic phase, containing the hydrophobic substrate to be converted, a continuous aqueous phase in which the biocatalyst is dissolved and the product is extracted. Lipases are the most used enzymes in synthetic organic chemistry, catalyzing the chemo-, regio- and/or stereo-selective hydrolysis of esters or the reverse reaction in organic solvents. Biotechnological applications of lipases in the synthesis of many organic molecules in non-aqueous media have rapidly increased. One of the most important characteristics of lipases is their activation by oil–water interfaces (interfacial activation) [86, 87]. Lipases exhibit recognition of enantiomeric molecules with high enantioselectivity. This important feature can be used in kinetic resolution of racemic mixtures achieving theoretical yield of 50% as well as for asymmetrization of a prochiral or *meso* compound with quantitative yield. Lipase catalyzed kinetic resolution and asymmetrization processes are well documented in the literature [88–90].

Some interesting examples were reported in which the racemic resolution is obtained in a biphasic enzyme membrane systems [91, 92]. In this case lipase was immobilized on polymeric membranes to selectively convert S-naproxen esters from racemic mixtures in S-naproxen acid.

The biocatalytic membranes thus created work as a membrane contactor, which keeps in contact the organic and aqueous phase at the membrane level avoiding dispersion of one phase into another. The organic solvent contains the substrate and the aqueous phase extracts the reaction product. This configuration regulates the interfacial activation of the enzyme, promotes the contact between the enzyme and the substrate and favours the separation of the product into the aqueous phase as it is formed. Therefore, the combination of enatioselective properties of lipase and separation properties of membrane permits the production and simultaneous separation of optically pure S-enantiomer.

A further improvement of the biphasic enzyme reactor concept using lipase for enantioselective transformation has been reported by Giorno and co-workers [92], i.e. an emulsion enzyme membrane reactor. Here, the enzyme distribution at the organic/water interface within the pores is achieved by stable oil-in-water emulsion, prepared by membrane emulsification. The two-separate phase membrane reactor was obtained by immobilizing an oil-in-water stable emulsion, containing the enzyme at the O/W interface, within the membrane. Therefore the emulsion+enzyme-loaded membrane was used as a contactor between an organic and an aqueous phase. In this way, a configuration like organic/emulsion+enzyme-loaded membrane/aqueous phases was obtained where the membrane contains the enzyme at the organic/aqueous interface of the immobilized oil-in-water emulsion.

The results showed that the observed reaction rate was much higher for the emulsion+enzyme-loaded membrane reactor compared to the enzyme-loaded membrane reactor (in absence of emulsion). The presence of the emulsion in the biocatalytic membrane modified the internal membrane liquid phase and improved reaction rate, enantioselectivity and mass transfer of ester substrate through the porous membrane matrix [91, 92].

10.7 Membrane Surface Modification

Today, the demand for membranes with specific properties is increasing in many industrial fields such as: bioseparation, gas separation, water treatment, biomedical, microelectronics, etc.. Zeman and Zydney reported that almost 50% of all microfiltration (MF) and ultrafiltration (UF) membranes on the market were surface modified [93]. Considering that the membrane performances also depend on the surface properties, more attention has been given to their modification. The goal of membrane surface modification is to introduce new properties, separation or catalytic, for improving the selectivity; for creating an entirely novel separation function; for minimizing the adsorption / adhesion so that to reduce fouling phenomena; or improving molecule attachment so that to modify surface energy and membrane wettability.

The different ways to obtain surface modifications proposed in the literature, e.g. coating, self- assembly, chemical treatment, plasma treatment and surface graft polymerization, are briefly described below. These methods can be useful to assist the preparation of membrane separations based on molecular recognition described in the previous paragraphs.

10.7.1 Coating

Coating is a physical modification method. The membrane surface can be modified depositing on it a different polymer. This deposition can be obtained via one of the following mechanisms:

1. Adsorption /adhesion: the membrane surface is modified by putting in contact one side of the membrane with a solution of a different polymer, then after solvent evaporation, a thin film of a different polymer forms a new surface. Physical adsorption can occur on the surface and inside the pores on the pore walls.
2. Interpenetration with interphase obtained mixing the membrane material with the new functional materials
3. Mechanical interpenetration of a new material and the pore structure of the membrane.

Membranes modified by coating method are employed in ultrafiltration process as antifouling membrane [94] in gas separation process to improve membrane separation performances [95] in the vapor separation of a wide variety of volatile compounds [96] Indeed, the PVDF hollow-fiber membrane coated with PDMS exhibited very high removal efficiency (>90%) for all the following VOCs, benzene, chloroform, acetone, ethyl acetate, and toluene.

The main limitation of this method is that it does not allow to prepare membranes with stable surfaces because the new material on the membrane surface runs away easily [97].

10.7.2 Self-assembly

Self-assembly technique permits to obtains supramolecular systems. It entails self-assembly monolayers (SAMs), in which there is the adsorption of an active surfactant on a solid surface [98] and layer-by-layer (LBL) assembly in which the assembly is obtained by means of alternative adsorption of linear polycations and polyanions. LBL technique has been initiated by Decher *et al.* [99, 100]. It is a coating method based on supramolecular assembly, very suitable for membrane surface modification. This technique not only creates a charged skin layer but also allows for a better control of the thickness, charge density and hydrophilicity of the active skin layer [101]. As a result, the membranes modified by LBL method have recently attracted significant attention for use in pervaporation [102], reverse osmosis [103] and nanofiltration [104] processes. The only condition for the use of the self-assembly method is that the base membrane must be able to adsorb the first polyelectrolyte layer by means of ionic bonds.

10.7.3 Chemical Treatment

The chemical treatment permits to obtain membrane surface modification with functional groups rather stable. The modified membrane retains the original mechanical properties while the interfacial properties are changed. The chemical modification is obtained using a chemical reaction such as oxidation, substitution, addition and hydrolysis [105, 106].

Chemical oxidation involves the use of oxidants such as chromic acid, nitric acid, and potassium permanganate to oxidize the membrane surface introducing oxygen-containing groups onto the surface of the membrane [107]. Redox initiators, such as ferric chloride can also oxidize the surface to create active sites where surface graft polymerization can occur.

A typical substitution reaction is the sulfonation reaction that is the addition of sulfonic groups to the (aromatic) backbone of PS/PES in which a hydrogen atom is replaced by sulfonic acid [108].

10.7.4 Plasma Treatment

Plasma is one of the four states of matter. It is generated by ionization of a gas and can be composed of energetic species such as ions, electrons, radicals and photons. This active species can activate the upper layer of membrane with the possibility to introduces functional groups on the membrane surface. Membrane surface with different properties can be obtained changing the plasma treatment parameter. Plasma gases are CF_4, Ar, H_2, He, Ne, N_2, O_2, CO_2 and the ionized gas can attack the C-C, C-H, C-S bonds, excluding the aromatic C-C, C-H bonds. Surface modification with CO_2 plasma, generally, allows to obtain surface oxidation and the formation of hydrophilic surface [109]. N_2 plasma makes amine, amide, imine and nitrile functional groups on the membrane surface [110]. O_2 plasma leads to incorporation of oxygen containing functional groups such as hydroxyl, carbonyl and carboxyl groups [111]. Instead, to improve hydrophobic property of the membrane surface, high degree of fluorinated compounds such as SF_6, CF_4, C_2F_6 are used as plasma gases [112].

10.7.5 Graft Polymerization

Surface grafting is a chemical modification method. Compared with the physical methods such as coating, the covalent bond between the polymer chain and the membrane surface avoids the desorption phenomena maintaining the chemical stability of modified surface for long-term. The grafting method is generally divided in two classes "*grafting to*" and "*grafting from*" based on different strategies of membrane functionalization. In the "*grafting to*" strategy the polymer containing the reactive groups at the end or side chains are covalently coupled to surfaces, while in the "*grafting from*" strategy, an initiation, due to active species on the membrane surface, permits the polymerization of monomer from the surface.

According to the different methods used for the surface activation this technique is classified in chemical, radiation, photochemical and plasma-induced grafting. The *chemical grafting* involves the activation of functional groups, by free radical or ion, on the membrane surface and after the reaction with monomers or macromolecules. A very effective method to generate radicals under mild conditions is the use of redox initiators. Redox initiation takes place when an oxidant, such as persulfate, and a reductant, such as sulfite, are present in the reaction mixture. Generally the used redox system is $K_2S_2O_8/Na_2S_2O_5$, which generates radicals on the membrane surface for subsequent grafting [113].

The *radiation grafting* can be performed in two ways: (1) pre-irradiation and (2) mutual irradiation method. In the pre-irradiation method, the membrane surface is first irradiated and then the formed radical is grafted with a monomer. In the mutual irradiation technique, membrane and monomer are irradiated simultaneously, to form free radicals and subsequent graft polymerization. Different parameters, such as, the properties of membrane and monomer, concentration of the monomer, the duration of radiation, reaction temperature, affect the surface modification by radiation grafting. Besides, high-energy radiation passes through the upper layer of the membrane, and may change the physical or chemical properties of the substrate [114].

The *photochemical grafting* is often used because of its simplicity and low cost. The mechanism of photochemical grafting is the following. When groups on the membrane surface absorb light, it enters in an excited state, which may generate reactive radicals and then initiate the grafting process. This technique does not change the properties of the original polymer. The grafting process by a photochemical technique can proceed in two ways: with or without a photo-initiator, which can promote the generation of reactive radicals. The mechanism 'without photo-initiator' involves the generation of free radicals on the backbone, which react with the monomer free radical to form a grafted co-polymer. On the other hand, in the mechanism 'with photo-initiator', [115] the photo initiator forms free radicals, which can abstract hydrogen atoms from the base polymer, producing the radical sites required for grafting.

The *Plasma grafting* is a combination of plasma treatment and grafting polymerization. Plasma-grafting polymerization is often employed to modify the surface hydrophilicity of a membrane. The creation of a plasma can be carried out in a microwave plasma generator or an induction coupled radiofrequency plasma generator [116]. When a polymeric membrane is exposed to plasma, radicals or peroxides are created on the surface. These radicals initiate polymerization reactions when they are put in contact with monomers in the liquid or gas phase. Then, grafting of monomers can be induced by using or an aqueous solutions at 50°C, through thermal decomposition of peroxides generated during plasma treatment or in gas phase, grafting in the gas phase, where after completion of plasma treatment, the plasma flow is interrupted, and the monomer gas is introduced in the reaction room. In the plasma polymerization procedure, both plasma and monomer gas are introduced at the same time. Grafting in the gas phase proved to yield superior results. Since the plasma and membrane surface produces radicals or peroxides near to the surface of the membrane, the polymerization by plasma grafting is restricted close to the surface. It is usually conducted by first exposing a membrane to a plasma such as argon, helium, or nitrogen for a short time (a few seconds), introducing many radicals onto the membrane surface. Afterwards, the membrane is brought into contact with the vapor of a monomer at an elevated temperature for a period of time. Oxygen in the monomer vapor or dissolved in the monomer solution inhibits the reactions and should be avoided.

These techniques permit to modify the physical-chemical properties of the membrane surface and in some cases the membrane pore walls. These properties can be used to improve the membrane functionalities, such as binding recognition molecules, affinity ligands, enzymes, functionalized nanoparticles (or the nanoparticle themselves can be functionalized using the mentioned techniques).

References

1. K. Mosbach, O. Ramstrom, *Biotechnology* Vol. 14 p163 1996.
2. S.A. Piletsky, I.Y. Dubey, D.M. Fedoryak, and V.P. Kukhar, *Biopolymer Kletka* Vol. 5 p55 1990.
3. F. Tasselli, L. Donato and E. Drioli, *Journal of Membrane Science* Vol. 320 p167 2008.
4. V. Kochkodan, *Advances in Colloid and Interface Science* Vol. 159 p180 2010.
5. C. K. M. Faizal and T. Kobayashi, *Polymer Engineering and Science* Vol 6 p1085 2008.
6. M. Ramamoorthy and M. Ulbricht, *Journal of Membrane Science* Vol 217 p207 2003.
7. L. Donato, A. Figoli and E. Drioli, *Journal of Pharmaceutical and Biomedical Analysis* Vol 37 p1003 2005.
8. F. Trotta, E. Drioli, C. Baggiani and D. Lacopo, *Journal of Membrane Science* Vol 201 p77 2002.

9. J. Yu, X. Hu, D. Li and C. Jiao, *Frontiers of Earth Science in China* Vol 3 p480 2009.
10. M.H. Lee, T.C. Tsai, J.L. Thomas and H.Y. Lin, *Desalination* Vol 234 p126 2008.
11. A.L. Hillberg, K. R. Brain and C.J. Allender, *Journal of Molecular Recognition* Vol 22 p223 2009.
12. L. Donato, F. Tasselli and E. Drioli, *Separation Science and Technology* Vol 45 p2273 2010.
13. T.A. Sergeyeva, O.V. Piletska, S.A. Piletsky, L.M. Sergeeva, O.O. Brovko, and G.V. El'ska, *Materials Science and Engineering C* Vol 28 p1472 2008.
14. R. Kalim, R. Schomäcker, S. Yüce and O. Brüggemann, *Polymer Bulletin* Vol 55 p287 2005.
15. S. G. Del Blanco, L. Donato and E. Drioli, *Separation and Purification Technology* Vol 87 p40 2012.
16. M. Yoshikawa, A. Tanioka and H. Matsumoto, *Current Opinion in Chemical Engineering* Vol 1 p18 2011.
17. I.A. Chernova, K.Y. Gurevich, *J. Membr. Sci* Vol 113 p161 1996.
18. M. Amounas, C. Innocent, S. Cosnier, P. Seta, *J. Membr. Sci* Vol 176 p169 2000.
19. E. Klein, *J. Membr. Sci*, Vol 179 p127 2000.
20. H. Zou, L. Quanzhou, D. Zhou, *J. Biochem.Biophys. Methods* Vol 49 p199 2001.
21. R. Axen, J. Porath, S. Ernback, *Nature* Vol 214 p1302 1967.
22. L. Sundenberg, J. Porath, *J. Chromatogr.* Vol 90 p87 1974.
23. J. Turkováa, J. Vajčnera, D. Vančurováa and J. Štamberg, *Collect. Czech. Chem. Commun* Vol 44 p3411 1979.
24. P.J. Neame, J. Parikh, *Applied Biochemistry and Biotechnology* Vol 7 p295 1982.
25. Scouten W.H., *Affinity Chromatography* New York: Wiley p42 1981
26. E. Klein, *Affinity Membranes* New York, Wiley 1991.
27. I.M. Chaiken, *Analytical affinity chromatography* Boca Raton FL: CRC Press 1987.
28. T.M. Philips, *Analytical techniques in immunochemistry* New York: Marcel Dekker pp50–58, 1992.
29. D.C. Kim, D.K. Kang, *Sensors* Vol. 8, p6605, 2008.
30. A. Bhardwaj, J. Lee, K. Glauner, S. Ganapathi, D. Bhattacharyya, D.A. Butterfield, *J. Membr. Sci.* Vol 119, p241 1996.
31. D.A. Butterfield, D. Bhattacharyya, S. Daunert, L. Bachas, *J. Membr. Sci.* Vol 181 p29 2001.
32. S. Dattaa, P.D. Rayb, A. Nathc, D. Bhattacharyya, *Journal of Membrane Science* Vol 280 p298, 2006.
33. S.R. Ahmed, A.T. Lutes, T.A. Barbari, *Journal of Membrane Science* Vol 282 p311 2006.
34. J. Romero, A. Zidney, *Biotech. Bioeng* 77 pp256–265 2001.
35. S. Poncet, J. Randon, J. Rocca, *Sep. Sci. Technol.* Vol 32 p 2029 1997.

36. J. Romero, A.L. Zydney, *J. Mem. Sci* 209 p107 2002.
37. P. Hadik, L.P. Szabò, E. Nagy, Z. Farkas, *Journal of Membrane Science* Vol 251 p223, 2005.
38. A. Higouchi, A. Hayashi, N. Kanda, K. Sanui, H. Kitamura, *Journal of Molecular Structure*. Vol 739 p145 2005.
39. A.M. Urmenyi, A.A. Poot, M. Wessling, M.H.V. Mulder, *Journal of Membrane Science* Vol 259 p91 2005.
40. K.C. Hou, R. Zaniewski, *J. Chomatogr*. Vol 525 p297 1990.
41. C. Weiner, M. SáRa, G. Dasgupta, U. B. and Sleytr, *Biotechnology and Bioengineering* Vol 44 p55 1994.
42. L.R. Castilho, F.B. Anspach, W. Deckwer, *J. Membr. Sci*. Vol 207 p253 2002.
43. N. Kubota, Y. Nakagawa, Y. Eguchi, *J. Appl. Polym. Sci*. Vol 62 p1153 1996.
44. N. Kubota, N. Tatsumoto, T. Sano, *Carbohydrate Polymers* Vol 40 p107 1999.
45. L. Hamilton, C. T. Kelly, W. M. Fogarty, *Enzyme and Microbial Technology* Vol. 26 p561 2000.
46. E. Fontananova, A. Basile, A. Cassano and E. Drioli, *Journal of Inclusion Phenomena and Macrocyclic Chemistry* Vol 47 p33 2003.
47. P. Vretblad, P. *FEBS Lett* Vol 47 p86 1974.
48. E. G. Yanes, S. R. Gratz, R. M. C. Sutton and A. M. Stalcup, *Fresenius Journal of Analytical Chemistry* Vol 369 p412 2001.
49. R. Li, C. Zhao, W. Zhang, L. Shi andJ. Tang, *Journal of Natural Sciences* Vol 14 p362 2009.
50. X. Ma, Z. Chen, R. Chen, X. Zheng, X. Chen, *Polym. Int*. Vol 60, p1455 2011.
51. H. Asanuma, T. Akiyama, K. Kajiya and T. Hishiya, M. Komiyama, Molecular imprinting of cyclodestrin in water for the recognition on nanometer-scaled guests, *Analytica Chimica Acta* 435 (2001) 25–33.
52. E. E. McLeary, J. C. Jansen and F. Kapteijn, *Microporous and Mesoporous Materials*, Vol 90 p198 2006.
53. P. Bernardo, C. Algieri, G. Barbieri and E. Drioli, *Catalysis Today*, Vol. 118, p. 90, 2006.
54. Y. Hasegawa, K.I. Sotowa, K. Kusakabe and S. Morooka, *Microporous and Mesoporous Materials*, Vol. 53, p. 37, 2002.
55. M.C. Lovallo, A. Gouzinis and M. Tsapatsis, *AICHE Journal*, Vol.44, p. 1903, 1998.
56. C. Algieri, P. Bernardo, G. Barbieri and E. Drioli, *Microporous and Mesoporous Materials*, Vol. 119, p. 129, 2009.
57. A. Huang, Y. S. Lin and W. Yang, *Journal of Membrane Science*, Vol 245 p41 2004.
58. H. Takahashi, B. Li, T. Sasaki, C. Miyazaki, T. Kajino and S. Inagaki, *Chemistry of Materials* Vol 12 p.3301 2000.
59. K. Sakaguchi, M. Matsui and F. Mizukami, *Applied Microbiology and Biotechnology* Vol 67 p306 2005.
60. J. E. Krown and M. Tsapatsis, *Langmuir* Vol 21 p8743 2005.
61. L. Donato, C. Algieri, V. Miriello, R. Mazzei, G. Clarizia and L. Giorno, *Journal of Membrane Science* Vol 407 p86 2012.

62. C. Liao, P. Yu, J. Zhao, L. Wang, and Y. Luo, *Desalination* Vol 272 p59 2011.
63. C. Algieri, E. Drioli and L. Donato, *Journal of Applied Polymer Science* Vol 128, p754 2013.
64. D. Zampino, T. Ferreri, C. Puglisi, M. Mancuso, R. Zaccone, R. Scaffaro and D. Bennardo, *Journal of Material Science* Vol 46 p6734 2011.
65. P. Kaali, E. Strömberg, R. E. Aune, G. Czél, D. Momcilovic and S. Karlsson, *Polymer Degradation and Stability* Vol 95 p1456 2010.
66. S. Sabbani, D. G. Perez, A. Nagy, W. J. Waldman, D. Hansford and P. K. Dutta, *Microporous and Mesoporus Materials*, Vol. 135, p. 131, 2010.
67. S. K. Kirdeciler, E. Soy, S. Öztürk, I. Kucherenko, O. Soldatkin, S. Dzyadevych and B. Akata, *Talanta* Vol 85 p1435 2011.
68. M. Vilaseca, J. Coronas, A. Cornet, J. R. Morante and J. Santamaria, *Catalysis Today* Vol 82 p179 2003.
69. W. L. Rauch and M. Liu, *Journal of Material Science* Vol 38 p4307 2003.
70. S. Mintova, and T. Bein, *Microporous and Mesoporous Materials* Vol 50 p159 2001.
71. M. Vilaseca, J. Coronas, A. Cirera, A. Cornet, J. R. Morante and J. Santamaria, *Sensors and Actuators B* Vol 124 p99 2007.
72. M. L. Hamlaoui, K. Reybier, M. Marrakchi, N. J. Renault, C. Martelet, P. Kherraqt and A. Walcarius, *Analytica Chimica Acta* Vol 466 p39 2002.
73. M. Lehmam, H. Brunner, *G.E.M. Tovar* Vol 149 p315 2002.
74. I. S. Chronakis, A. Jakob, B. Hagström, L. Ye, *Langmuir* Vol 22 p8960 2006.
75. R. Büttiker, J. Ebert, C. Hinderling, C. Adlhart, *Chimia* Vol 65 p182, 2011.
76. D. Silvestri, N. Barbani, C. Cristallini, P. Giusti, G. Ciardelli, *J Membrane Sci* Vol 282 p284 2006.
77. G. Ciardelli, C. Borrelli, D. Silvestri, C. Cristallini, N. Barbani, P. Giusti, *Biosens Bioelectron* Vol 21 p2329 2006.
78. H. Fu, T. Kobayashi, *J Mater Sci* Vol 45 p6694 2010.
79. M. Avramescu, M. Gironès, Z. Borneman, M. Wessling, *Journal of Membrane Science* Vol 218 p219 2003.
80. M. Avramescu, Z. Borneman, M. Wessling, *Journal of Chromatography A* Vol 1006 p171 2003.
81. K. Zhang, X. Y. Wu, *Journal of Controlled Release* Vol 80 p169 2002.
82. M.S.L. Tijink, M. Wester, J. Sun, A. Saris, L. A. M. Bolhuis-Versteeg, S. Saiful, J. A. Joles, Z. Borneman, M. Wessling, D. F. Stamatialis, *Acta Biomaterialia* Vol 8 p2279 2012.
83. S. Verbych, M. Bryk, M. Zaichenko, *Desalination* Vol 198 p295 2006.
84. R. Bade, L. S. Hwan, *Journal of Water Sustainability* Vol 1 p85 2011.
85. K. Volchek, E. Krentsel, Y. Zhilin, G. Shtereva and Y. Dytnersky, *J. Membr. Sci.* Vol 79 p253 1993.
86. H. Stamatis, A. Xenakis, F.N. Kolisis, *Biotechnology Advances* Vol 17 p293 1999.
87. P. Reis, K. Holmberg, H. Watzke, M.E. Leser, R. Miller, *Advances in Colloid and Interface Science* Vols 147–148 p237 2009.
88. K. Sakaki, L. Giorno, E. Drioli, *Journal of Membrane Science* Vol 184 p27 2001.

89. S.Takaç and D. Mutlu, *Applied Biochemistry and Biotechnology* Vol 141, p15 2007.
90. H.Y. Lin and S.W. Tsai, *J Mol Catal B Enzym* Vol 24–25 p111 2003.
91. L. Giorno, N. Li, E. Drioli, *Biotechnology and Bioengineering*, Vol. 84, p. 677, 2003.
92. L. Giorno, E. D'Amore, R. Mazzei, E. Piacentini, J. Zhang, E. Drioli, R. Cassano, N. Picci, *Journal of Membrane Science* Vol 295 p95 2007.
93. L. J. Zeman, A. L. Zydney, *Microfiltration and Ultrafiltration: Principles and Applications;* Marcel Dekker: New York, Vol. 69, p. 1479, 1996.
94. M. L. Luo, J. Q. Zhao, W. Tang, C. S. Pu, *Appl Surf Sci,* Vol. 249, p. 76, 2005.
95. N. Widjojo, S. D.Zhang, T. S. Chung, Y.Liu, *J Membr Sci,* Vol. 306, p. 147, 2007.
96. H. Zhen, S. M. Jang, W. K. Teo, K. Li, *J Appl Polym Sci,* Vol. 99, p. 2497, 2006.
97. J.M. Dickson, R.F. Childs, B.E. Mc Carry, D.R. Gagnon, *J Membr Sci,* Vol. 16, p. 25, 1998.
98. E. Ruckenstein, Z.F. Li, *Adv Colloid Interf Sci,* Vol. 113, p. 43, 2005.
99. G. Decher, J. D. Hong, B. Bunsen-Ges, *Phys. Chem.,* Vol. 95,p. 1430,1991.
100. G. Decher, Science, Vol. 277, p. 1232, 1997.
101. T. Kouketsu, S. Duan, T. Kai, S. Kazama, K. Yamada, *J. Membr. Sci.*, Vol. 287, p. 51, 2007.
102. Z. Zhu, X. Feng, A. Penlidis, *Mater. Sci. Eng. C,* Vol. 27, p. 612, 2007.
103. A. Toutianoush, W. Jin, H. Deligoz, B. Tieke, *Appl. Surf. Sci.,* Vol. 246, p. 437, 2005.
104. M.D. Miller, M.L. Bruening, *Langmuir,* Vol. 20 p. 11545, 2004.
105. A.H. Lu, W.C. Li, N. Muratova, B. Spliethoff, F. Schuth, *Chem Commun,* Vol. 51, p.84, 2005.
106. S.S. Barton, M.J.B. Evans, E. Halliop, J.A.F. MacDonald, *Carbon,* Vol. 35, p.1361, 1997.
107. M.G. Carneiro-da-Cunha, J.M.S. Rocha, F.A.P. Garcia, M.H. Gil, *Biotechnology Techniques,* Vol. 13, p. 403, 1999.
108. B. Van der Bruggen, *Journal of Applied Polymer Science*, Vol. 114, p. 630, 2009.
109. D. S. Wavhal, E. R. Fisher, *Desalination,* Vol. 172, p. 189, 2005.
110. M. Bryjak, I. Gancarz, G.Pozniak, W. Tylus, *Eur Polym J,* Vol. 38, p. 717, 2002.
111. D. Tyszler, R. G. Zytner, A. Batsch, A. Brugger, S. Geissler, H. Zhou, D. Klee, T. Melin, *Desalination,* Vol. 189, p. 119, 2006.
112. N. De Geyter, R. Morent, L. Gengembre, C. Leys, E. Payen, S. Van Vlierberghe, E. Schacht. *Chem Plasma Process* (2008) 28:289–298
113. S. Belfer, R. Fainshtain, Y. Purinson, J. Gilron, M. Nyström, M. Mänttäri, *J Membr Sci,* Vol. 239, p. 55, 2004.
114. R. Chennamsetty, I. Escobar, *Langmuir,* Vol. 24, p. 5569, 2008.
115. M. Ulbricht, H. Matuschewski, A. Oechel, H. G. Hicke, J *Membr Sci,* Vol. 115, p. 31, 1996.
116. K. Kato, E. Uchida, E. T. Kang, Y. Uyama, Y. Ikada, *Prog Polym Sci,* Vol. 28, p. 209, 2003.

Part 4

MEMBRANE SENSORS AND CHALLENGED APPLICATIONS

11

Electrospun Membranes for Sensors Applications

Pierangiola Bracco*, Dominique Scalarone and Francesco Trotta

Department of Chemistry and NIS Centre of Excellence, University of Torino, Torino, Italy

Abstract
This chapter summarizes the basic principles of the electrospinning technology and the recent advances in its application to the preparation of one dimension nanostructures. The main advantages of using electrospun membranes to fabricate ultrasensitive sensors are highlighted. The applications of electrospinning in the production of resistive, optical, acoustic wave and amperometric sensors are presented.

Keywords: Electrospinning, nanofibers, nanotubes, core-shell fibers, metal oxide fibers, polymer fibers

11.1 Introduction

Chemical sensors are essential to many applications such as environmental monitoring, food inspection, medical diagnosis, defense and security, etc.

In response to the increasing demand for more timely and reliable detections, considerable advances have been made in the development of ultrasensitive sensors [1–3]. In particular, greater control over the micro- and nanostructure of materials has resulted in sensor materials with increased sensitivities, multiplexing capabilities, or both [4, 5]. Recently, one-dimension (1D) nanostructures have attracted much attention, due to their fascinating properties and applications.

**Corresponding author*: pierangiola.bracco@unito.it

Annarosa Gugliuzza (ed.) Smart Membranes and Sensors, (303–336) 2014 © Scrivener Publishing LLC

There are several techniques to produce nanofibers, from high-volume to highly precise production methods. However, their applicability is often restricted by combinations of limited material ranges, possible fiber assembly, cost and production rate.

Electrostatic spinning or, more commonly, electrospinning has recently emerged as an inexpensive, extremely powerful technique to produce nanofibers membranes with a broad variety of characteristics [6–9]. In the electrospinning process, a polymer solution or melt is subjected to a high voltage electric field and ejected through a nozzle, resulting in 1D nano or micron sized fibrous structures deposited on a collector

The remarkable specific surface area and high porosity make electrospun membranes highly attractive to many applications, including catalysis, filtration, fuel cells, tissue engineering, drug delivery and, notably, ultrasensitive sensors.

Although the first patents on the subject appeared in the US already at the beginning of the last century [10–12], the electrospinning process was not commercially exploited until the early 1990s. At that time, extensive research activities, mainly by Reneker and co-workers [13–16], renewed the interest in this technique by demonstrating its enormous potential in fabricating attractive nanostructures. From then on, electrospinning has experienced an upsurge of interest, attracting both the academia and the industry in a continuously increasing number of theoretical studies and practical applications.

In particular, the distinctive features of the electrospun materials has attracted their application as potential sensing elements and many recent studies have been devoted to explore this possibility. Some useful reviews published on the subject are cited in Ref. [5, 17–23]-

The present work provides an overview of the electrospinning process and summarizes recent advances in the field of electrospun membranes for sensing applications, with a particular focus on the sensing approaches that have taken more advantage of the electrospinning technology.

11.2 Basic Principles of Electrospinning

Almost any soluble or fusible polymer with sufficiently high molecular weight can be electrospun. Nanofibers made of synthetic or natural polymers, polymer blends, nanoparticle-loaded or drug-impregnated polymers and ceramic precursors can be fabricated in a controlled manner with dimensions down to a few nanometers. The small diameter provides a high surface area to volume ratio, and a high length to diameter ratio. Fibers

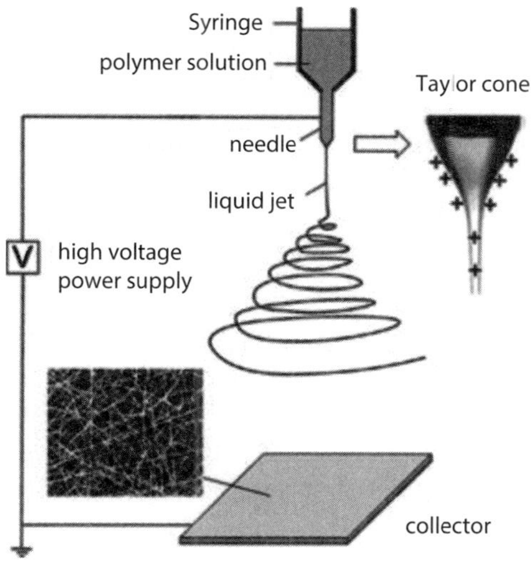

Figure 11.1 Scheme Of A Basic Electrospinning Setup. The Insets Show A Drawing Of The Electrified Taylor Cone And A Typical SEM Image Of Randomly Oriented Fibers Deposited On The Collector (Reprinted With Permission From [6] © 2004, Wiley).

with a variety of cross sectional shapes and morphology such as beaded, ribbon, porous, and core-shell may also be produced [6–8].

A typical laboratory electrospinning setup consists of three major components: a high-voltage power supply, a spinneret (usually a metallic needle), which also serves as an electrode, and a collector (a grounded conductor or another substrate brought into contact with the counter electrode) (Figure 11.1).

A polymer solution (or melt), hosted in a syringe, is usually fed through the spinneret at a constant and controllable rate, with the use of a syringe pump.

The apparatus is often vertically aligned (i.e. the spinneret is placed over the collector, at a distance normally ranging from 10 to 25 cm), but horizontal or even vertical "bottom to top" alignments have also been successfully used.

The application of a high voltage (usually in the range of 1 to 30 kV) causes a cone-shaped deformation (commonly known as Taylor cone) of the pendent drop of polymer solution at the nozzle of the spinneret. The distortion in the direction of the counter electrode is the result of a combination between the electrostatic repulsion between the surface charges and the Coulombic force exerted by the external electric field. If the voltage exceeds a critical value, a jet is formed from the deformed drop, which is ejected towards the counter electrode, undergoing a stretching and whipping process. As the jet travels towards the counter electrode, the solvent evaporates (or the melt solidifies), while the entanglements of the polymer chains prevent the jet from breaking up, and continuous solid fibers with diameters ranging from micrometers to nanometers are precipitated on the counter electrode, as a randomly oriented, nonwoven mat.

To date, the use of this relatively simple technique has allowed to process more than 100 different types of synthetic and natural polymers into fibers with diameters ranging from tens of nanometers to a few micrometers and lengths of kilometers.

A complex set of parameters controls the feasibility of the process, along with the shape and the dimensions of the formed fibers: the properties of the polymer itself (molecular weight, molecular-weight distribution, glass-transition temperature and solubility), those of the polymer solution (concentration, viscosity, electrical conductivity and surface tension), instrumental variables (feed rate of the solution, applied voltage and distance between the electrodes) and ambient parameters (temperature and humidity in the electrospinning chamber).

The electrospinning of polymers from the melt avoids the use of solvents and is, therefore, attractive from the perspective of productivity and environmental considerations. However, as a result of the high melt viscosities of the polymers, fibers with average diameters lower than 1μm and narrow diameter distributions are difficult to be obtained, thereby limiting the applicability of the technique [8].

A brief overview on the influence of the most relevant parameters on the quality of the final product will be given in the next section.

11.3 Control of the Electrospinning Process

11.3.1 Fibers Morphology and Diameter

Fibers with circular cross-section are usually obtained by electrospinning. The control of the diameter has been discussed in a number of publications [8, 9, 24, 25]. It has been proved that the major factors affecting the diameter of electrospun fibers include the concentration and the electrical conductivity of the solution and its feeding rate.

The concentration also determines the spinnability of a solution, that is whether fibers forms or not. For a specific polymer or blend, an optimum range of concentrations exists in which fibers can be electrospun, when all other parameters are held constant (Figure 11.2). The solution

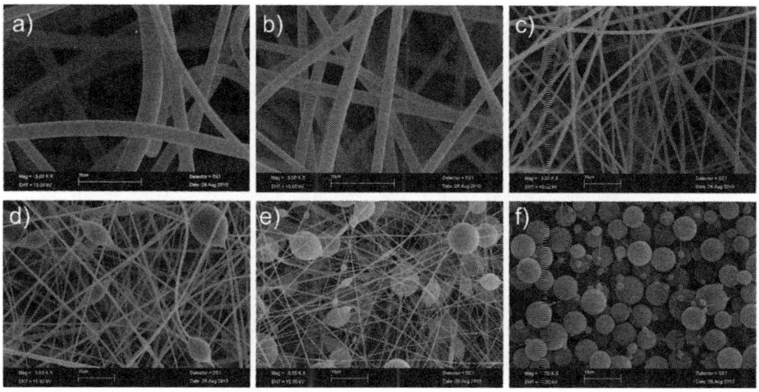

Figure 11.2 Effect Of The Solution Concentration On The Quality And Diameter Of The Electrospun Fibers. PMMA Fibers Spun From a) 30%, b) 28%, c) 26%, d) 22%, e)20%, f) 16% w/w PMMA/Chloroform Solutions.

must be concentrated enough for chain entanglements to occur. In general, the fibers become more uniform and assume a cylindrical shape with increasing the solution concentration; fiber diameters also increase significantly with increasing concentration. However, if the solution is too concentrated, the fibers cannot be formed due to the high viscosity, which makes the control of the solution flow rate through the capillary difficult. On the other hand, at lower concentrations, increasingly thinner fibers are formed. If the solution is too dilute, beads are created along the fiber axis. At very high dilution, the polymer fiber will break up into droplets before reaching the collector, due to the effects of surface tension, and fiber formation no longer takes place.

The presence of beads in electrospun fibers is a common problem. The density of beads usually decreases or beads disappear if more viscous solutions (i.e., higher concentrations of polymer) are used. The formation of beads can also be prevented by adding some salts, thereby increasing the electrical conductivity of the solution, so that fibers with diameters down to a few nanometers can be produced. Conversely, it has been observed that increasing voltages might lead to an increase in the density of beads in the electrospun fibers.

The selection of an appropriate solvent is also critical as to whether fibers are capable of forming, as well as influencing fiber porosity. A volatile solvent must be used, in order to allow sufficient solvent evaporation between the capillary tip and the collector.

By using suitable combination of the mentioned parameters or by appropriately modifying the electrospinning setup, a broad variety of fiber shapes and morphology, as well as ordered fiber arrangements can also be obtained.

11.3.2 Fibers Arrangement, Composition and Secondary Structure

In order to control the structure, morphology and alignment of the fibers, a number of modifications to the basic electrospinning setup have been proposed, mostly involving either the spinneret or the collector.

For example, two nozzles in a side-by-side configuration have been successfully used as spinneret to overcome the problems arisen when trying to electrospin polymer blends in which the polymers of interest are not soluble in a common solvent [26].

Arrays composed of multiple needles have also been proposed [27, 28], mainly to improve the low productivity of conventional electrospinning in scaled-up processes. Indeed, to obtain ultrathin fibers, the polymer solution must be fed at relatively low rates. The use of several needles in parallel alignment, as electrodes in multijet electrospinning, can overcome this drawback and gives an opportunity for the fabrication of electrospun tissues consisting of different fiber materials. The configuration of the arrayed needles needs to be carefully designed, since electric field interference between adjacent nozzles may arise.

Coaxial electrospinning have also been demonstrated as a powerful technique to fabricate more complex architectures, from core-shell to hollow fibers, by using a wide range of materials [29, 30]. In the coaxial configuration, two separate polymer solutions flow through two different capillaries and two concentrically aligned nozzles are used for spinning, so that a smaller fiber can essentially be encapsulated in a larger fiber, leading to what is known as a core-shell morphology (Figure 11.3). In this case, the immiscibility of the core and shell solution plays a crucial role in determining the quality of the final product. This technique has received great interest and has been largely employed in a variety of applications [31–33].

Using a standard apparatus for electrospinning, equipped with a flat counter electrode, the fibers are deposited in a statistical orientation and a nonwoven mat is fabricated by layer-by-layer deposition.

Highly ordered, aligned nanofibers can be beneficial or even essential in some applications. A rapidly rotating drum has been introduced [34] as an alternative collector to obtain aligned electrospun fibers as relatively uniform mats. Parallel fibers can also be produced by modifying the design of

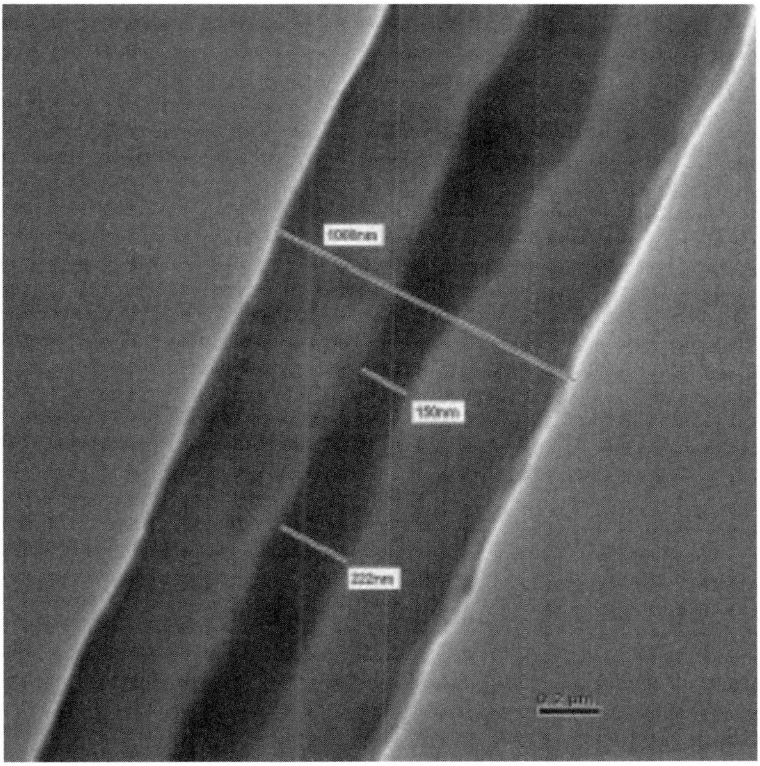

Figure 11.3 TEM Micrograph Of A Core-Shell Fiber Of Co-Electrospun Polyethyleneoxide (Shell) And Poly(dodecylthiophene) (Core). (Reprinted With Permission From [29] © 2003, Wiley WCH).

the collector electrodes, for example with special electrode arrangements consisting of two parallel flat bars [35]. By using a collector consisting of two conductive strips separated by a void gap of variable widths, electrospun fibers could be uniaxially aligned over long length scales during the spinning process. A very high degree of orientation can be achieved with this method.

In general, electrospinning generates smooth fibers with a circular cross section and a solid interior. Nevertheless, it has been observed that modified system designs or appropriate selection of processing parameters can enable the production of nanofibers with specific secondary structures (i.e. hollow, core-shell) and/or with porous surfaces, which may be greatly advantageous for a variety of applications.

Besides by coaxial electrospinning, core-shell nanofibers can also be fabricated from a conventional electrospinning setup, by using a solution containing two polymers that will phase separate as the solvent is evaporated.

The fabrication of hollow nanofibers has been obtained by various techniques. One approach involves the electrospinning of a polymer template, which is then coated with a precursor material, from which the hollow fibers are to be prepared by various deposition methods. Subsequently, the inner section of the electrospun nanofibers is removed by selective dissolution or thermal degradation, and the hollow fibers are obtained [36, 37].

Alternatively, core-shell composite nanofibers, directly produced either by coaxial or by phase-separation electrospinning, undergoes selective removal of the core component, yielding the hollow fibers [30].

The fabrication of porous nanofibers results in a further increase in surface areas which may be beneficial in many applications. It is possible to generate different fiber topologies during the electrospinning process by choosing particular solvents or solvent mixtures, by varying the humidity or by using block copolymers or polymer mixtures. Most commonly, porous nanofibers have been prepared by selective removal of a component from nanofibers made of a composite or blend material [38].

Another broad field of research is the electrospinning of polymer composites, loaded with a variety of organic and inorganic particles of different forms and dimensions. For example, carbon nanotubes (CNTs) have been successfully immobilized in electrospun polymer fibers, pursuing advantages in terms of electrical conductivity or mechanical strength of the resulting product [39, 40]. The benefits of electrospun nanofibers loaded with various metal nanoparticles have also been reported in numerous applications [41–44].

Moreover, electrospun nanofibers have been extensively used in biomedical applications as carriers to encapsulate functional materials, such as drugs, enzymes, viruses and bacteria [24, 31].

Nanofibrous structures of pure inorganic or organic-inorganic hybrid systems were also obtained by electrospinning. A large number of reports describes the preparation of polymer/metal oxide composite solid or hollow nanofibers by electrospinning, in combination with sol–gel processes or by following a polymer-based precursor route [37, 45–50]. In many cases, the composite fibers was converted into metal oxide fibers by subsequent pyrolysis. A broad range of metal oxides, including for example TiO_2 [51–54], SnO_2 [55–57], ZnO [58–60] and various mixtures[61–63] have been employed. Again, different fiber morphologies can be obtained by appropriately controlling the processing conditions, enabling the production of solid, as well as hollow, porous, and core-shell structures.

11.4 Application of Electrospun Materials to Ultrasensitive Sensors

Recent advances on the development of novel sensors have been focused on nanostructured materials to achieve increased surface-to-volume ratios and reduced cross sections. Most approaches are based on conventional three dimensional (3D) and two-dimensional (2D) material architectures, such as thick mesoporous layers or thin films, but new types of one dimensional (1D) nanostructures have became extremely popular in recent years.

The electrospun nanofiber mats exhibit a high surface area to volume ratio (approximately 1 to 2 orders of magnitude higher than that of continuous films [64]), providing more reaction/absorption sites on the sensing surface; in addition, the small diameter provides a short diffusion length, which could enhance the charge transfer and electron conduction along the length direction. Moreover, high porosity, and good interconnectivity are additional features that makes electrospun membranes highly attractive for potential application in ultrasensitive sensors.

The following paragraphs will be devoted to illustrate some of the last advances in the field of electrospun nanofibres based sensing devices.

11.4.1 Metal-Oxide-Based Resistive Sensors

The use of metal oxide semiconductors (MOS) has been very popular in the field of ultrasensitive resistive sensors in the last 40 years. High response value, fast response, quick recovery, excellent stability and simplicity in fabrication are some of the characteristics which made MOS the ideal candidates for high performance sensors. MOS are known for their ability to detect trace concentrations of various analytes via charge transfer interactions between the sensor and chemisorbed species that modify the sensor's resistance.

Rapid and accurate determination of gases is of great practical importance in many branches of chemistry, medicine and in environmental applications. In recent years, many efforts have been devoted to prepare ultrasensitive sensors to detect moisture, O_2, NH_3, HCl, NO_2, CO_2, H_2S, etc.

MOS nanoparticles have been widely investigated due to their high surface area, which provides high response to the detected specie. Yet, one common drawback is represented by their facile aggregation that leads to a significant degradation of the sensing properties. Recently, MOS nanofibers and nanotubes have become very popular because they do not suffer of such degradation. Electrospinning has been largely employed as a versatile method to prepare nanostructured sensing membranes of MOS, such as TiO_2 [65–69], ZnO [70, 71], SnO_2 [72, 73], WO_3 [74], etc .

Zhang *et al* [75] demonstrated the high efficiency for humidity detection of Na^+ doped ZnO nanofiber membranes, prepared via electrospinning and calcination. A solution of polyvinyl alcohol (PVA), Zinc acetate ($Zn(Ac)_2 \cdot 2H_2O$) and NaCl was electrospun to obtain nanofiber membranes, which were then calcined at 550°C for 5 hours to remove the polymer template. The resulting product showed remarkable sensing performances, with high sensitivity and linear response in a range of relative humidity from 11% to 95%; good reproducibility, fast response and recovery time (about 3 and 6 s, respectively), were also obtained.

Hollow ZnO nanofibers were fabricated by electrospinning of a polyacrylonitrile (PAN), polyvinylpyrrolidone (PVP), and zinc acetate composite solution via a facile single capillary technique [76]. Phase separation between the electrospun composite materials occurred during electrospinning and the obtained precursor nanofibers exhibited a core-shell structure (PAN as the core and PVP/zinc acetate composite as the shell). Calcination of the as-electrospun composite eliminated the polymers and resulted in hollow ZnO nanofibers (Figure 11.4 A). The hollow nanofibers displayed excellent ethanol sensing characteristics (Figure 11.4 B), in a wide range of concentration (10–3000 ppm), with notably fast response and recovery times (3 and 5s, respectively).

Hollow ZnO nanofibers were also successfully fabricated by sputtering ZnO onto an electrospun polyvinyl-acetate (PVAc) fibers template, deposited on silicon or alumina substrates [58]. In this case, alignment of the PVAc fibers during electrospinning was obtained by using a modified counter electrode, as discussed in section 11.3.2. Two stripes of aluminum wires were placed along opposite edges of the substrate, connected to the ground terminal of the power supply that applied the electrical field between the nozzle and the substrate. This imposed a directional distribution of the electrical field flux lines between the nozzle and the aluminum wires, facilitating alignment of segments of the fibers across the aluminum wires.

Calcination of the ZnO/PVAc composite fibers resulted in hollow fibers of nanocrystalline ZnO. The inner diameter of the hollow fibers ranged between 100 and 400 nm and their wall thickness varied from 100 to 40 nm from top to bottom. The NO_2 sensing behavior of the hollow quasi-aligned and nonaligned ZnO nanofibers was compared to that of reference ZnO thin films (200 nm thick) directly deposited onto interdigitated alumina.

The hollow fibers were considerably more sensitive than their thin film counterparts, displaying well resolved response signals down to 2 ppm

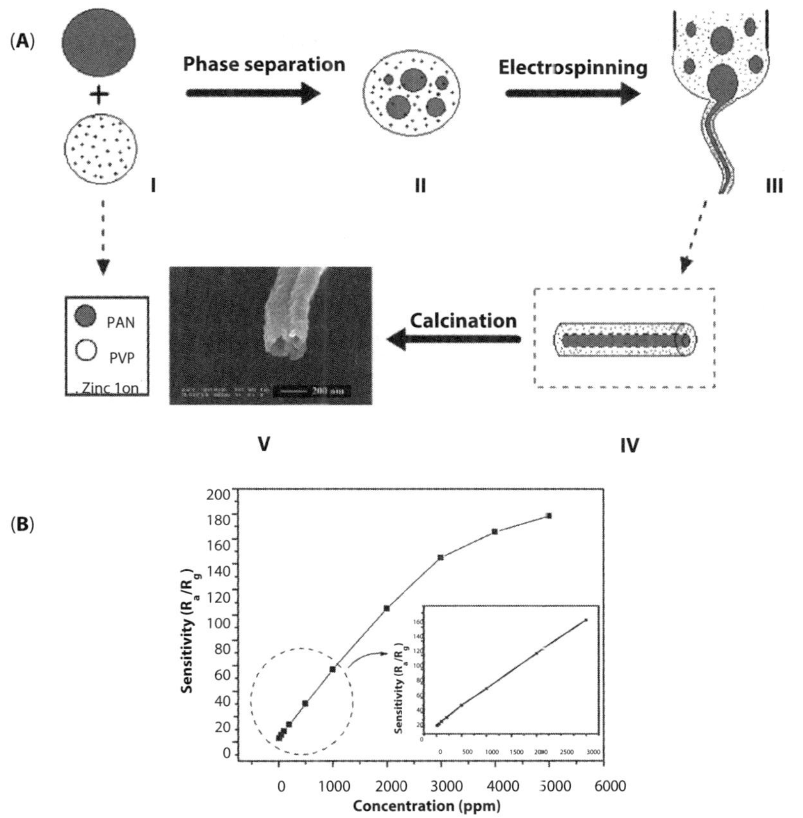

Figure 11.4 A) Scheme Of The Formation Mechanism Of ZnO Hollow Nanofibers. B) Hollow ZnO Nanofibers-Based Sensor Sensitivity Versus Ethanol Concentration; The Inset Shows The Linear Dependence Of The Sensitivity To The Ethanol Concentration. (Adapted With Permission From [76] © 2009, American Chemical Society)

NO_2 concentration. In contrast, the thin film reference sensors did not show a detectable response below 8 ppm, and their sensitivity at higher concentrations was considerably lower compared to the hollow fiber sensors.

The high sensing properties of the hollow MOS fibers were attributed, in part, to their high surface to volume ratio, which is twice as large as that of planar films of the same thickness. Further enhancement of the sensitivity may also result from effects connected with their unique morphology, which yields unusual electronic transport properties. Such effects may include, for instance, constriction of the current to the conductive cores of the fibers, giving rise to 1D transport phenomena. The 1D character of the electronic transport was further enhanced in the latter case for the quasi-aligned fibers, which accordingly demonstrated approximately 2-fold higher sensitivity compared to their randomly oriented counterparts.

Among the MOS, SnO_2 nanostructures are particularly popular due to their high transparency, wide-bandgap semi-conductivity, elevate sensitivity to a wide range of gases and vapors, lower working temperature and adaptability to sense different gases. Qi and coworkers have developed nanofibrous SnO_2 membranes for detection of toluene, ammonia, ethanol and acetone. First [77], they have synthesized SnO_2 nanofibers with a conventional procedure, including electrospinning of a solution of tin dichloride and poly(vinyl pyrrolidone) (PVP) in N,Ndimethylformamide/ethanol, followed by calcination at 600°C to remove the polymer and convert the Sn chloride into SnO_2. The sensor fabricated from these fibers exhibited excellent toluene sensing properties at 350°C, with a linear and fast (1s) response in the range of 10–300ppm. In addition, fast recovery (5 s) and good selectivity to common interference gases were also demonstrated.

In another study[78], the same authors have implemented the SnO_2 nanofibers morphology by adding a block copolymer in the precursor to obtain highly porous nanofibers with enhanced sensing properties. It was found that these procedure dramatically increases the surface area of the fibers (BET surface area was 111.261 vs. 18.025 m^2/g) and creates porous nanofibers with a relative narrow pore size distribution centered at about 5.3 nm. This resulted in more target molecules to be adsorbed on the fiber surface, with an increase in both sensitivity and saturated-detection concentration, as was proved for sensing of ammonia, ethanol and acetone.

In addition, K^+ ion-doped, electrospun SnO_2 nanofibers were also demonstrated to be good candidates to prepare humidity sensors with higher sensitivity, more rapid response and recovery, smaller hysteresis, and better linearity and stability compared to most of the conventional electrical conductivity-based SnO_2 humidity sensors [79].

A number of successful attempts to prepare TiO_2 and TiO_2 composite based resistive sensors by electrospinning for detecting a broad range of analytes was also reported [65, 67–69].

Kim *et al.* [80] fabricated TiO_2 nanofiber mats with high sensitivity to NO_2 by electrospinning of a precursor solution, followed by hot pressing and calcinations (Figure 11.5 A). A solution of dimethylformamide, poly(vinyl acetate) (PVAc), titanium(IV) propoxide and acetic acid was directly electrospun on alumina substrates with interdigitated Pt electrode arrays. The as-spun TiO_2/PVAc composite fiber mats were then hot pressed (120 °C, for 10 min) and calcined. The purpose of hot pressing was two-fold: enhancing the adhesion between the TiO_2 fiber mats and

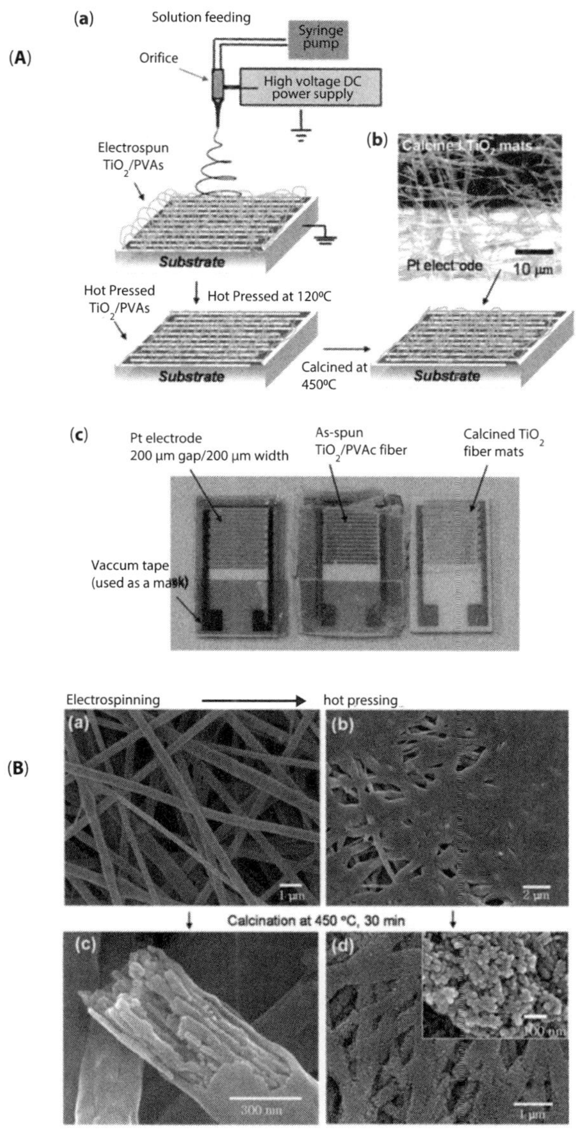

Figure 11.5 [A] (a) Schematic Diagram Of The Processing Steps Used To Fabricate Tio_2 Nanofiber Mats On Al_2O_3 Substrates With Interdigitated Pt Electrode Arrays; (b) Scanning Confocal Laser Micrograph Of A Calcined Tio2 Nanofiber Mat On Top Of The Al_2O_3 Substrate (Dark Region) And Pt Electrode (Bright Region); (c) Optical Micrographs Of Gas Sensor Test Devices (10 Mm × 15 Mm) With Tio_2 Nanofiber Mats After Different Processing Steps. [B] SEM Images Of (a) As-spun TiO_2/PVAc Composite Fibers; (b) TiO_2/PVAc Composite Fibers After Hot Pressing At 120 °C For 10 Min;. (c) Unpressed Tio_2 Nanofibers After Calcination At 450 °C. & (d) Hot Pressed Tio2 Nanofibers After Calcination At 450 °C. (Adapted With Permission From [80] © 2006, American Chemical Society).

the substrates and creating an accessible large active area for the gas-sensing interactions, by exposing the complex aligned fibrillar structure, created by phase separation during the electrospinning step (Figure 11.5 B). The resulting sensors were tested against traces of NO_2, CO, H_2, CH_4, and dimethyl methylphosphonate (DMMP) vapors. The sensitivity against NO_2 was found to be more than two order of magnitude higher than that reported for conventional TiO_2-based gas sensors, with a detection limit estimated to be well below 1 ppb. Encouraging results were also obtained with H_2, CO, and DMMP vapors, with detection limits in the order of few parts per million, while the response to CH_4 was considerably smaller.

11.4.2 Conducting Polymer Based Resistive Sensors

Conducting polymers, such as polyaniline (PANI), polythiophene, polypyrrole, etc. are successfully employed in various applications, including solar cells, light emitting diodes, electromagnetic radiation shielding and actuators. Distinctive features such as mechanical flexibility, ease of processing and modifiable electrical conductivity and chemical selectivity, make them particularly attractive also for sensing applications.

Among conducting polymers, Polyaniline (PANI) is one of the most widely investigated because of its relative low cost, simple preparation, high environmental stability and high conductivity. PANI and its composites have been proposed as a gas sensing materials for the detection of a number of gases and vapours, including ammonia, NO_2, CO, HCl, $CHCl_3$, N_2H_4, methanol, etc [81, 82]

Ji and co-workers [83] have prepared a composite of coaxial PANI/PMMA nanofibers, via electrospinning of poly(methyl methacrylate) (PMMA), followed by in situ polymerization of polyaniline on the surface of the electrospun nanofibres, in the presence of a suitable doping acid. The nanostructured composite was transferred to an interdigitated gold electrode to construct a gas sensor that was proposed as an alternative to metal oxide-based sensors to detect triethylamine (TEA) vapor. The electrospinning conditions were adjusted as to obtain thin PMMA fibers (avg. diameter ca. 250 nanometers), whose composite showed a markedly increased sensitivity towards TEA, compared with that of an in-situ prepared PANI thin film. A high, linear response was observed in the range of 20 to 500 ppm concentration, while the average response and recovery times were estimated in 131 and 600 s, respectively, thus comparable with that the thin film counterpart.

Gao and co-workers [84] have demonstrated a successful route to synthesize PANI nanotubes for gas sensing purposes, using the inner eggshell membrane as a template. Meanwhile, they have proposed electrospun poly(vinyl alcohol) (PVA) fiber mats as an alternative template to obtain PANI nanotubes with similar high sensing properties. It was demonstrated that PANI with nanotubular structure has higher sensitivity and more rapid, reversible, response to TEA than that prepared without a template. Again, the high surface area of NTs favors an easy access of the gas vapor. This leads to a much greater extent of dedoping over short times for the PANI nanotubes.

An interesting alternative approach [85] used a scanned-tip electrospinning method for depositing isolated and oriented polyaniline/poly-(ethylene oxide) (PANI/PEO) nanowires on lithographically defined microelectrodes. In this case, the electrospinning apparatus included a non conventional spinneret formed by a microfabricated, arrow-shaped tip, which was directly dipped in the PANI/PEO solution [86]. The application of high voltage to the droplet deposited on the tip resulted in deposition of oriented polymeric nanowires, with diameters in the range of 100–500 nm, on gold electrodes positioned on the counter electrode plate. A benefit of this single-step fabrication method is that it does not include chemical vapor deposition or etching processes. The resulting device was found to be able to detect ammonia at concentrations as low as 0.5 ppm, with rapid response and recovery time. The controllable geometry and high surface-to-volume ratio of the nanowires yielded improved response and increased sensitivity compared to similar polyaniline sensors based on films and random fiber networks.

Among the other conducting polymers, poly(3,4-ethylenedioxythiophene) (PEDOT) is regarded as an attractive alternative to PANI, due to a superior environmental stability and excellent properties in term of high conductivity, optical transparency in its doped state, low band gap, and moderate redox potential.

Researchers at the University of Connecticut [87] fabricated conductive core-shell TiO_2- Poly(3,4-ethylenedioxythiophene) (PEDOT) nanocables using electrospun TiO_2 nanofibers as a template, followed by vapor phase polymerization of 3,4-ethylenedioxythiophene.

TiO_2 nanofibers were prepared by electrospinning a solution of titanium isopropoxide ($Ti(OiPr)_4$) and PVP in acetic acid and ethanol. The collected nanofibers were then calcined at 500 °C for 3 h to remove PVP and generate TiO_2 nanofibers. The fibers mat was treated with $FeCl_3$ and exposed to saturated EDOT vapor at 95 °C for 2 days. The product characterization showed the formation of a core-shell structure where the TiO_2 core had an average diameter of ~78 nm while the PEDOT sheath had a uniform thickness of ~6 nm. Such ultra-thin, highly homogeneous layer of PEDOT, its porous mesh structure and highly specific surface to volume ratio, favor the adsorption and desorption of gas molecules and suppress the noises generated from structural defects. The TiO_2-PEDOT nanocables displayed a fast and reversible response to gaseous NO_2 and NH_3 with a limit of detection as low as 7 and 675 ppb, respectively.

Kwon *et al.* [88] have successfully fabricated coaxial PVA/PEDOT nanocables (NCs) and PEDOT nanotubes (NTs) flexible membranes for ammonia detection. PVA nanofibers were electrospun as a core part from a solution of PVA in distilled water. Polyvinyl alcohol (PVA)/PEDOT coaxial nanocables (NCs) were then prepared by vapor deposition polymerization, via 3,4-ethylenedioxythiophene adsorption onto the electrospun PVA nanofibers template. To obtain the PEDOT NTs membrane, the core polymer nanofibers were removed from PVA/PEDOT coaxial NCs with distilled water (Figure 11.6).

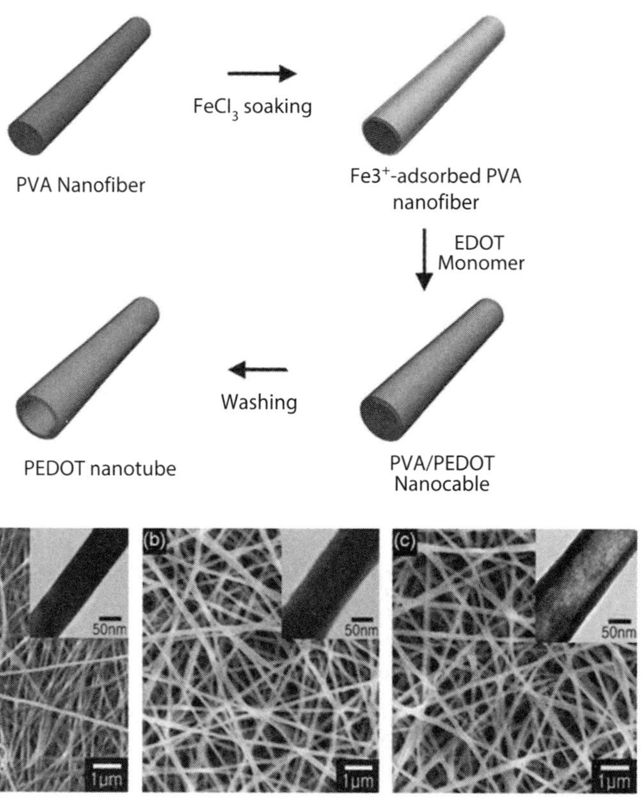

Figure 11.6 Scheme Of The Reaction Steps For The Fabrication Of A PEDOT Nanotubes Membrane And SEM And TEM (Inset) Images Of (A) The Pristine PVA Nanofibers; (b) PVA/PEDOT Nanocables, & (c) PEDOT Nanotubes (Adapted With Permission From [88] © 2010, Elsevier B.V.).

The diameter of PVA nanofibers was ca. 100nm and the PEDOT sheath thickness of both PVA/PEDOT coaxial NCs and PEDOT NTs was ca. 20nm; in addition, it was demonstrated that the PEDOT NTs did not collapse after washing (Figure 11.6c).

The minimum detectable level of these membrane sensors for NH_3 was found to be ca. 5ppm, which is significantly lower than the recommended threshold limit value of 25 ppm for human exposure [85]. It was also observed that the sensitivity of PEDOT NTs membrane sensor was higher than that of the PVA/PEDOT coaxial NCs and this was explained by the higher surface area of the former (31.6m^2 g^{-1} vs. 18.1m^2 g^{-1}). Improved reproducibility, rapid response time (ca. 1 s) and fast recovery time (ca. 30s) were observed in particular for PEDOT NTs, again due to faster diffusion of NH_3 gas molecules in and out of transducer.

Although conducting polymers have shown promising results in the field of chemical sensing, their commercial applications are somewhat limited by issues of thermal and environmental stability.

An alternative approach has focused on tailoring the electrical properties of more stable and easily processable non-conducting polymers, by loading with electrically active fillers, for example carbon black, carbon fibers and, more recently, carbon nanotubes.

In a recent study [89], multi-walled carbon nanotubes (MWCNTs) have been employed as conductive fillers in Nylon-6. Two fabrication routes were explored: nanofiber composites have been prepared by adsorbing the MWCNTs on the surface of electrospun Nylon-6 with the aid of Triton® X-100 as surfactant or a 1 wt % MWCNTs was dispersed in a Nylon-6/hexafluoroisopropanol solution and directly electrospun. Both MWCNTs/Nylon-6 nanofiber composite based devices were exposed to a wide variety of organic analytes (dichloromethane, trichloromethane, tetrahydrofuran, ethyl acetate, ethanol, acetone, hexane and toluene) and the magnitude of responsiveness has been found to be dependent on the dipole moment, vapour pressure and nature of polar functional groups of the analytes. Interconnectedness of the MWCNTs allows the charge transport, while their accessibility to the analyte is crucial for the sensing properties. The composite resulting from direct electrospinning of the polymer/MWCNTs solution showed much lower sensitivity and this was attributed to the MWCNTs being "embedded" in the polymer fibers and thus less accessible to the analyte.

11.4.3 Optical Sensors

Recently, optical sensors based on one-dimensional nanostructures, have been attracting considerable attention [5, 19]. Optical fiber sensors have many advantages compared to traditional types of sensors such as high sensitivity, fast response, immunity to electromagnetic interference, etc.

The majority of the recently developed electrospun optical sensors are fluorescent sensors, whose sensing mechanism is based on the quenching of fluorescence due to the interaction between an appropriate fluorophore and the target analyte, acting as a quencher. The degree of quenching depends on the amount of analyte.

The quantitative measure of fluorescence quenching is generally described by the Stern-Volmer constant, K_{sv}:

$$I_0/I = 1 + K_{sv}[Q] \tag{11.1}$$

where I_0 and I are the intensities of fluorescence in the absence and in the presence of the quencher, respectively, and [Q] is the quencher concentration. K_{sv} define the efficiency of quenching and can be expressed as:

$$K_{sv} = k_2\tau_1 \tag{11.2}$$

where τ_1 is the luminescence decay time of the fluorophore in the absence of the quencher, and k_2 is the bimolecular quenching rate constant.

Eq 11.1 indicates that, when all other variables are held constant, the higher K_{sv}, the lower the concentration of quencher required to quench the fluorescence. Eq.11. 2 implies that the sensitivity of the quenching process is enhanced by employing fluorophores with long luminescence decay times (τ_1) and that it can be tailored by controlling the quencher diffusion rate to fluorophores via the microstructural properties of the sensing film.

To this effect, nanostructured assemblies with high porosity and surface area showed considerable advantages.

Wang and co-workers paved the way to an extensive work on the use of fluorescence quenching-based optical sensors. In an early study [90], they prepare electrospun membranes sensitive to 2,4-dinitrotoluene (DNT). A fluorescent polymer was newly synthesized via covalent attachment of the fluorescent indicator pyrene methanol (PM) onto polymethylmethacrylate (PMMA) and quenching-based optical sensors were then fabricated by electrospinning. The resulting membranes showed much higher sensitivities than conventional continuous thin films to the tested analyte and this was attributed to the higher surface-to volume ratio of the nanofiber mat, which increases the rate of quenching. In addition, lower concentration of fluorophore were found to be advantageous to increasing sensitivity, since the change of fluorescence intensity due to small amount of quencher is more significant if the fluorescence intensity without quencher is lower.

In another work [91], the same group tested the performance of sensing membranes prepared by simultaneous electrospinning of polyacrylic acid (PAA) – PM and poly(allylamine hydrochloride), used as a thermal cross-linker, in a dual-electrode configuration. This water-insoluble, fluorescent membrane showed effective quenching in the presence of both electron-deficient metal cations, such as Fe^{3+} and Hg^{2+}, and nitro aromatic compound, such as DNT and TNT.

A further improvement to this approach involved the fabrication of membranes by direct electrospinning of polyacrylic acid (PAA) – PM in the presence of a thermally cross-linkable polyurethane [64]. It was found that the fluorescence intensity decreases with increasing the analytes concentration and Stern- Volmer constants (K_{sv}) of these electrospun films for Fe^{3+}, Hg^{2+} and DNT were found to be 2 to 3 orders of magnitude higher than that obtained previously from thin film sensors. The sensitivities of the electrospun sensors were in the range of a few to tens parts per billion. The significant enhancement of the sensitivities of the sensors was again attributed to the higher surface area of the electrospun membranes.

Porous nanofiber membranes were successfully fabricated by electrospinning of cellulose acetate (CA) doped with 9-chloromethylanthracene (9-CMA) in the presence of Polyethylene glycol (PEG) as a porogen [92] and their fluorescence quenching sensing behavior was tested using methyl violet as model compound. It was demonstrated that the nanofiber membranes exhibited a much higher sensitivity than that of the corresponding 9-CMA doped, CA continuous film; the sensitivity was further improved by the secondary porous structures, which provided increased surface areas and facilitated the quenchers diffusion into nanofibers, accelerating the contact with fluorophores.

Electrospun membranes prepared by electrospinning of porphyrinated polyimide showed both chromo- and fluorogenic responses upon exposure to HCl vapors [93]. Bonding of the porphyrin fluorophores into polyimide main chains overcomes the disadvantage of porphyrin aggregation, while the polyimide frame provides good mechanical properties, as well as superior thermal and chemical stabilities, which are strongly required for many applications, such as monitoring HCl emissions in incinerators of wastes. The nanofibrous membrane showed high sensitivity, fast response and good recovery time and thus a good potential in HCl sensing application.

Nanofibrous, oxygen sensing optical devices have also been prepared by electrospinning of phosphorescent complexes of transition metals incorporated into a polystyrene matrix [94, 95]. Cu(I) ([Cu(POP)(ECI-Phen)]

BF_4, POP = bis[2-(diphenylphosphino)-phenyl]ether; ECI-Phen=1-ethyl-2-(N-ethyl-carbazole-yl-4-)imidazo[4,5-f]1,10-phenanthroline) or Eu(III) ([Eu(TTA)$_3$(phencarz)] TTA=2-thenoyltrifluoroacetonate, phencarz=2-(*N*-ethylcarbazolyl-4)imidazo[4,5-*f*]1,10-phenanthroline) phosphorescent complexes were investigated. Both membranes showed good linearity of response in a whole 0–100% range of oxygen concentrations, with a high sensitivity and reasonably short response-recovery times.

Besides gas and vapours sensors, highly sensitive fluorescence quenching-based sensors for determination of trace amounts of metal ions were also fabricated by electrospinning. PMMA and ethyl cellulose electrospun membranes, doped with methoxy azomethine ionophore as a fluorescent indicator, were demonstrated to give a selective response to femtomolar concentrations of silver ion, with a one to two order of magnitude sensitivity increase compared to thin film based sensors [96].

The same polymer matrix doped with N'-3-(4-(dimethylaminophenly) allylidene)isonicotinohydrazide showed high selectivity and sensitivity towards Cu(II) ions, with a large concentration working range (10^{-12} to 10^{-2} M), good equilibrium response time (<1 min) and satisfying recovery properties [97].

11.4.4 Acoustic Wave Sensors

Sensors based on the effects of acoustic waves occupy a prominent place among chemical gas and vapors sensors. The distinctive characteristic of acoustic wave sensors is that a gas-adsorbing support (usually a thin film) serves as a receptor, while a piezoelectric device is the transducer. Because of this, sensors of this class are often referred to as piezoelectric sorption or resonance sensors [1].

Among the sensing techniques based on the piezoelectric effect, surface acoustic wave (SAW) and quartz crystal microbalance (QCM) are by far the most popular.

Although a few attempts have been reported of depositing nanostructured coatings on SAW *via* electrospray or electrospinning techniques to improve the sensor sensitivity [98], much more extensive efforts have been focused on QCM [99–102]

The quartz crystal microbalance (QCM) is a simple, cost effective, high-resolution mass sensing technique [3, 99]. The QCM device is a resonating polymer-coated disk with metal electrodes, each side of which is connected to a lead wire. In operating mode, the device resonates at a characteristic frequency via excitation by a specific oscillating signal.

When an analyte is adsorbed by the polymer, the resulting increase in the mass of the device reduces the frequency of resonance. Up to now, a variety of materials such as metals, ceramics, polymers, self-assembled monolayers, etc, have been employed as sensitive coatings on QCM to improve the sensor sensitivity and selectivity for chemical analytes. It is widely accepted that the sensitivity of QCM sensors towards a specific analyte is enhanced by increasing the specific surface area of the sensing materials. Recent efforts have been focused on the development of nano-structured coatings on QCM to improve the sensor sensitivity [102]. Thanks to their large specific surface area, high porosity and good inter-connectivity, the electrospun fibrous membranes represent an attractive alternative to replace continuous membranes to enhance the sensitivity of the QCM based sensors.

Ding *et al* [100, 101] demonstrated highly enhanced sensitivity of a quartz crystal microbalance coated with electrospun nanofibrous membranes as gas sensor for NH_3 detection, compared to continuous films coated QCM sensors. PAA, a weak anionic polyelectrolyte, is regarded as ideal sensing material for coating on QCM because of the interaction between ammonia and the carboxyl groups of PAA. However, pure PAA solutions are difficult to electrospin due to the formation of strong hydrogen bond. To overcome this drawback two different approaches were attempted. In a first study [100], water-soluble PVA was blended with PAA for preparing the electrospun membranes. The electrospun nanofibers with diameter of 100–400 nm were directly deposited on the surface of QCM by electrospinning homogenous blends solutions of different proportions of PAA and PVA. A heat treatment was then employed to promote crosslinking between PAA and PVA, thus providing water insoluble nanofibrous membranes.

It was observed that the average resonance frequency shifts were gradually increased on increasing the content of PAA in the nanofibrous membranes, because of the increase in the amount of absorption sites at higher PAA content. Relative humidity and NH_3 concentration were also found to positively influence the sensitivity. A drift in the signal baseline was observed upon subsequent exposures to the analyte, likely due to an irreversible absorption of NH_3 originated from strong interactions with the carboxyl groups of PAA.

In a subsequent study [101] the fibrous membranes were deposited on the electrode of QCM by electrospinning of pure PAA from H_2O/ethanol mixtures in different proportion. The fibers morphology was found to be strongly dependent on the solvent composition. Although those obtained from pure water were beaded and showed a poor morphology, the fibers obtained from pure water or pure ethanol solutions also exhibited smaller average diameters and thus higher sensing response, compared to those obtained from solvent mixtures.

The sensors exhibited high sensitivity towards concentration of ammonia as low as 130 ppb at the relative humidity of 40%. The sensitivity of these fibrous membranes coated QCM sensors was proved to be four times higher than that of a continuous films coated QCM sensor.

It was found that the sensors have cross sensitivity to ammonia and H_2O and pre-sorbed water in the fibrous membranes was shown to be the key factor affecting the sensitivity.

Following this observation, PAA electrospun membranes were also proposed as sensitive coatings on a quartz crystal microbalance (QCM) to fabricate humidity sensors with good performance [103].

Another successful application of nanofibrous membranes coated QCM was demonstrated by Wang and co-workers [102] in the fabrication of ultrasensitive formaldehyde sensors based on Polyethyleneimine (PEI), a cationic polyelectrolyte.

Again, due to the inherent difficulties encountered in electrospinning pure PEI, PVA was mixed with PEI for preparing the sensing fibrous membranes and various H_2O/ethanol mixtures were used as solvents. The specific surface area of the fibrous membranes decreased with increasing the PEI content (Figure 11.7 A), while it increased in the presence of ethanol as co-solvent.

The combination of PEI content and specific surface area of the nanofibrous mat was recognized as the key factor affecting the sensitivity (Figure 11.7 B).

As already observed [101], the responses of the QCM sensors were reinforced with increasing the relative humidity. This was explained by the increase in conductivity of PEI at increasing relative humidity and by the absorption of water on the hydrophilic membrane, facilitating the concurrent absorption of formaldehyde through the formation of hydrogen bond between formaldehyde and H_2O.

In addition, the sensor response was found to be reversible, reproducible, and showed a linear relationship with increasing formaldehyde concentrations and a good selectivity to formaldehyde rather than other conventional volatile organic compounds.

Alternative approaches were also recently proposed by using electrospun inorganic or organic-inorganic hybrid sensitive coatings on QCM, such as ZnO fibermats [104] and nanoporous TiO_2 fibers functionalized with PEI [105] for humidity and formaldehyde detection, respectively.

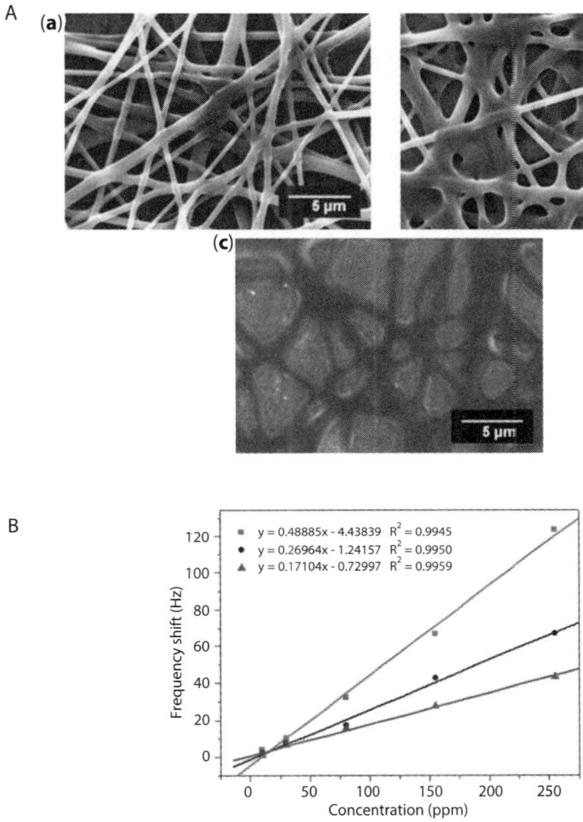

Figure 11.7 [A] FE-SEM Images Of Fibrous Membranes With Various Weight Ratios Of PEI/PVA: (a) 0.8/1; (b) 1.6/1; & (c) 3.2/1. [B] Dependence Of The Frequency Shift Of QCM Sensors Coated With: (a) Flat Membrane; (b) Fibrous Membrane Formed From A PEI/PVA Solution In Water, & (c) Fibrous Membrane Formed From PEI/PVA Solution In Water/Ethanol, On Formaldehyde Concentration (Adapted With Permission From [102] © 2009, Elsevier B.V.).

11.4.5 Amperometric Biosensors

Nanofibers, nanowires, and nanotubes have been widely investigated for use within biosensors [23].

In the past decades, intensive research has been focused on the development of fast and sensitive devices for monitoring blood-glucose levels, because of their practical applications in treating diabetes mellitus, one of the most prevalent metabolic disorders worldwide. Among the various techniques commonly applied in glucose determination, electrochemical methods, especially amperometry, have shown great potential and attracted much attention.

Most of the recent research on the development of glucose sensors has been based on the immobilization of enzymes, which catalyzes the oxidation of glucose, on the electrode surfaces. Electrospun nanofibers have been shown to be effective supports to this purpose, providing large specific surface areas for highly efficient immobilization, as well as stabilizing enzymes [106].

Electrospinning of sacrificial polymer templates containing metal oxide precursors have been used to produce nanofibrous metal oxides coated electrodes, on which glucose oxidase (GOx) has been immobilized. Ahmad at al.[107] have explored the performance of a gold electrode coated with a single, GOx-functionalized, ZnO nanofibers (Figure 11.8). The study revealed that the nanofiber improved the electrocatalytic activity of the enzyme, which in turn enhanced the sensitivity of the biosensor allowing a high, fast and reproducible sensitivity, with a linear response in a range of glucose concentration from 0.25 to 19 mM, and a lower limit of detection of 1 μM. The sensor also exhibits good selectivity and favorable stability over relatively long-term storage (more than 4 months).

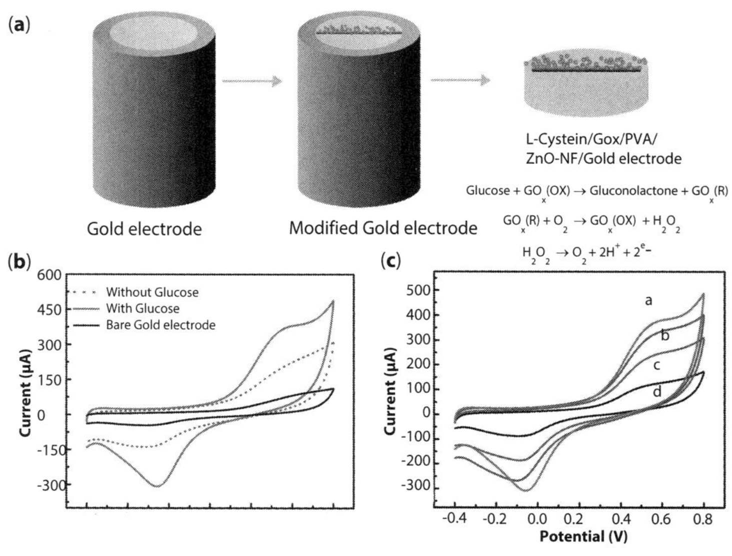

Figure 11.8 (*a*) Schematic Illustration Of A Gox Functionalized, Zno Nanofiber Modified Gold Electrode And Its Mechanism Of Glucose Sensing; (*b*) Cyclic Voltammograms Of The Bare And Modified Gold Electrode With And Without 100 Mm Glucose In Ph 7.0 PB Solution; And (*c*) Cyclic Voltammograms Of The Biosensor In PB Solution (Ph 7.0) Containing 100 Mm Glucose At A Scan Rate Of (A) 100 Mv, (B) 80, (C) 50, And (D) 20 Mv S-1. (Reprinted With Permission From [107] © 2010, American Chemical Society).

Similarly, GOx-functionalized TiO_2 [108] and Mn_2O_3-Ag nanofibers [109] were also found to be effective coatings in enhancing the sensitivity of glucose sensors.

Polymeric electrospun membranes have also been used for direct immobilization of enzymes. Ren and co-workers [110] demonstrated a facile method to prepare a glucose sensing device by direct electrospinning of PVA and GOx on a gold electrode, followed by cross-linking of PVA with glutharaldehyde. It was shown that the immobilized enzyme remains active inside the biocomposite membrane, through the electrospinning process, and that the resulting electrode exhibited very good performance, in term of rapid response time (1 s), linear response and lower detection limit (0.05 mM).

Scampicchio *et al* [111], proposed a polyamide-6 electrospun membrane as a support for immobilization of GOx with the aid of an enzyme-immobilizing solution, containing glutaraldehyde and bovine serum albumin. They demonstrated that the nanofibrous environment had only a minimal effect on the kinetic of the biocatalysis and the resulting coated electrode showed promising performances as a glucose sensor.

Despite the encouraging performances of these enzymatic sensors, some inherent problems, such as the chemical and thermal instability originated from the intrinsic nature of enzymes, may limit their analytical applications [112]. Efforts have been made for direct determination of glucose at enzyme-free electrodes in which glucose is electrocatalytically oxidized at metal (Pt, Au, Cu, Ni, etc.) or metal oxide surfaces.

For example, Liu and co-workers [113] have presented a novel non-enzymatic glucose sensor, based on a Ni nanoparticle-loaded carbon nanofibers (CFs) paste electrode. The nanocomposite was prepared by carbonizing electrospun PAN/Ni acetylacetonate (NiAA) composite fibers, via a carefully controlled thermal treatment, which yielded the Ni nanoparticle-loaded CFs (Figure 11.9 A and B). The electrode was fabricated by packing a glass tube with a paste, prepared by mixing the nanocomposite with mineral oil. The resulting device combines the good conductivity of CFs with the electrocatalytic activity of the Ni nanoparticles towards glucose oxidation. It exhibited excellent sensing performances (Figure 11.9 C), with a lower detection limit of 1 μM, linear response in the range 2 μM – 2,5mM and high resistance to surface fouling.

The good performances of Ni and NiO-based glucose sensors were also demonstrated by a number of other studies [114–116].

Also Co_3O_4 nanofibers were applied to construct a non-enzymatic sensor for glucose detection in alkaline solutions [117]. The MOS

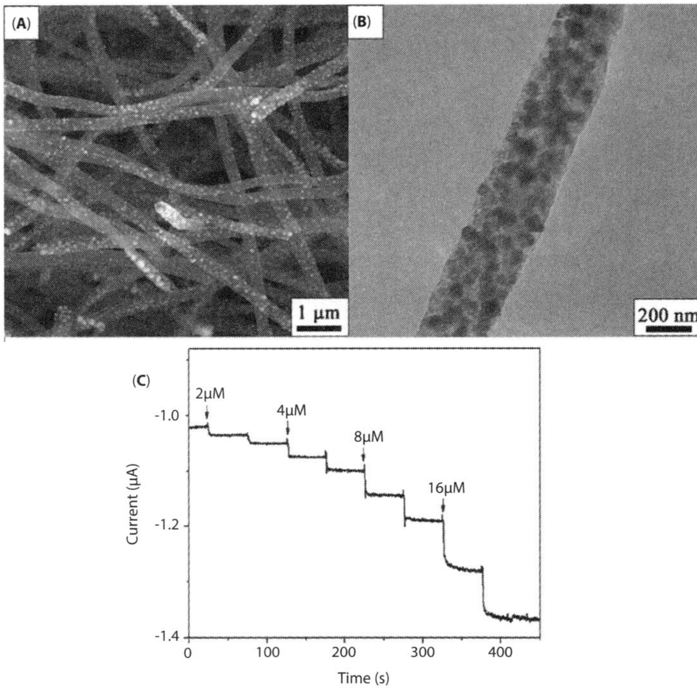

Figure 11.9 SEM Image (A) And TEM Image (B) Of A NiCF Nanocomposite. (C) Current-Time Responses Of A Ni-Carbon Fiber-Paste Electrode Upon Successive Addition Of Glucose At Low Concentrations. (Adapted With Permission From [113] © 2009, Elsevier B.V.)

nanofibers were fabricated by a conventional two-step procedure consisting of electrospinning and subsequent calcination. Then a modified GC electrode was prepared by casting a nanofiber suspension on the surface of the electrode and entrapping them into a Nafion membrane. The developed sensor showed a fast response time (less than 7 s), high sensitivity, good reproducibility and selectivity, and a lower detection limit of 0.97 µM.

Concerns regarding the selectivity and the relatively easy poisoning of metal and metal oxide based enzymeless sensors stimulated the research of alternative materials for preparing enzymeless sensors. Manesh and coworkers [118], fabricated a sensor electrode based on a composite electrospun nanofibrous membrane of poly(vinylidene fluoride) (PVdF) and poly(aminophenylboronic acid) (PAPBA), in which the boronate groups in PAPBA preferentially sense the glucose, while PVdF provides mechanical strength. A linear response to the detection of glucose in a broad range of concentrations (1–15mM) was demonstrated, together with a short time of response (<6 s), high selectivity, reproducibility and excellent storage stability.

Beside glucose, amperometric biosensors based on nanofiborus biomaterials, have also been prepared to detect hydrogen peroxide [119, 120], urea, nitrite [121], etc.

As an example, a hydrogen peroxide biosensor was recently fabricated by immobilizing hemoglobin on the surface of electrospun gold nanoparticle-chitosan-poly(vinyl alcohol) composite nanofibers [122]. The electrode showed a linear response to H_2O_2 in a wide concentration range ($5.6 \cdot 10^{-7}$ - $5.2 \cdot 10^{-2}$ M), with a low detection limit of $1.98 \cdot 10^{-7}$M and a short response time of 4 s. Moreover, it exhibited superior properties in terms of long-term stability, good reproducibility, and high selectivity.

Similarly, a nanocomposite fibermat prepared by electrospinning of a solution of urease and PVP has been demonstrated to show a promising potential for urea detection down to 0,5 mM concentration [123].

11.5 Conclusions

The interest on the electrospinning technology has experienced a tremendous increase in the last ten years that resulted in enormous progresses in terms of variety and type of nanostructures produced by this technique.

Electrospun nanofibers have been proved to be excellent materials, intermediates or supports to produce sensors with highly enhanced capability. The high surface area to volume ratio and low cross section of the nanofibers have been shown to dramatically increase the available sensing surface, facilitating the analyte adsorption and the mass and electric charge transport. This generally results in higher sensitivity, larger linear ranges and lower limits of detection, when compared to other sensors.

In addition, recent strategies for the fabrication of electrospun nanofibers with hierarchical structures resulted in novel, more complex nanostructures, which demonstrated even more promising performances.

Nevertheless, further development steps are required: first, more fundamental studies would be beneficial in exploring new sensing materials and providing further control over the size, morphology and architecture of the electrospun fibers. Then, nanofibers are still difficult to fabricate in batches, although multijet systems have been developed, and this limits their large-scale application. Moreover, many of the recently developed sensors have been tested in well-controlled laboratory conditions, but they still need to be validated in real-case conditions.

However, given the continuous and fast progresses of both electrospinning and sensors technologies, it may be expected that new and expanded uses will become available in the near future.

Abbreviations

1D	One-dimension
9-CMA	9-Chloromethylantracene
BET	Brunauer-Emmett-Teller
CA	Cellulose acetate
CFs	Carbon fibers
CNTs	Carbon nanotubes
DNT	2,4-dinitrotoluene
GC	Glassy carbon
MOS	Metal oxide semiconductors
NCs	Nanocables
NTs	Nanotubes
PAA	Poly(acrylic acid)
PAN	Poly(acrylonitrile)
PANI	Polyaniline
PAPBA	Poly(aminophenylboronic acid)
PEDOT	Poly(3,4-ethylenedioxythiophene)
PEG	Poly(ethylene glycol)
PEI	Polyethyleneimine
PEO	Polyethylenoxide
PM	Pyrene methanol
PMMA	Poly(methyl methacrylate)
PVA	Poly(vinyl alcohol)
PVAc	Poly(vinyl acetate)
PVdF	Poly(vinylidene fluoride)
PVP	Poly(vinyl pyrrolidone)
QCM	Quartz crystal microbalance
SAW	Surface acoustic wave
TEA	Triethylamine
TNT	2,4,6-trinitrotoluene

References

1. L.M. Dorozhkin and I.A. Rozanov, *Journal of Analytical Chemistry*, Vol. 56, p. 399, 2001.
2. C. McDonagh, C.S. Burke and B.D. MacCraith, *Chemical Reviews*, Vol. 108, p. 400, 2008.

3. S.K. Pandey, K.H. Kim and K.T. Tang, *TrAC - Trends in Analytical Chemistry*, Vol. 32, p. 87, 2012.
4. J.W. Kemling, A.J. Qavi, R.C. Bailey and K.S. Suslick, *Journal of Physical Chemistry Letters*, Vol. 2, p. 2934, 2011.
5. X. Chen, C.K.Y. Wong, C.A. Yuan and G. Zhang, *Sensors and Actuators, B: Chemical*, Vol. 177, p. 178, 2013.
6. D. Li and Y. Xia, *Advanced Materials*, Vol. 16, p. 1151, 2004.
7. S. Ramakrishna, K. Fujihara, W.E. Teo, T. Yong, Z. Ma and R. Ramaseshan, *Materials Today*, Vol. 9, p. 40, 2006.
8. A. Greiner and J.H. Wendorff, *Angewandte Chemie - International Edition*, Vol. 46, p. 5670, 2007.
9. G.C. Rutledge and S.V. Fridrikh, *Advanced Drug Delivery Reviews*, Vol. 59, p. 1384, 2007.
10. J.F. Cooley, Apparatus for electrically dispersing fluids, US Patent 692,631, 1902..
11. W.J. Morton, Method of Dispersing Fluids, US Patent 705,691, 1902.
12. A. Formhals, Process and apparatus for preparing artificial threads, US Patent 1,975,504, 1934.
13. J. Doshi and D.H. Reneker, *Journal of Electrostatics*, Vol. 35, p. 151, 1995.
14. D.H. Reneker and I. Chun, *Nanotechnology*, Vol. 7, p. 216, 1996.
15. X. Fang and D.H. Reneker, *Journal of Macromolecular Science - Physics*, Vol. 36, p. 169, 1997.
16. H. Fong and D.H. Reneker, *Journal of Polymer Science, Part B: Polymer Physics*, Vol. 37, p. 3488, 1999.
17. C. Kriegel, A. Arrechi, K. Kit, D.J. McClements and J. Weiss, *Critical Reviews in Food Science and Nutrition*, Vol. 48, p. 775, 2008.
18. B. Ding, M. Wang, J. Yu and G. Sun, *Sensors*, Vol. 9, p. 1609, 2009.
19. B. Ding, M. Wang, X. Wang, J. Yu and G. Sun, *Materials Today*, Vol. 13, p. 16, 2010.
20. S. Chigome and N. Torto, *Analytica Chimica Acta*, Vol. 706, p. 25, 2011.
21. M. Scampicchio, A. Bulbarello, A. Arecchi, M.S. Cosio, S. Benedetti and S. Mannino, *Electroanalysis*, Vol. 24, p. 719, 2012.
22. M.M. Arafat, B. Dinan, S.A. Akbar and A.S.M.A. Haseeb, *Sensors (Switzerland)*, Vol. 12, p. 7207, 2012.
23. L. Matlock-Colangelo and A.J. Baeumner, *Lab on a Chip - Miniaturisation for Chemistry and Biology*, Vol. 12, p. 2612, 2012.
24. T.J. Sill and H.A. von Recum, *Biomaterials*, Vol. 29 p. 1989, 2008.
25. D.H. Reneker and A.L. Yarin, *Polymer*, Vol. 49, p. 2387, 2008.
26. P. Gupta and G.L. Wilkes, *Polymer*, Vol. 44, p. 6353, 2003.
27. B. Ding, E. Kimura, T. Sato, S. Fujita and S. Shiratori, *Polymer*, Vol. 45, p. 1895, 2004.
28. S.A. Theron, A.L. Yarin, E. Zussman and E. Kroll, *Polymer*, Vol. 46, p. 2889, 2005.
29. Z. Sun, E. Zussman, A.L. Yarin, J.H. Wendorff and A. Greiner, *Advanced Materials*, Vol. 15, p. 1929, 2003.

30. D. Li and Y. Xia, *Nano Letters*, Vol. 4, p. 933, 2004.
31. S. Agarwal, J.H. Wendorff and A. Greiner, *Polymer*, Vol. 49, p. 5603, 2008.
32. A.L. Yarin, *Polymers for Advanced Technologies*, Vol. 22, p. 310, 2011.
33. X. Lu, C. Wang and Y. Wei, *Small*, Vol. 5, p. 2349, 2009.
34. J.S. Kim and D.H. Reneker, *Polymer Engineering and Science*, Vol. 39, p. 849, 1999.
35. D. Li, Y. Wang and Y. Xia, *Nano Letters*, Vol. 3, p. 1167, 2003.
36. H. Hou, Z. Jun, A. Reuning, A. Schaper, J.H. Wendorff and A. Greiner, *Macromolecules*, Vol. 35, p. 2429, 2002.
37. R.A. Caruso, J.H. Schattka and A. Greiner, *Advanced Materials*, Vol. 13, p. 1577, 2001.
38. M. Bognitzki, W. Czado, T. Frese, A. Schaper, M. Hellwig, M. Steinhart, A. Greiner and J.H. Wendorff, *Advanced Materials*, Vol. 13, p. 70, 2001.
39. F. Ko, Y. Gogotsi, A. Ali, N. Naguib, H. Ye, G. Yang, C. Li and P. Willis, *Advanced Materials*, Vol. 15, p. 1161, 2003.
40. Y. Dror, W. Salalha, R.L. Khalfin, Y. Cohen, A.L. Yarin and E. Zussman, *Langmuir*, Vol. 19, p. 7012, 2003.
41. G.M. Kim, A. Wutzier, H.J. Radusch, G.H. Michler, P. Simon, R.A. Sperling and W.J. Parak, *Chemistry of Materials*, Vol. 17, p. 4949, 2005.
42. L. Chen, S. Hong, X. Zhou, Z. Zhou and H. Hou, *Catalysis Communications*, Vol. 9, p. 2221, 2008.
43. X. Xu, Q. Yang, J. Bai, T. Lu, Y. Li and X. Jing, *Journal of Nanoscience and Nanotechnology*, Vol. 8, p. 5066, 2008.
44. X. Xu, Q. Wang, H.C. Choi and Y.H. Kim, *Journal of Membrane Science*, Vol. 348, p. 231, 2010.
45. N.A.M. Barakat, M.S. Khil, F.A. Sheikh and H.Y. Kim, *Journal of Physical Chemistry C*, Vol. 112, p. 12225, 2008.
46. C. Drew, X. Liu, D. Ziegler, X. Wang, F.F. Bruno, J. Whitten, L.A. Samuelson and J. Kumar, *Nano Letters*, Vol. 3, p. 143, 2003.
47. Q. Fan and M.S. Whittingham, *Electrochemical and Solid-State Letters*, Vol. 10, p. A48, 2007.
48. S. Madhugiri, B. Sun, P.G. Smirniotis, J.P. Ferraris and K.J. Balkus Jr, *Microporous and Mesoporous Materials*, Vol. 69, p. 77, 2004.
49. K.M. Sawicka, A.K. Prasad and P.I. Gouma, *Sensor Letters*, Vol. 3, p. 31, 2005.
50. W. Zheng, X. Lu, W. Wang, Z. Li, H. Zhang, Y. Wang, Z. Wang and C. Wang, *Sensors and Actuators, B: Chemical*, Vol. 142, p. 61, 2009.
51. A. Kumar, R. Jose, K. Fujihara, J. Wang and S. Ramakrishna, *Chemistry of Materials*, Vol. 19, p. 6536, 2007.
52. J.A. Lee, K.C. Krogman, M. Ma, R.M. Hill, P.T. Hammond and G.C. Rutledge, *Advanced Materials*, Vol. 21, p. 1252, 2009.
53. W. Nuansing, S. Ninmuang, W. Jarernboon, S. Maensiri and S. Seraphin, *Materials Science and Engineering B: Solid-State Materials for Advanced Technology*, Vol. 131, p. 147, 2006.

54. S. Zhan, D. Chen, X. Jiao and C. Tao, *Journal of Physical Chemistry B*, Vol. 110, p. 11199, 2006.
55. N. Dharmaraj, C.H. Kim, K.W. Kim, H.Y. Kim and E.K. Suh, *Spectrochimica Acta - Part A: Molecular and Biomolecular Spectroscopy*, Vol. 64, p. 136, 2006.
56. Q. Qi, T. Zhang, L. Liu and X. Zheng, *Sensors and Actuators, B: Chemical*, Vol. 137, p. 471, 2009.
57. Y. Zhang, X. He, J. Li, Z. Miao and F. Huang, *Sensors and Actuators, B: Chemical*, Vol. 132, p. 67, 2008.
58. S.H. Choi, G. Ankonina, D.Y. Youn, S.G. Oh, J.M. Hong, A. Rothschild and I.D. Kim, *ACS Nano*, Vol. 3, p. 2623, 2009.
59. H. Wu and W. Pan, *Journal of the American Ceramic Society*, Vol. 89, p. 699, 2006.
60. X. Yang, C. Shao, H. Guan, X. Li and J. Gong, *Inorganic Chemistry Communications*, Vol. 7, p. 176, 2004.
61. C. Shao, X. Yang, H. Guan, Y. Liu and J. Gong, *Inorganic Chemistry Communications*, Vol. 7, p. 625, 2004.
62. C. Wang, C. Shao, X. Zhang and Y. Liu, *Inorganic Chemistry*, Vol. 48, p. 7261, 2009.
63. S. Zhan, D. Chen, X. Jiao and Y. Song, *Chemical Communications*, Vol. p. 2043, 2007.
64. X. Wang, C. Drew, S.H. Lee, K.J. Senecal, J. Kumar and L.A. Samuelson, *Nano Letters*, Vol. 2, p. 1273, 2002.
65. Z. Li, H. Zhang, W. Zheng, W. Wang, H. Huang, C. Wang, A.G. MacDiarmid and Y. Wei, *Journal of the American Chemical Society*, Vol. 130, p. 5036, 2008.
66. Q. Qi, Y. Feng, T. Zhang, X. Zheng and G. Lu, *Sensors and Actuators, B: Chemical*, Vol. 139, p. 611, 2009.
67. J. Moon, J.A. Park, S.J. Lee, T. Zyung and I.D. Kim, *Sensors and Actuators, B: Chemical*, Vol. 149, p. 301, 2010.
68. J.A. Park, J. Moon, S.J. Lee, S.H. Kim, T. Zyung and H.Y. Chu, *Materials Letters*, Vol. 64, p. 255, 2010.
69. J.Y. Park, S.W. Choi, J.W. Lee, C. Lee and S.S. Kim, *Journal of the American Ceramic Society*, Vol. 92, p. 2551, 2009.
70. W. Wang, Z. Li, L. Liu, H. Zhang, W. Zheng, Y. Wang, H. Huang, Z. Wang and C. Wang, *Sensors and Actuators, B: Chemical*, Vol. 141, p. 404, 2009.
71. W.Y. Wu, J.M. Ting and P.J. Huang, *Nanoscale Research Letters*, Vol. 4, p. 513, 2009.
72. J.K. Choi, I.S. Hwang, S.J. Kim, J.S. Park, S.S. Park, U. Jeong, Y.C. Kang and J.H. Lee, *Sensors and Actuators, B: Chemical*, Vol. 150, p. 191, 2010.
73. D.J. Yang, I. Kamienchick, D.Y. Youn, A. Rothschild and I.D. Kim, *Advanced Functional Materials*, Vol. 20, p. 4258, 2010.
74. G. Wang, Y. Ji, X. Huang, X. Yang, P.I. Gouma and M. Dudley, *Journal of Physical Chemistry B*, Vol. 110, p. 23777, 2006.

75. H. Zhang, Z. Li, W. Wang, C. Wang and L. Liu, *Journal of the American Ceramic Society*, Vol. 93, p. 142, 2010.
76. Z. Zhang, X. Li, C. Wang, L. Wei, Y. Liu and C. Shao, *Journal of Physical Chemistry C*, Vol. 113, p. 19397, 2009.
77. Q. Qi, T. Zhang, L. Liu and X. Zheng, *Sensors and Actuators B: Chemical*, Vol. 137, p. 471, 2009.
78. Q. Qi, T. Zhang, L. Liu, X. Zheng and G. Lu, *Sensors and Actuators B: Chemical*, Vol. 141, p. 174, 2009.
79. X. Song, Q. Qi, T. Zhang and C. Wang, *Sensors and Actuators B: Chemical*, Vol. 138, p. 368, 2009.
80. I.D. Kim, A. Rothschild, B.H. Lee, D.Y. Kim, S.M. Jo and H.L. Tuller, *Nano Letters*, Vol. 6, p. 2009, 2006.
81. H. Bai and G. Shi, *Sensors*, Vol. 7, p. 267, 2007.
82. D. Nicolas-Debarnot and F. Poncin-Epaillard, *Analytica Chimica Acta*, Vol. 475, p. 1, 2003.
83. S. Ji, Y. Li and M. Yang, *Sensors and Actuators B: Chemical*, Vol. 133, p. 644, 2008.
84. Y. Gao, X. Li, J. Gong, B. Fan, Z. Su and L. Qu, *Journal of Physical Chemistry C*, Vol. 112, p. 8215, 2008.
85. H. Liu, J. Kameoka, D.A. Czaplewski and H.G. Craighead, *Nano Letters*, Vol. 4, p. 671, 2004.
86. J. Kameoka, D. Czaplewski and H.G. Craighead, *Journal of Photopolymer Science and Technology*, Vol. 16, p. 423, 2003.
87. Y. Wang, W. Jia, T. Strout, Y. Ding and Y. Lei, *Sensors*, Vol. 9, p. 6752, 2009.
88. O.S. Kwon, E. Park, O.Y. Kweon, S.J. Park and J. Jang, *Talanta*, Vol. 82, p. 1338, 2010.
89. N.L. Lala, V. Thavasi and S. Ramakrishna, *Sensors*, Vol. 9, p. 86, 2009.
90. X. Wang, S.H. Lee, B.C. Ku, L.A. Samuelson and J. Kumar, *Journal of Macromolecular Science - Pure and Applied Chemistry*, Vol. 39 A, p. 1241, 2002.
91. X. Wang, C. Drew, S.H. Lee, K.J. Senecal, J. Kumar and L.A. Samuelson, *Journal of Macromolecular Science - Pure and Applied Chemistry*, Vol. 39 A, p. 1251, 2002.
92. Y. Yang, X. Fan, Y. Long, K. Su, D. Zou, N. Li, J. Zhou, K. Li and F. Liu, *Journal of Materials Chemistry*, Vol. 19, p. 7290, 2009.
93. Y.Y. Lv, J. Wu and Z.K. Xu, *Sensors and Actuators, B: Chemical*, Vol. 148, p. 233, 2010.
94. L.Y. Wang, Y. Xu, Z. Lin and N. Zhao, *Journal of Luminescence*, Vol. 131, p. 1277, 2011.
95. Y. Wang, B. Li, L. Zhang, Q. Zuo, P. Li, J. Zhang and Z. Su, *ChemPhysChem*, Vol. 12, p. 349, 2011.
96. S. Kacmaz, K. Ertekin, A. Suslu, M. Ozdemir, Y. Ergun, E. Celik and U. Cocen, *Sensors and Actuators, B: Chemical*, Vol. 153, p. 205, 2011.

97. M.Z. Ongun, K. Ertekin, M. Gocmenturk, Y. Ergun and A. Suslu, *Spectrochimica Acta - Part A: Molecular and Biomolecular Spectroscopy*, Vol. 90, p. 177, 2012.
98. X. He, R. Arsatb, A.Z. Sadekb, W. Wlodarskib, K. Kalantar-zadehb and J. Li, *Sensors and Actuators B: Chemical*, Vol. 145, p. 674, 2010.
99. K.A. Marx, *Biomacromolecules*, Vol. 4, p. 1099, 2003.
100. B. Ding, J. Kim, Y. Miyazaki and S. Shiratori, *Sensors and Actuators, B: Chemical*, Vol. 101, p. 373, 2004.
101. B. Ding, M. Yamazaki and S. Shiratori, *Sensors and Actuators, B: Chemical*, Vol. 106, p. 477, 2005.
102. X. Wang, B. Ding, M. Sun, J. Yu and G. Sun, *Sensors and Actuators, B: Chemical*, Vol. 144, p. 11, 2010.
103. X. Wang, B. Ding, J. Yu, M. Wang and F. Pan, *Nanotechnology*, Vol. 21, p. 2010.
104. N. Horzum, D. Tascioglu, S. Okur and M.M. Demir, *Talanta*, Vol. 85, p. 1105, 2011.
105. X. Wang, F. Cui, J. Lin, B. Ding, J. Yu and S.S. Al-Deyab, *Sensors and Actuators, B: Chemical*, Vol. 171–172, p. 658, 2012.
106. Z.G. Wang, L.S. Wan, Z.M. Liu, X.J. Huang and Z.K. Xu, *Journal of Molecular Catalysis B: Enzymatic*, Vol. 56, p. 189, 2009.
107. M. Ahmad, C. Pan, Z. Luo and J. Zhu, *Journal of Physical Chemistry C*, Vol. 114, p. 9308, 2010.
108. H. Tang, F. Yan, Q. Tai and H.L. Chan, *Biosensors and Bioelectronics*, Vol. 25, p. 1646, 2010.
109. S. Huang, Y. Ding, Y. Liu, L. Su, R. Filosa and Y. Lei, *Electroanalysis*, Vol. 23, p. 1912, 2011.
110. G. Ren, X. Xu, Q. Liu, J. Cheng, X. Yuan, L. Wu and Y. Wan, *Reactive and Functional Polymers*, Vol. 66, p. 1559, 2006.
111. M. Scampicchio, A. Arecchi, N.S. Lawrence and S. Mannino, *Sensors and Actuators, B: Chemical*, Vol. 145, p. 394, 2010.
112. P. Si, Y. Huang, T. Wang and J. Ma, *RSC Advances*, Vol. 3, p. 3487, 2013.
113. Y. Liu, H. Teng, H. Hou and T. You, *Biosensors and Bioelectronics*, Vol. 24, p. 3329, 2009.
114. F. Cao, S. Guo, H. Ma, D. Shan, S. Yang and J. Gong, *Biosensors and Bioelectronics*, Vol. 26, p. 2756, 2011.
115. Y. Ding, Y. Liu, J. Parisi, L. Zhang and Y. Lei, *Biosensors and Bioelectronics*, Vol. 28, p. 393, 2011.
116. W. Wang, Z. Li, W. Zheng, B. Dong, S. Li and C. Wang, *Journal of Nanoscience and Nanotechnology*, Vol. 10, p. 7537, 2010.
117. Y. Ding, Y. Wang, L. Su, M. Bellagamba, H. Zhang and Y. Lei, *Biosensors and Bioelectronics*, Vol. 26, p. 542, 2010.
118. K.M. Manesh, P. Santhosh, A. Gopalan and K.P. Lee, *Analytical Biochemistry*, Vol. 360, p. 189, 2007.

119. J. Huang, D. Wang, H. Hou and T. You, *Advanced Functional Materials*, Vol. 18, p. 441, 2008.
120. J. Shen, X. Yang, Y. Zhu, H. Kang, H. Cao and C. Li, *Biosensors and Bioelectronics*, Vol. 34, p. 132, 2012.
121. F. Cao, S. Guo, H. Ma and J. Gong, *Electroanalysis*, Vol. 24, p. 418, 2012.
122. J.P. Wu and F. Yin, *Analytical Letters*, Vol. 46, p. 818, 2013.
123. K. Sawicka, P. Gouma and S. Simon, *Sensors and Actuators, B: Chemical*, Vol. 108, p. 585, 2005.

12

Smart Sensing Scaffolds

Carmelo De Maria[1,2], Yudan Whulanza[1,3], Giovanni Vozzi[1,2] and Arti Ahluwalia[1,2,*]

[1]*Research Center E. Piaggio, University of Pisa, Pisa, Italy*
[2]*Department of Information Engineering, University of Pisa, Pisa, Italy*
[3]*Department of Mechanical Engineering, University of Indonesia, Depok, Indonesia*

Abstract
There is great interest in measuring cell functions on three dimensional scaffolds without using invasive and destructive methods. Here we describe how scaffolds for tissue engineering can also be used as sensors for monitoring cellular activity such as adhesion and spreading.

Carbon nanotube polyester-polymer composites were fabricated into membranes and scaffolds with electro-conductive properties. Several nanotube concentrations were introduced aiming at understanding their influence on mechanical properties, impedance features and electric percolation threshold of the polymer matrix. It was observed that a concentration of 0.3% was able to transform an insulating matrix into a conductive one. Experimental results were compared with theoretical models.

Impedance techniques were used to measure the effects of media and cell cultures on composite membranes and the results were analysed using lumped parameter models. Impedance changes can be correlated with protein or cell adhesion, spreading and changes in cell density.

Keywords: Smart scaffolds, CNTs, impendence measurements, mechanical properties, percolation theory

12.1 Introduction

Currently, there is great interest in designing smart materials capable not only of directing cellular activity, but also of monitoring cell adhesion

**Corresponding author*: arti.ahluwalia@centropiaggio.unipi.it

Annarosa Gugliuzza (ed.) Smart Membranes and Sensors, (337–366) 2014 © Scrivener Publishing LLC

and function [1]. At present, most methods for measuring cell function on three dimensional (3D) scaffolds are invasive and destructive as they involve measurement of gene expression or labelling of cytoplasmic proteins. In principles, if one could transform a conventional scaffold into a "smart scaffold" which acts as a sensor for monitoring cell behaviour, it would be possible to perform long term experiments for the study of chronic disease models, or for repeat dose experiments.

Scaffolds serve as a synthetic structure able to induce cell organization into a three-dimensional architecture and to provide the right stimuli in directing cell growth and proliferation. One of principal requirements that a scaffold has to fulfil is mechanical stability, which is crucial for maintaining the predesigned tissue structure. Mechanical stability mainly depends on the selection of biomaterial and architectural design of the matrix. Although polymers, such as PLLA (polyl-L-lactide acid), PLGA (poly(lactic-coglycolic acid)) or PCL (poly-caprolactone), are the most commonly used biopolymers for realisation of scaffolds in various tissue engineering applications, they do not often have the necessary mechanical strength. Moreover, although they are useful for creating tissue constructs, it is usually difficult to assess cell function using non-destructive methods.

Giaever and Keese [2, 3] have shown that as the cells attach and spread on a small electrode surface, they alter the effective area available for current flow causing an increase in the impedance of the system. After these initial changes, the impedance fluctuates with time. These fluctuating impedance characteristics, after careful interpretation can be used for examining the attachment, spreading and motion of cell populations. This method was then adapted by Connelly's group to better understand extracellular action potential measurements [4]. Wegener *et al.* [5] also used the same method to monitor the proliferation of epithelial cells. Ehret *et al.* [6] used an interdigitated electrode structure to monitor the spreading, attachment and morphology of fibroblasts in culture. In summary, these reports have shown that the impedance of microelectrodes of various sizes and structures can give useful information on cell behaviour.

The unique properties of carbon nanotubes (CNTs) such as extremely high strength, low density and high electric conductivity, offer advantages over other nano-fillers [7, 8]. The extraordinary mechanical properties of CNTs make them very attractive and promising as reinforcing fillers for the production of a new generation of tissue substitutes. Experimental data reveal that CNTs dispersed in a polymer matrix significantly improve its mechanical properties [9–12].

CNTs have dimensions which are comparable to extracellular matrix (ECM) molecules such as collagen and laminin, and are reported to support cell adhesion [13–16]. CNTs are in fact extensively explored for biomedical applications [17–19], particularly those involving scaffolds for neural and bone tissue. For example neurons grown on a CNT network are reported to exhibit better signal transmission [20], possibly due to the fact that CNTs form tight contacts with neuron membranes leading to electrical shortcuts [21].

Moreover, as CNTs behave like an inert matrix, combined with other natural or synthetic materials in biocomposites, they can be effective in bone tissue engineering applications. CNT-based substrates have been shown to support the growth of osteoblastic cells which can be expected to become functional bone [19, 22, 23].

However, the effective utilization of the excellent properties of nanotubes in composite applications strongly depends on the ability to disperse CNTs homogeneously throughout the matrix, as well as on the interfacial bonding and the content of nanotubes in the matrix. Producing well-dispersed CNTs in a composite is difficult because the addition of solid 'powder' (carbon) in a liquid polymer in the early mixing stages often leads to phase separation between the carbon and the polymer matrix due to low interfacial bonding between the two. Several methods for enhancing interfacial adhesion between CNTs and the polymer matrix are described [24]. Following a simple two-solvent mixing method, a tolerable dispersion with low content of CNT in a polymer matrix has been reported by the authors [25].

CNTs have also been used to create electrically conductive polymers and tissue scaffolds with the capacity to provide controlled electrical stimulation. It has been reported that current-conducting CNT/polymer composites promoted various osteoblast cell functions. By applying alternating current to these nanocomposites, an increase in osteoblast proliferation by 46%, and calcium deposition by 307% has been observed [26]. This result has suggested that CNT-based composites may be used to stimulate bone formation. In a previous work, we have used impedance techniques to measure the effects of media and cell cultures on our CNT composite films. Considering recent publications, it can be assumed that the use of CNTs for tissue engineering appears to be challenging, but on the other hand, potentially rewarding perspective to develop a novel generation of engineered biomaterials for monitoring cellular activities, as explained in the following sections [27, 28].

12.2 Composite Sensing Biomaterial Preparation

Polymer (PLLA, PCL, PLGA) was solved in chloroform (Sigma Aldrich, Italy) to a concentration of 0.2 g/mL. SWNTs (single-walled nanotubes, kind gift from Prof. Daraio of Caltech University, USA) with a density of 1.8 g/cm^3 were used as prepared without any chemical pre-treatment or modification. The SWNTs were sonicated in benzene (Sigma-Aldrich, Italy) for 2 minutes to obtain an homogenous CNTs dispersion with a concentration of 1, 2, 3, 4, 5, 8 mg/mL. Various composites solutions were obtained adding in a volume ratio of 1:3 Polymer solution:CNT dispersion and sonicating them 2 minutes. The resulting SWNTs/polymer weight ratios were of 0.2, 0.3, 0.5, 0.7, 0.8 and 1.3%. Composite films were obtained by spin coating (BLE Equipment, Germany) onto a 12 mm glass slide and then they were dried in a silica tank for seven days to allow complete evaporation of solvents.

12.3 Composite Sensing Biomaterial Characterisation

In order to visualise the formation of carbon nanotube clusters in the polymer matrix, the samples were analysed under an optical microscope (Olympus AX70, Olympus, Italy) and pictures of different areas of each sample were taken.

The mechanical properties of SWNT composites were measured using a Zwick/Roell testing device (Z005 series, Genova, Italy). In this test the sample was pulled until 10% elongation. The load was applied in the axial direction (in-plane) of the sample. The specimens were stretched at a constant load speed of 0.05 mm/min. Load F and strain ε were recorded by a computer connected to the control unit of tensile apparatus. From the stress–strain graphs the tensile strength of the samples was estimated and the elastic modulus calculated.

Samples were also characterised in terms of their electromechanical behaviour.

Samples of 1 cm × 2 cm were prepared from spun cast films. Two copper wires were glued to opposite ends of the CNTs composite films. A tensile test was performed in parallel with impedance measurement to characterise them mechanically and also electrically. The impedance of SWNT/polymer films was measured using Agilent E4980 LCR Meter (Agilent, Milan, Italy). Voltage amplitudes of 1 V AC and frequencies ranged from 20 Hz to 2 MHz were applied. From the results of the AC impedance, the specific conductivity σ of the samples was calculated according to Equation 12.1:

$$\sigma(f) = \frac{1}{|Z(f)|} \cdot \frac{1}{t} \tag{12.1}$$

where t and |Z(f)| are the sample thickness and the complex impedance of the sample as a function of frequency, respectively.

12.4 SWNTs-Based Composite Films Structural Properties

Before performing mechanical and electrical tests of SWNT-based composites, the structure and morphology of samples were characterized by using of optical microscopy. Composites solutions of SWNTs and polymers (PLLA, PLGA and PCL) were found to be relatively homogeneous without any evident phase segregation of the components. The mixtures were black and uniform in colour, stable for days, indicating efficient dispersion of CNTs in polymer solution. The high shear forces introduced by ultrasonic energy were sufficient to disperse the SWNTs uniformly in the polymer solution. Only the composites of SWNTs/PLGA at a concentration higher than 0.8 wt.% revealed some phase segregation after sonication: high concentrations of carbon nanotubes tended to aggregate during the mixing process. The stability of the dispersions could be attributed to electrostatic stabilisation, a well-known phenomenon in colloidal dispersions [29]. The results obtained by optical microscopy analysis of 0.2, 0.5 and 0.8 wt.% of CNTs in PLLA respectively is reported in Figure 12.1 a-b-c. In Figure 12.2, images of cast films at concentration of 0.5 wt.% CNTs in different polymer matrix are reported. Comparing the images, it can be concluded that CNTs/PLGA composites showed

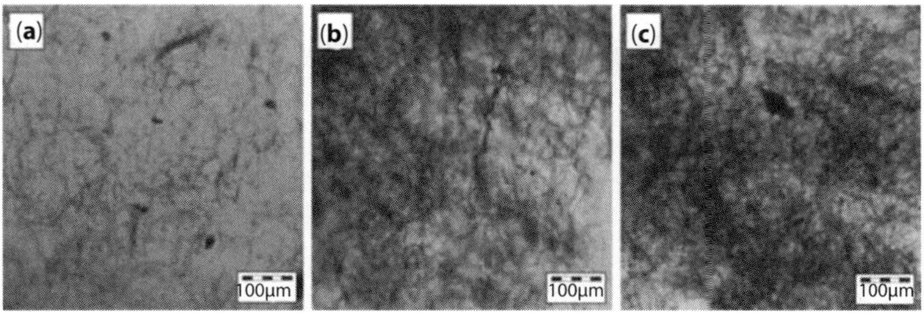

Figure 12.1 Optical microscope image for plla/swnts composite with (a) 0.2 wt.%, (b) 0.5 wt.% and (c) 0.8% cnts concentration (Reprinted with permission from [27], Whulanza Y et al. *J Nanosci Nanotechnol. 2013 Jan;13(1):188-97.* Copyright ©American Scientific Publishers).

Figure 12.2 Optical microscope image for composite with 0.5 wt.% swnts in : (*a*) plla, (*b*) pcl and (*c*) plga matrix (Reprinted with permission from [27], Whulanza Y *et al. J Nanosci Nanotechnol.* 2013 Jan;13(1):188-97. Copyright © American Scientific Publishers).

significantly more massive aggregation than that of PLLA or PCL system (Fig. 12.2).

Because of strong intrinsic van der Waals forces, individual carbon nanotubes were interwoven with each other, forming an interconnected network within the host material. In general, SWNTs were fairly dispersed across the matrix. Bigger black spots were observed in figure 12.1c. These observations clearly indicated that the agglomeration process of nanotubes in the composite was concentration-dependent. These observations agree well with former results on carbon black particles in the PLLA matrix [30, 31]. Kim *et al.* [30] showed that the application of low shear forces could induce agglomeration of initially well-dispersed carbon black particles, explaining that the shear forces provided particles with sufficient kinetic energy to overcome the repulsive interactions of the electric double layers. At the same time, agglomerates could be disrupted by high shear forces. Therefore, the cast films also presented uniformly distribution of CNTs within polymer matrix. As described above, the CNTs-PLGA solutions tend to heavily aggregate in the 0.5 wt% addition of CNTs. On the other hand, at the same blend composition, PLLA and PCL solutions still show good dispersion. The difference in aggregation between the different polymers (Fig. 12.2) might due to the property of the PLGA material which is composed of PGA and PLA. In fact PGA has relatively high portion of crystallinity (45–55%) which made it hard to disperse CNTs in polymer. On the contrary, PLLA is an amorphous polymer so is less problematic to distribute CNTs inside it [32]. On the basis of this analysis, we found a maximum value of 0.5 wt.% CNTs to form a uniform dispersion in PLGA.

12.5 Tensile Properties of SWNTs-Based C omposite Films

The tensile modulus and ultimate strength were: 65.2± 5.3 MPa, 2.1±0.3 MPa, for the pure PLLA; and 223.8±5.3 MPa, 6.3±0.5 MPa, for a PLLA composite with 0.5% CNT content, respectively. The tensile strength of the neat polymers increased roughly around 3 times; the elastic modulus increased 3.6 times, when SWNTs were incorporated into the host material. As shown in Figure 12.3a, a decrease in ultimate strength of the PLLA/SWNTs composites was evident. This confirmed that composites with greater agglomeration of nanotubes were far weaker than films with finer blocks of SWNTs. In Figure 12.4 we have reported the elastic modulus measured experimentally as a function of SWNTs concentration for PLLA, PCL and PLGA. The mechanical properties of the spun films were essentially the same, suggesting that the three-dimensional distribution of nanotubes controlled the failure mechanism. To explain this behaviour we have used different mechanical models (Voigt, Reuss and Halpin–Tsai) as reported in Figure 12.5, as explained later in the text.

In general, the elastic moduli and ultimate strengths of CNT-based composites are reported to increase compared to the neat polymer. The

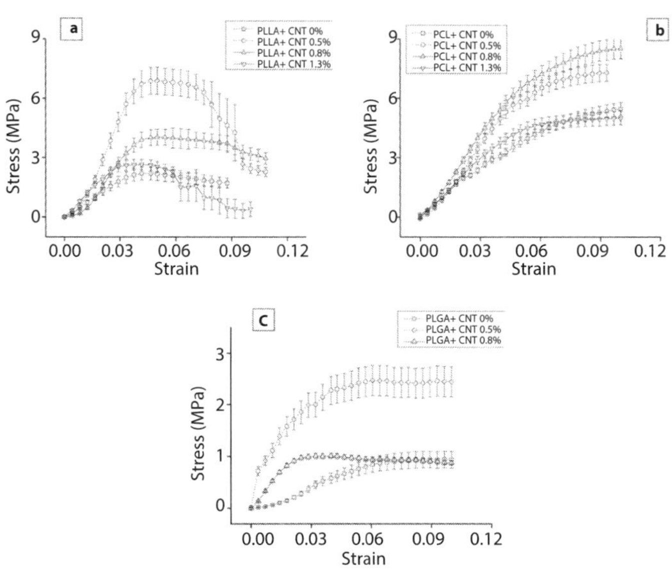

Figure 12.3 Stress-strain graphs of (*a*)PLLA, (*b*) PCL and (c) PLGA composite films with various concentration of CNTS (Reprinted with permission from [27], Whulanza Y *et al. J Nanosci Nanotechnol.* 2013 Jan;13(1):188-97. Copyright©American Scientific Publishers).

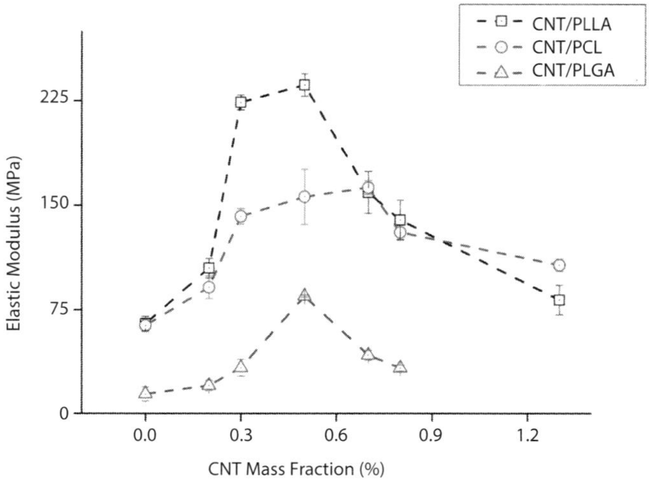

Figure 12.4 Elastic modulus of swnts based composite films as function of SWNTS contents (Reprinted with permission from [27], Whulanza Y *et al. J Nanosci Nanotechnol.* 2013 Jan;13(1):188-97. Copyright ©American Scientific Publishers).

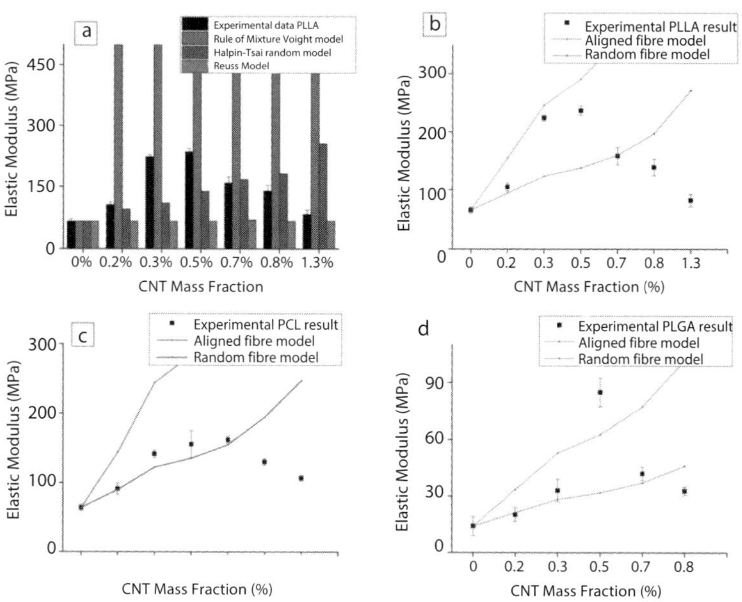

Figure 12.5 (*a*) comparative graphs of experimental and modelled elastic modulus using Halpin-Tsai model, Voigt and Reuss model prediction of elastic modulus using Halpin-Tsai both aligned and random model for (*b*) PLLA; (*c*) PCL and (*d*) PLGA composites. (Reprinted with permission from [27], Whulanza Y *et al. J Nanosci Nanotechnol.* 2013 Jan;13(1):188-97. Copyright ©American Scientific Publishers).

assessment of the data accumulated in numerous studies on CNTs composites revealed that effective reinforcement of these materials strongly depends on several factors such as: a high aspect ratio of the CNTs, good dispersion of the nanotubes in a matrix, good interfacial bonding, interactions and mechanical anchoring between CNTs and polymer molecules. These simple design guidelines for carbon nanotube composites could permit substantial advances in the composites' mechanical properties [33–36]. As it is possible to see in Figure 12.3, the massive aggregation of carbon nanotubes apparently reduced the mechanical performance of composites due to the disruption of interfacial interaction between CNTs and polymer matrix. It was also observed that increasing the loading amount of nanotubes in these composites caused a significant increase of stiffness, which eventually led to brittle fracture, as indicated by lower elongation at break. Moreover, similar reinforcing behaviour also was present in PCL and PLGA composites. The increase of the elastic modulus and tensile strength of these composites has been shown to change by a factor in the range of 2.5–3.6 and 2–3, respectively for PCL composites. On other hand, PLGA composites have shown the greatest increase of elastic modulus and ultimate strength at around 6 and 6.6 times, respectively. These results demonstrated that the presence of carbon nanotubes was crucial for achieving improved mechanical properties of polymer composite.

The tensile analysis (Fig. 12.4) led to the following observations:

- Elastic modulus increased with increasing SWNT concentration up to optimum CNT content then decreased. The optimum CNTs additional content was 0.5 wt.% for PLLA and PLGA composite. On the other hand, PCL composite had an optimal value of 0.7 wt.%. It has been suggested that these optimum CNTs contents were corresponding to the optimal dispersion of CNTs in the polymer matrices.
- The trends for tensile strength and elastic modulus were similar for the 3 polymer composites.

To explain this behaviour we used different mechanical models. Initially, we have modelled the composite in a simple way as an isotropic and elastic matrix filled with aligned elastic fibres that span the full length of specimen. It was assumed that the matrix and fibres were very well-bonded. When a stress was applied along the fibre alignment direction, the matrix and fibres were equally strained and total stress was the sum of contributions. Under these circumstances, the composite tensile modulus in the alignment direction, E_{cV}, was given by Equation 12.2:

$$E_{cV} = \left(E_f - E_m\right)V_f + E_m \tag{12.2}$$

where E_f was the fibre modulus, E_m was the matrix modulus and V_f was the fibre volume fraction, on the basis of the well-known Voigt's rule of mixtures [37].

We also compared the results with Reuss model, where the two components were assumed as a series connection:

$$E_{cR} = \frac{E_f E_m}{\left(1 - V_f\right)E_f + V_f E_m} \tag{12.3}$$

These parallel and series models have been used to give an estimate of the upper and lower bounds of the blend modulus. However, they could describe a rather idealised situation: in fact the above models do not because of the fibres dimension and orientation.

Tucker [38] have been provided a good review of the application of several classes of micromechanical models to discontinuous fibre-reinforced polymers. He noted that, of the existing models, the widely used Halpin–Tsai equations [39–41] gave reasonable estimates for effective stiffness. On the other hand, the Mori–Tanaka type models [42, 43] have reported the best results for large- aspect-ratio fillers. Therefore, we focused on prediction based on the Halpin-Tsai model. This prediction was originally developed for continuous fibre composites which had two approximations for fibre orientation in the polymer matrix: aligned and random fibre orientation. For aligned fibre composites, the Halpin–Tsai model reported the composite modulus to be (Eqs. 12.4a & 12.4b):

$$E_c = E_m \frac{1 + 2\left(L/D\right)V_f\eta}{1 - V_f\eta} \tag{12.4a}$$

$$\eta = \frac{(E_f / E_m) - 1}{(E_f / E_m) + 2\left(L/D\right)} \tag{12.4b}$$

For randomly orientated composites the expression became (Eqs 12.5a,b &c):

$$\frac{E_c}{E_m} = \frac{3}{8}\left[\frac{1 + 2\left(L/D\right)\eta_L V_f}{1 - \eta_L V_f}\right] + \frac{5}{8}\left[\frac{1 + 2\eta_T V_f}{1 - \eta_T V_f}\right] \tag{12.5a}$$

where

$$\eta_L = \frac{(E_f / E_m) - 1}{(E_f / E_m) + 2(L / D)} \tag{12.5b}$$

$$\eta_T = \frac{(E_f / E_m) - 1}{(E_f / E_m) + 2} \tag{12.5c}$$

In this model L and D were the length and the diameter of fibers respectively.

Both rule of mixtures and the Halpin–Tsai equations were taking E_m = 500 GPa and L/D = 500. Several studies have been reported values in this range [9, 44, 45]. It should be emphasized here that L was not the true or actual length of the nanotubes in the composite, but it can be considered as the effective nanotube length responsible for reinforcement in the composite system. Any difference in modelled and experimental values in different composites could be attributed to differences in interfacial strength in different systems or in same system at different nanotube loading. As predicted, the Voigt model which takes into account total stress of fibres and matrix correspondingly, showed a much higher values than experimental data (Figures 12.5). On the other hand, the Reuss model, as the lower bound of the rule of mixtures, gave results closer prediction. However, the Reuss model failed in following the trend of elastic modulus with increasing CNTs. The Reuss' estimates were quasi-constant during the addition of 0.2–1.3% CNT (Figures 12.5). On the other hand, the Halpin-Tsai random model gave closer prediction to experimental data. Therefore, the Halpin-Tsai random and aligned models were then taken into account in estimating the composite systems.

Moreover, the results of these modelling phase showed (Figures 12.5) that the real values were in the range between aligned and random fibre composites model, except for PLGA +0.5 wt.% CNT. However, at higher nanotube volume fractions (more than 0.8 wt.%) the deviations were greater. The reason for this difference was that the models were not suitable for massive CNT agglomeration in polymer matrix. the Halpin-Tsai equation was derived based on the assumption that stress was uniformly transferred between polymer matrix and nanotubes. The interfacial shear strength is an important parameter for any fibre-reinforced composite and many recent studies are devoted to it. The first thing to determine is whether any stress is transferred to nanotubes at all. Hence, non-uniform dispersion of CNTs leads to poor stress transfer to the matrix.

12.6 Electrical Properties of SWNTs-Based Composites Films

The electrical conductivity of the composites was determined by using Eq. (12.1). The conductivity values were found in range of 2×10^{-3} – 2 S/m, which is close to the range of SWNTs/polymer system that has been reported by Deng [46]. The results from the electrical conductivity measurements of composites with various concentrations of SWNTs were reported in Figure 12.6:

We analysed the conductivity spectra as a function of frequency for composites with different concentrations of SWNTs. Conductivity of PLLA/SWNTs composites increased as SWNTs concentration increased

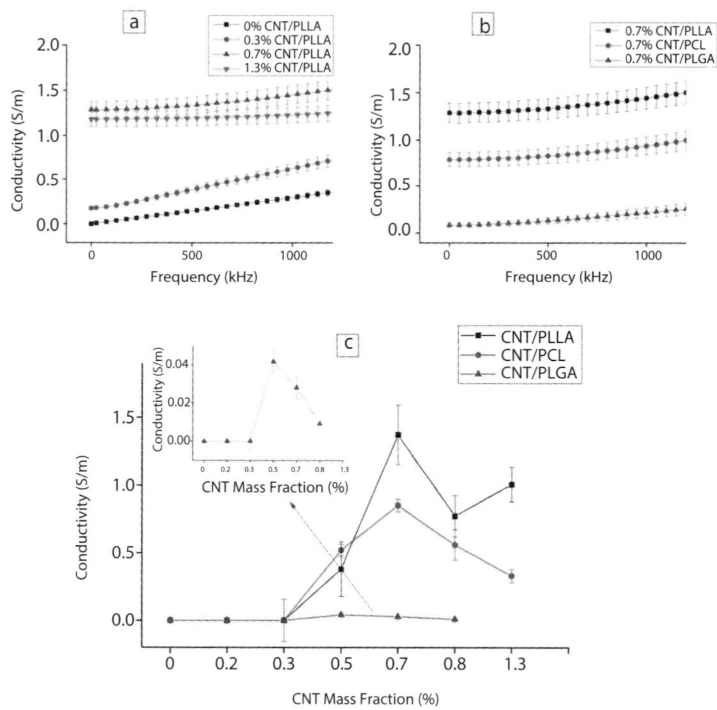

Figure 12.6 (*a*) Conductivity of PLLA/SWNTS-based composite films as function of frequency; (*b*) conductivity of 0.7% SWNTS-based composite films as function of frequency. (*c*) conductivity of SWNTS based composite films as function of SWNTS concentration, with a zoom of conductivity of SWNTS/PLGA based composite films as function of swnts concentration (Reprinted with permission from [27], Whulanza Y *et al. J Nanosci Nanotechnol.* 2013 Jan;13(1):188-97. Copyright ©American Scientific Publishers).

(fig.12.6a). In particular under 0.5% SWNTs concentration the conductivity has not reached a plateau value; this could indicate that a perfect electric connection between the carbon nanotubes embedded in polymer matrix was not obtained and the capacitor effect was dominant in these composites. Instead from 0.7% SWNTs concentration the plateau was reached, sign that the electrical connection between the CNTs was reached at low frequencies and the capacitor effect acted at only low frequencies. This results was observed in all different analysed composites (Fig. 12.6b).

The conductivity of composites sharply increased in order of 10,000 (the pure polymer had value around 2×10^{-6} S/m) at around 0.5 SWNTs wt.% indicating the formation of a percolating network. These results have been highlighted the importance of nanotube clustering for the formation of a conductive network.

At low percentage of carbon nanotubes (figure 12.6c), the conductivity of the compounds was very low. This was because the CNTs particles were isolated from each other, and there was no conducting path. As the loading of CNTs was increased, the conductivity increased rapidly up to a particular concentration, thereafter, the increase in conductivity was not so rapid. This is because, when the amount of CNTs was increased, cluster or aggregates of CNTs were formed and a conducting path was already formed. Thus, the internal contact resistance between the aggregates decreased as the loading level increased. The composite became conductive above a critical value percolation threshold that defines the insulator-conductor transition [47, 48].

As described previously, SWNTs were dispersed in various polymer matrices (PLLA, PCL and PLGA) always in the same way; however, PLGA/SWNTs composites showed a lower conductivity level. Such differences in electrical properties can be explained by a difference of tube–tube contact electrical properties between PLLA/PCL and PLGA composites. An explanation can be related to the higher amount of crystallinity in the case of PLGA polymer. The organisation of this tube-polymer network in the film was highly uncertain due to the low solubility. This kind of network could strongly modify electrical properties of tube–tube contacts. Therefore, correspondingly to mechanical property explanation, the conductivity analysis led also to observation:

- conductivity increased with increasing SWNTs concentration up to optimum CNTs content then decreased.
- the optimum CNTs additional contents were 0.7 wt.% for PLLA and PCL system. Alternatively, PLGA system had optimum value at 0.5 wt.%.

Table 12.1 Result of fitting experimental data with percolation scaling law.

Polymer system	Volume filler critical (%)	Mass filler critical (%)	Power constant	Error (%)
PLLA	0.20	0.32	1.72	11.7
PCL	0.20	0.31	1.71	13.5
PLGA	0.18	0.29	1.86	16.2

The relationship between the composite conductivity σ and the concentration above the percolation threshold (p_c) could be described by a scaling law [48–50]:

$$\sigma \propto (p - p_c)^{\gamma} \tag{12.6}$$

where p was a CNTs concentration and γ was the critical exponent.

The experimental data were fitted to the scaling law to obtain the p_c and γ. The concentrations would most commonly be expressed as volume fraction (p) of filler particles, however, in the current case, since the density of the nanotubes can only be estimated approximately (1.7–1.9 g/cm^3), weight fractions have been used. Volume fraction based percolation thresholds may be estimated by multiplying the weight fraction results by the approximate matrix/filler density ratio (e.g. 1.127/1.8 for PLLA).

Since PLLA is an insulating polymer, the percolation threshold was observed at 0.3 wt.%, where the conductivity jumped from 10^{-6} S/m (insulating region) to ~10^{-3} S/m (conducting region). By fitting the plot of log (σ) versus log (p-p_c), the values of correlation coefficient R and γ were estimated to be 0.96 and 1.72 respectively. In Table 12.1 we have summarised the percolation fitting for three composite systems. The SWNTs critical percolation occurred at around 0.3 wt.% which applied for all our polymer systems (Table 12.1). These results were interesting because power constants were lower than would be theoretically predicted for a 3D percolation occurred to be about 2. This discrepancy could be explained in terms of the SWNTs starting to aggregate in much lower concentration (about 0–0.2%).

12.7 Electromechanical Characterisation and Strain-Dependence Measurement

The resistivity of a conductive polymer composite can be sensitive to various external stimuli [51, 52], such as gas, pressure, temperature, and deformation. Resistivity of CNTs/polymer composites with strain deformation dependence was studied. The resistivity was measured using a two-point measurement system under a constant applied voltage of 1 V. The rectangular samples were stretched in an Zwick/Roell universal test machine, and the resistances were measured by an Agilent impedance analyszer. The resistivity measurement setup and tensile test machine were both interfaced with the same computer. Hence, resistivity-strain dependence could be measured.

The effect of CNTs concentration on the resistivity-strain dependence was measured by observing the relative change in resistance as a function of strain. The result of 0.7% CNTs composites is plotted in Figure 12.7.

Electromechanical characterisation (Figure 12.7) showed that the change in impedance magnitude was proportional to the change in length. The resistivity-strain dependence can be quantitatively expressed by a gauge factor G or sensitivity factor which is commonly used in strain gauge measurement [53]:

$$GF = \frac{\Delta R / R_0}{\varepsilon} \tag{12.7}$$

where $\Delta R/Ro$ is normalized resistance with respect to its initial time and ε is corresponding strain.

It was found that the change in impedance magnitude was proportional to the change in length. The impedance profile was increased significantly from value of 25 kΩ to 75 Ω at the end of tensile test (fig.

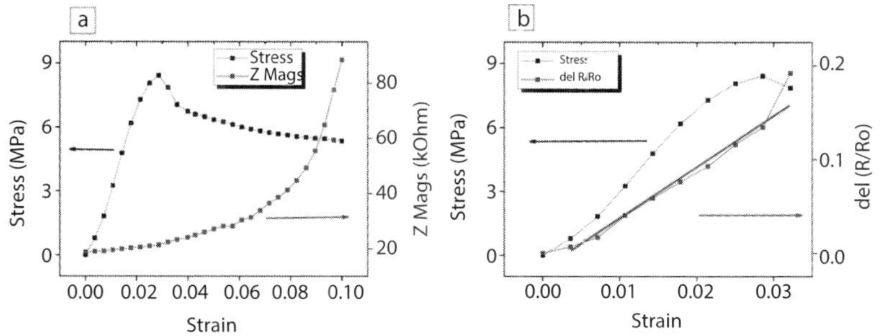

Figure 12.7 (a) Plot of strain vs stress and impedance magnitude and (B) strain vs normalised resistance at elastic region for 0.7 Wt% SWNTs (Reprinted with permission from [27], Whulanza *Y et al. J Nanosci Nanotechnol.* 2013 Jan;13(1):188-97. Copyright ©American Scientific Publishers).

Table 12.2 Calculation of Gauge factor as function of SWNTs content.

CNT concentration	Gauge factor/strain sensing		
	PLLA	PCL	PLGA
0	0.06 ± 0.02	0.05 ± 0.01	0.04 ± 0.05
0.2	0.10 ± 0.05	0.23 ± 0.03	0.34 ± 0.07
0.5	16.5 ± 6.7	1.21 ± 0.35	9.0 ± 3.1
0.7	20.4 ± 2.4	2.82 ± 0.29	0.69 ± 0.07
0.8	8.5 ± 0.6	1.96 ± 0.9	0.8 ± 0.56
1.2	6.7 ± 0.5	2.18 ± 0.91	–

12.7(a). This result suggested that the composite film was still in conducting region after the tensile test. As the definition of gauge factor, the normalised resistance, ΔR/Ro should be reversible with strain and compression. Therefore, only the elastic region was accounted for calculation of gauge factor. In this case, the elastic region was observed until an elongation of 0.03 and plotted in Figure 12.7b. The gauge factor was easily determined by calculating the slope of curve fitting from Figure 12.7b. In Table 12.2 we summarised the gauge factor of CNTs/polymer composites as a function of CNTs weight content. It can be observed that the gauge factor for PLLA and PLGA got higher with a more CNTs addition up to some point and then decreased. On the other hand, gauge factor for PCL composites showed a rather constant value in the range of 0.7–1.2 wt.% CNTs concentration.

As a reference, gauge factor of a stiff material such as metal is around 2 [53]. Another aspect can be deduced from the view of CNTs conducting network. An higher CNTs content have been developed a more connected networks to allow conductive percolation. With the same strain applied, a film with higher CNTs content still established this percolation network. Hence normalised resistance tended to be lower compared to a low CNTs content film.

12.8 Cell Sensing Scaffolds

12.8.1 Preparation

The 1.3% CNT/PLLA composite, prepared as described in the previous section, was chosen to fabricate cell sensing scaffolds. because 2% or higher

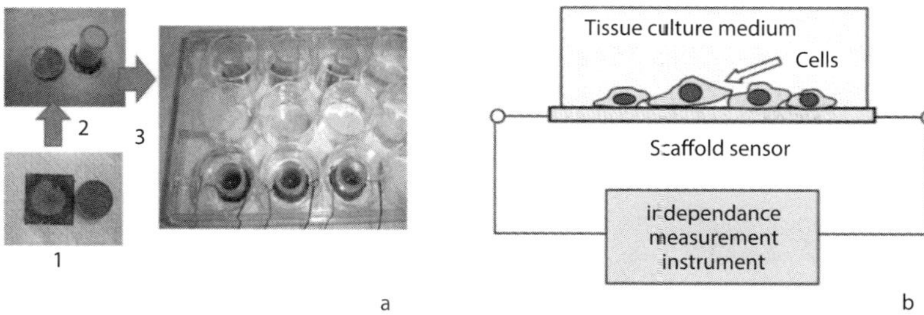

Figure 12.8 (*a*) Membrane electrode preparation and (*b*) membrane electrode measurement.

concentrations are phase separated with CNT agglomerates visible to the naked eye, while lower concentrations are less conductive. 300ul of the mixture was spin coated at 1500rpm (BLE Equipment, Germany) onto a 12mm glass slide and dried in a silica tank for seven days to evaporate the solvents. Control membranes of PLLA were also fabricated using the same spinning parameters. Two wires were glued to opposite ends of the CNT/PLLA membranes and a silicon tube of ID 10mm was glued over the centre of the membrane to realise a cell culture chamber (Fig. 12.8a). Finally the membranes were placed in a 12 well plate and sterilised using H_2O_2 gas plasma.

12.8.3 Cell Testing

The human hepatoma-derived cell line C3A (ATCC Culture, USA) was used as a model. They were cultured in Dulbecco's Modified Eagle Medium (DMEM, Lonza, Milan, IT) with 10% FBS, 1% Penicillin/Streptomycin/Amphotericin B, 1% l- Glutamine 200mM(all from Lonza); 1% non-essential Amino acids 100× (EuroClone, Milan, IT) and 1% MEM vitamins solution (Sigma, Milan, IT). This medium was used for all experiments. Sterilised membranes were first were coated with 0.3 mg/ml of collagen extracted from rat tails according to standard procedures [54] (Beken *et al.*, 1998) and then equilibrated in complete medium for 24 h. Subsequently membranes were seeded with 160 000 cells/cm^2 in 250μl of media for up to 2 days. As a control group, collagen coated membranes without cells were placed in the same volume of media and in the incubator for the same time. To confirm the presence of cells on the membrane, cells were fixed and stained with DAPI and imaged using a fluorescent microscope (Olympus IX81, Olympus Italia). Viability was assessed using the Cell Titer-BlueTM Cell Viability Assay (Promega, Madison, USA) and compared with controls cultured in collagen coated tissue culture plates (24 well plates with the same number of cells) as well as with collagen coated PLLA membranes.

12.8.4 Membrane Impedance Measurement

Impedance measurements were performed using an Agilent 4086 RCL meter over the range 0 kHz to 2MHz with 10 kHz increments. The experimental scheme is shown in Figure 12.8b.

The following protocol was used:

- measurement of dry CNT/PLLA membrane impedance;
- measurement of dry PLLA membrane impedance;
- measurement of CNT/PLLA and PLLA collagen coated membrane impedance in culture medium over 44 h. A sterile membrane was immersed in medium at 37 °C and its impedance was monitored at regular intervals. This was done to evaluate the time dependence of wetting and swelling phenomena in the membrane;
- measurement of CNT/PLLA and PLLA membrane impedance with cells over 44 h, at 37 °C. Cells were only added after a 24 h equilibration period in the medium, as initial observations showed that the membranes did not attain a stable impedance value until at least 6–24 h in medium.

To evaluate the impedance change over the observation time, we calculated the fractional impedance change $|Z_p|$ by using the initial impedance $|Z_r|$ of a wet membrane just before seeding cells (t = 0 h) as a reference point and the successive time points as $|Z_t|$.

$$\left|Z_p\right| = \frac{\left|Z_t\right| - \left|Z_r\right|}{\left|Z_r\right|} \cdot 100 \tag{12.8}$$

Figure 12.9a shows the measured impedance of one scaffold sensor as function of frequency measured at 0, 6, 24 and 44 h while Fig. 12.9b reports the corresponding value of |Zp|. The measurement was carried out by immersing samples in complete medium for up to 44 h. The 24 h and 44 h responses are coincident, indicating that the membranes were completely equilibrated with medium at 24 h. Since the medium contains serum proteins, it is likely that the equilibration process involves dynamic protein adsorption as described by Vroman [55]. After equilibration the

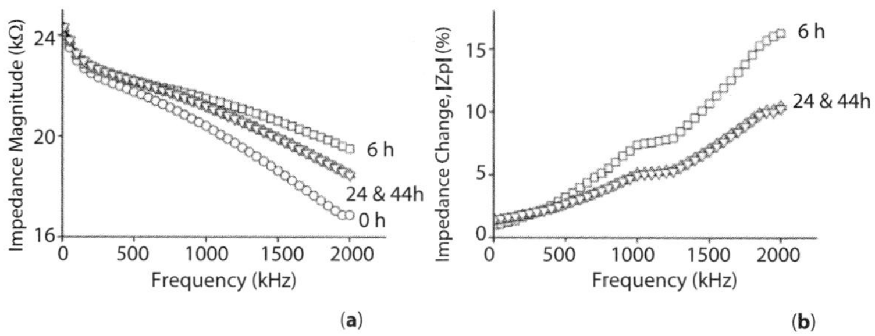

Figure 12.9 Plot of impedance magnitude over frequency for scaffold without cell suspension. (Reprinted with permission from [28], Whulanza Y *et al. Biosens Bioelectron.* 2011 Mar 15;26(7):3303-8).

sensors were extremely stable in medium with a maximum of 4% deviation in $|Z_p|$ for up to 7 days. In the presence of cell culture media the impedance decreases with frequency up to 2MHz; this is rather different than the results reported by Lo *et al.* [57] and Huang *et al.* [58] where between 50 kHz and 1MHz the measured impedance without cells is constant and very low in magnitude. It should be noted that the impedance measurement set up is slightly different here because the conductive substrate completely covers the surface and there are no insulating islands to contribute to large capacitive effects.

Figure 12.10a shows the impedance change over 44 h for a CNT/PLLA membrane after adding cells to a previously equilibrated membrane. We observed a peak of impedance magnitude in first 6 h. Moreover, Fig. 12.10b shows the fractional impedance change, $|Z_p|$. High values of $|Z_p|$ are observed in the high frequency domain (>1000 kHz). After 6 h, the peak value started to fluctuate reaching the lowest point after 44 h. Similar behaviour was observed in all samples. C3A cells typically require around 3–6 h for complete adhesion [59], therefore the initial rise in the curve is probably due to cell attachment which increases the resistive portion of the impedance at 1000 kHz by about 30% with respect to the initial value. After a further 12 h the magnitude of impedance decreases as the cells flatten out and spread, forming a high capacitance layer, confirmed also by the increase in impedance phase after 6 h. The gradual rise in $|Z_p|$ at 24 h could be due to the initiation of cell proliferation. C3A cells are known to have a high proliferative capacity with a typical doubling time of 24 h [59].

Figure 12.11 shows DAPI stained cells on the CNT/PLLA membranes after 44 h. A large number of intact nuclei were observed indicating that the membranes were able to support cell adhesion.

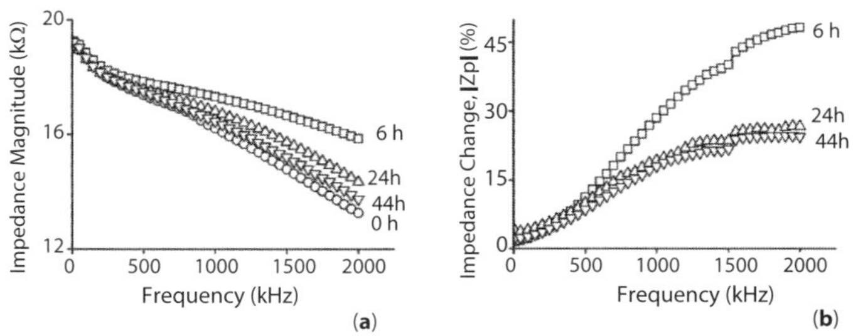

Figure 12.10 (*a*) Plot of impedance magnitude and (*b*) impedance change (Z_p) over frequency for scaffold with cell suspension (Reprinted with permission from [28], Whulanza *et al. Biosens Bioelectron.* 2011 Mar 15;26(7):3303-8).

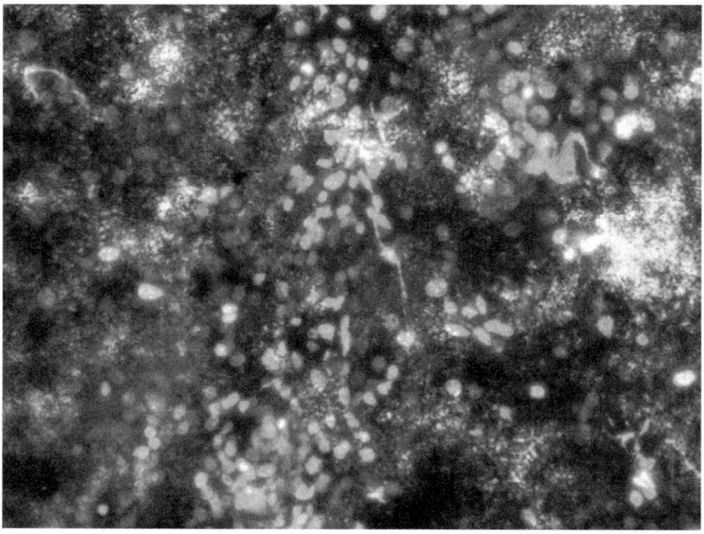

Figure 12.11 Micrograph of DAPI stained cells on CNT/PLLA membrane 44 H after seeding (10× magnification) (Reprinted with permission from [28], Whulanza Y *et al. Biosens Bioelectron.* 2011 Mar 15;26(7):3303-8).

Cell viability was also measured, and compared with viability in controls. The viability on the CNT/PLLA membranes remained within 5% of those of controls, but then fell by about 20% at 44 h, indicating a small degree of cell loss. This decrease in cell density could explain the decrease in $|Z_p|$ at 44 h. On PLLA membranes, cells viability remains at about 80% of the controls.

Figure 12.12 shows average $|Z_p|$ values of different membranes with and without cells. Pure dry PLLA membranes have an impedance value in the

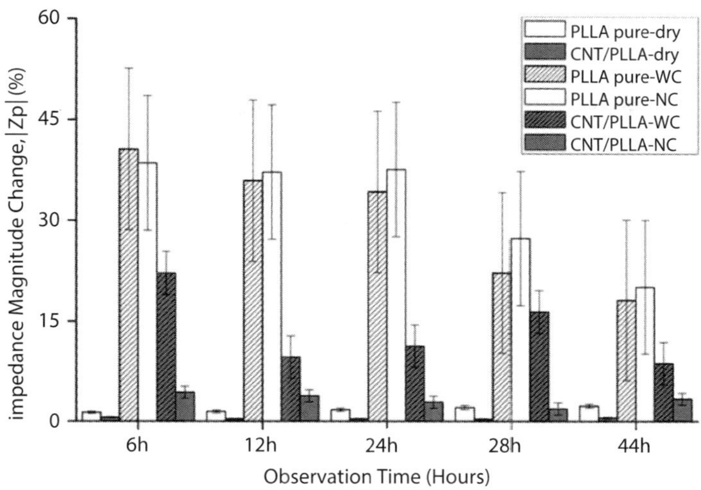

Figure 12.12 $|Z_p|$ At 1000khz for CNT/PLLA and control membranes (dry, medium wetted, pure PLLA membrane and CNT/PLLA membrane) with cells (WC) and without cells (NC) (Reprinted with permission from [28], Whulanza Y *et al. Biosens Bioelectron.* 2011 Mar 15;26(7):3303-8).

range of 10–12MΩ. Upon wetting with medium, PLLA probably swells and proteins may adsorb onto the membrane surface. Although we observed a high change in impedance ($|Z_p|$ = 15–50%) in medium wetted PLLA membranes, exactly the same range of values was noted for the membranes with cells. Given the high impedance of PLLA, the measurement of $|Z_p|$ was subject to large errors and any observed variations in $|Z_p|$ over time can be considered insignificant (for example comparing the impedance change in PLLA at 6 h with that at 44 h, $p = 0.219$ two tailed t-test). Therefore pure PLLA membranes do not change their impedance characteristics specifically as a function of cell activity. On the other hand, in CNT/PLLA membranes cell adhesion was well distinguishable from protein adhesion as the value of $|Z_p|$ was consistently higher ($p = 0.015$, two tailed t-test) in the presence of cells throughout the experiment.

12.8.5 Modelling Sensing Scaffold

In order to understand the physical basis of the impedance changes, a lumped parameter model analysis was employed. The result of the impedance monitoring was compared with a variety of different equivalent circuit models using the Matlab SimPowerSystem® toolbox (The Mathworks,

Incorporated). The toolbox is widely used for analysis of electrical circuit equivalents and employs a least squares fitting process to determine the most suitable values of circuit elements based on an input experimental transfer function.

We used a step-wise approach in which the starting point was a dry CNT/PLLA membrane. Following this the medium wetted membrane and cell-seeded membranes were modelled. Reasonable values of parameters were first chosen and then used to calculate values of impedance. The calculated and experimental values were then compared and the model parameters altered to improve data fitting. This iterative process was continued until an acceptable fit was obtained. These parameters were then used as starting values for analysis of the next time step. A similar process was used to fit the time varying impedance data in the presence of cells. Fitting was performed from 0 kHz to 2MHz with increments of 10 kHz (same as the measurement setup). The relative standard errors (RSE) reported in Table 12.1 are expressed as:

$$RSE = 100\frac{\sqrt{\left(\frac{1}{n}\right)\cdot\sum_{j=1}^{n}\left(y_{data} - y_{model}\right)}}{\mu_{data}} \tag{12.9}$$

where n = number of data, y_{data} = values from experimental data, y_{model} = values from circuit equivalent model, μ_{data} =mean values of experimental data.

In the first phase, dry membranes were measured and the resistance and capacitance were used to build the first circuit equivalent. Although a numbers of models were investigated, the circuit shown in Fig. 12.13a provides a good fit to the data with a minimum number of circuit elements. The RSE, over all membranes, of the model in Figure 12.13a is 3.25% and 6.33% for impedance magnitude and phase respectively.

We first followed the resistance and capacitance of the membrane sensors during the immersion in media for 24 h. The circuit in Fig. 12.13b contains the block RMain–CMain which was taken from the RC circuit in Fig. 12.13a to represent the dry membrane; the values remain unchanged.

Parameters R1–C1, R2–C2 and R3–C3 were added and iterated to produce the best fit with experimental impedance measurement. Block R1–C1 is in series with the dry membrane block and modifies the value of resistance whereas the parallel block R2–C2 plays role in modifying the capacitance. Block R3–C3 was introduced later to better fit the experimental data

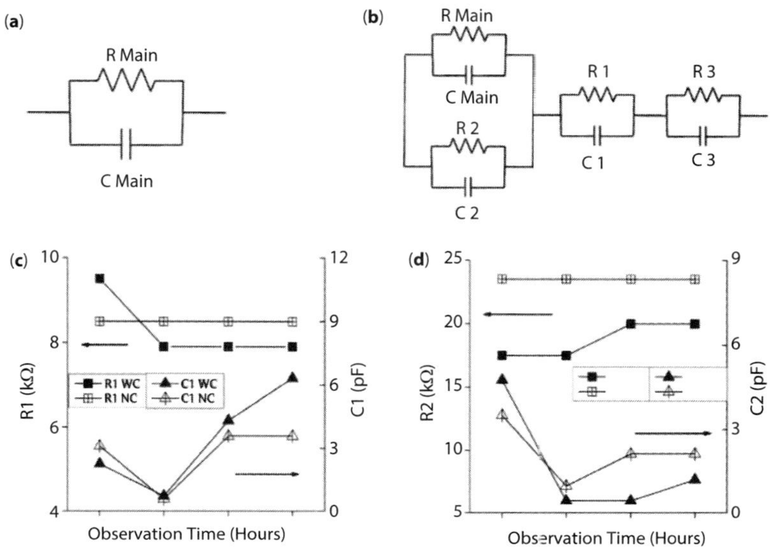

Figure 12.13 (*a*) Proposed equivalent circuit for dry CNT/PLLA Membrane, (*b*) Membrane immersed in media. Fitted parameters values of (*c*) R1 C1 and (*d*) R2 C2 as a function of time with cells (WC) and without cells (NC) (Reprinted with permission from [28], Whulanza Y *et al. Biosens Bioelectron.* 2011 Mar 15;26(7):3303-8).

which has a slight inflexion at low frequencies (<10 kHz). Relative standard errors are reported in Table 12.3.

To fit the impedance data over time in the presence of cells a number of models were investigated. However a reasonable fit was obtained for the same circuit used for wet membranes, and this was used for subsequent analysis. In the fitting, RC main was kept constant at the values obtained from the dry film fitting. Variations in impedance were shown to be primarily controlled by R1–C1 and R2–C2, with R3–C3 maintaining the same values as in the wet membrane. Fig. 12.13c-d shows how these four elements change over the 44 h observation period in the presence and absence of cells. The results show that both R1 and R2 remain constant in the absence of cells, while in the presence of cells R1 decreases whereas R2 increases. On the other hand both C1 and C2 have similar trends independent of the presence of cells. While it is not possible to identify specific cellular behaviour in correspondence with these variations, clearly the changes in capacitance are related to protein adsorption while the resistance is principally modulated by cell adhesion and spreading, and to a small extent protein expression by the cells themselves. It should be noted the electrical

Table 12.3 Relative standard error from the fitting process.

Obsevation time (h)	Relative standard error(%)			
	Wet membrane with cells		Wet membrane without cells	
	Impedance magnitude	Impedance Phase	Impedance magnitude	Impedance phase
0	0.678	2.415	2.658	2.521
6	2.854	4.237	2.875	5.421
24	3.183	2.645	3.092	3.425
44	1.960	2.556	3.376	3.856

configuration used here is quite different to that used by Giaever and Keese [2]. The authors used microelectrodes with extremely small surface areas (10^{-4} cm^2) such that only a few cells were responsible for the observed impedance changes. Our measurements represent a sort of integral impedance response of several thousand cells over a surface area of about $0.5cm^2$. In fact our aim was to develop a sensing scaffold for monitoring the overall activity or function of cells rather than a close observation of single cell behaviour or of cell density.

12.9 Processing of CNT Composite: Microfabrication of Sensing Scaffold

An additional issue is that of CNT composite processing. Microfabrication techniques which do not compromise CNT dispersion are essential to the realisation of reproducible scaffolds. Scaffolds should have high porosity and proper pore size, desirable mechanical integrity and promote cell adhesion. Here, to demonstrate that sensing scaffolds can be microfabricated using rapid prototyping, well-defined CNT/polymer scaffolds were fabricated using the Pressure-Activated Microsyringe (PAM). PAM is a rapid prototyping microfabrication system which has been purposely developed to extrude polymer solutions for biomedical and biosensing applications. It is a syringe-based CAD/CAM (Computer Aided Design and Manufacturing) method for fabricating polymer scaffolds [60, 61]. It consists of a stainless-steel syringe barrel with a glass capillary tip mounted on a 3-axis micropositioner with a precision of 1μm. Compressed air is

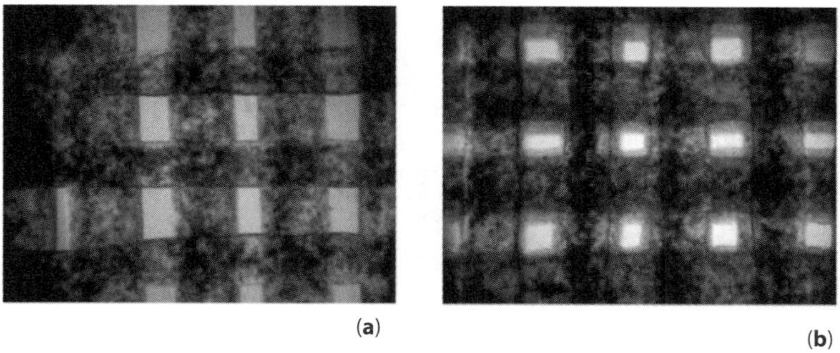

Figure 12.14 Scaffold realisation using PAM System: (*a*) Single layer and (*b*) two layers (Fig. 12.14a reprinted with permission from [28], Whulanza Y. *et al Biosens Bioelectron.* 2011 Mar 15;26(7):3303-8).

used to extrude polymer solutions, and both x, y and z positioning and the applied pressure are controlled using purpose designed software. The tip is pulled using a patch-clamp pipette puller (ESF Electronic, Goettingen, Germany), and in this study the tip inner diameter was 150μm. The CNT/PLLA dispersion was extruded through the capillary tip at a pressure of 20mmHg, while the deposition plane (x, y) was moved at 4mm/s to produce a simple square geometry (unit cells 300μm×300μm). Finally, to demonstrate the feasibility of processing the CNT/PLLA composite for Tissue Engineering two and three dimensional scaffolds were microfabricated using PAM. Square grids with a line width of $250 \pm 12 \mu m$ were obtained, as shown in Figure 12.14. As discussed by Tirella *et al.* [62] the resolution of the system depends on several parameters including the diameter of the deposition needle. Owing to the particulate nature of the CNT/PLLA suspension, small diameter needles tend to clog up, so the line width of the scaffolds is much lower than that usually achieved with the PAM for pure PLLA scaffolds [61].

12.10 Conclusions

The fabrication of "smart scaffolds" which act as a sensor for monitoring cell behaviour can allow long term experiments for the study of chronic disease models, or for repeat dose experiments. In this chapter we described a possible solution aiming at developing them by exploiting physical properties of CNTs.

The incorporation of CNTs into polymeric systems can result in segregation or inhomogeneous dispersion. In the studies here presented SWNTs were dissolved in organic solvent and incorporated into PLLA/PCL/PLGA matrices through simple mixing. The result obtained showed that CNTs/polymer composites exhibited not only a good dispersion, but also an improvement in mechanical and electrical properties.

CNTs were effective reinforcements, in terms of both mechanical strengthening in biodegradable scaffold applications. Addition of 0.5–0.7 wt.% CNTS gave a better dispersion in the PLLA/PCL/PGL matrix, whereas more than 0.8 wt.% CNTs addition have been led to their massive agglomeration. The dispersion of CNTs in polymer matrices resulted in substantial decreases in the electrical resistivity of the composite material.

CNT/PLLA composite membranes have been tested as cell sensor. The matrices are nontoxic and adhesive for cells. The processing of CNT/polymers system with microfabrication techniques requires a tight control over fabrication parameters, for maintaining the CNTs suspension and to avoid aggregates. The PAM system, with an air-pressure driven extrusion, was demonstrated to be an effective technique to fabricate structures with micrometric features.

From these analyses it is possible to select the optimal composite composition of SWNTs/polymer to realise scaffolds for tissue engineering applications such as bone, muscle or neural regeneration.

References

1. Tran PA, Zhang L, Webster TJ. Carbon nanofibers and carbon nanotubes in regenerative medicine. *Adv Drug Deliv Rev.* 2009 Oct 5;61(12):1097–114.
2. Giaever I, Keese CR. Monitoring fibroblast behavior in tissue culture with an applied electric field. *Proc Natl Acad Sci U S A.* 1984 Jun;81(12):3761–4.
3. Giaever I, Keese CR. Micromotion of mammalian cells measured electrically. *Proc Natl Acad Sci U S A.* 1991 Sep 1;88(17):7896–900.
4. Lind R, Connolly P, Wilkinson CDW, Breckenridge LJ, Dow JAT. Single cell mobility and adhesion monitoring using extracellular electrodes. *Biosens. Bioelectron.* 1991;6:359–367.Lind, R., Connolly, P., Wilkinson, C.D.W., Breckenridge, L.J., Dow, J.A.T., 1991. *Biosens. Bioelectron.* 6, 359–367.
5. Wegener J, Keese CR, Giaever I. Electric cell-substrate impedance sensing (ECIS) as a noninvasive means to monitor the kinetics of cell spreading to artificial surfaces. *Exp Cell Res.* 2000 Aug 25;259(1):158–66.
6. Ehret R, Baumann W, Brischwein M, Schwinde A, Stegbauer K, Wolf B. Monitoring of cellular behaviour by impedance measurements on interdigitated electrode structures. *Biosens Bioelectron.* 1997;12(1):29–41.

7. An LN, Xu WX, Rajagopalan S, Wang CM, Wang H, Fan Y, Zhang LG, Jiang DP, Kapat J, Chow L, Guo BH, Liang J, Vaidyanathan R, Carbon-Nanotube-Reinforced Polymer-Derived Ceramic Composites 2004. *Adv. Mater.* 16 (22), 2036–2040.
8. Meyyappan, M., 2005. *Carbon Nanotubes Science and Applications.* CRC Press, Boca Raton, FL.
9. Coleman JN, Cadek M, Blake R, Nicolosi V, Ryan KP, Belton C, *et al.* High performance nanotube-reinforced plastics: understanding the mechanism of strength increase. *Advance Functional Mater* 2004;14:791–8.
10. Coleman JN, Khan UM, Gun'ko YK. Mechanical reinforcement of polymers using carbon nanotubes. *Adv Mater* 2006;18: 689–706.
11. Liu T, Phang IY, Shen L, Chow SY, Zhang WD. Morphology and mechanical properties of multiwalled carbon nanotubes reinforced nylon-6 composites. *Macromolecules* 2004; 37: 7214–7222
12. Mattioli-Belmonte M, Vozzi G, Whulanza Y, Seggiani M, Fantauzzi V, Orsini G, *et al.* Tuning polycaprolactone–carbon nanotube composites for bone *Tissue en*gineering scaffolds. *Materials Science and Engineering*: C 2012;32: 152–159.
13. Freire E, Gomes FC, Linden R, Neto VM, Coelho-Sampaio T. Structure of laminin substrate modulates cellular signaling for neuritogenesis. *J Cell Sci.* 2002 Dec 15;115(Pt 24):4867–76.
14. Kleinman HK, Cannon FB, Laurie GW, Hassell JR, Aumailley M, Terranova VP, Martin GR, DuBois-Dalcq M. Biological activities of laminin. *J Cell Biochem.* 1985;27(4):317–25.
15. Luckenbill-Edds L. Laminin and the mechanism of neuronal outgrowth. *Brain Res Brain Res Rev.* 1997 Feb;23(1–2):1–27.
16. He W, Bellamkonda RV. Nanoscale neuro-integrative coatings for neural implants. *Biomaterials.* 2005 Jun;26(16):2983–90.
17. Balasubramanian K, Burghard M. Biosensors based on carbon nanotubes. *Anal Bioanal Chem.* 2006 Jun;385(3):452–68.
18. Harrison BS, Atala A. Carbon nanotube applications for *Tissue en*gineering. *Biomaterials.* 2007 Jan;28(2):344–53.
19. Zhang X, Prasad S, Niyogi S, Morgan A, Ozkan M, Ozkan C. Guided neurite growth on patterned carbon nanotubes. 2005. *Sens. Actuators* 106, 843–850.
20. Lovat V, Pantarotto D, Lagostena L, Cacciari B, Grandolfo M, Righi M, Spalluto G, Prato M, Ballerini L. Carbon nanotube substrates boost neuronal electrical signaling. *Nano Lett.* 2005 Jun;5(6):1107–10.
21. Cellot G, Cilia E, Cipollone S, Rancic V, Sucapane A, Giordani S, Gambazzi L, Markram H, Grandolfo M, Scaini D, Gelain F, Casalis L, Prato M, Giugliano M, Ballerini L. Carbon nanotubes might improve neuronal performance by favouring electrical shortcuts. *Nat Nanotechnol.* 2009 Feb;4(2):126–33.
22. Li B, Ma Y, Wang S, Moran PM. Influence of carboxyl group density on neuron cell attachment and differentiation behavior: gradient-guided neurite outgrowth. *Biomaterials.* 2005 Aug;26(24):4956–63.

23. Pankratz MT, Li XJ, Lavaute TM, Lyons EA, Chen X, Zhang SC. Directed neural differentiation of human embryonic stem cells via an obligated primitive anterior stage. *Stem Cells*. 2007 Jun;25(6):1511–20.
24. Coleman JN, Khan U, Blau WJ, Gun'ko YK. Small but strong: A review of the mechanical properties of carbon nanotube-polymer composites 2006. *Carbon* 44, 1624–1652.
25. Pioggia G, Di Francesco F, Marchetti A, Ferro M, Leardi R, Ahluwalia A. A composite sensor array impedentiometric electronic tongue Part II. Discrimination of basic tastes. *Biosens Bioelectron*. 2007 May 15;22(11):2624–8.
26. Supronowicz PR, Ajayan PM, Ullmann KR, Arulanandam BP, Metzger DW, Bizios R. Novel current-conducting composite substrates for exposing osteoblasts to alternating current stimulation. *J Biomed Mater Res* 2002; 5: 499–506.
27. Whulanza Y, Battini E, Vannozzi L, Vomero M, Ahluwalia A, Vozzi G. Electrical and mechanical characterisation of single wall carbon nanotubes based composites for *Tissue eng*ineering applications. *J Nanosci Nanotechnol*. 2013 Jan;13(1):188–97.
28. Whulanza Y, Ucciferri N, Domenici C, Vozzi G, Ahluwalia A. Sensing scaffolds to monitor cellular activity using impedance measurements. *Biosens Bioelectron*. 2011 Mar 15;26(7):3303–8.
29. Hunter RJ. *Foundations of colloid science*, Oxford: Clarendon Press,1987: vol. 1.
30. Kim KS, Chae YS, Park BH, Yoon JS, Kang MS, Jin HJ. Thermal and electrical conductivity of poly(l-lactide)/multiwalled carbon nanotube nanocomposites. *Curr App Phys* 2008;8:803–806.
31. Kuan CF, Kuan HC, Ma CC, Chen CH. Mechanical and electrical properties of multi-wall carbon nanotube/poly(lactic acid) composites. *Journal of Physics and Chemistry of Solids* 2008;69:1395–1398
32. Gilding DK, Reed AM. Biodegradable polymers for use in surgery- polyglycolic/poly(actic acid) homo- and copolymers: 1. *Polymer* 1979; 20: 1459–1554.
33. Zhang XF, Liu T, Sreekumar TV, Kumar S, Moore VC, Hauge R H, *et al.* Poly(vinyl alcohol)/SWNT composite film. *Nano Letters* 2003; 3: 1285–1288.
34. Dalton AB, Collins S, Munoz E, Razal JM, Ebron VH, Ferraris JP, *et al*, Super-tough carbon-nanotube fibres. These extraordinary composite fibres can be woven into electronic textile. *Nature* 2003; 423: 703–704.
35. Geng HZ, Rosen R, Zheng B, Shimoda H, Fleming L, Liu J, *et al.*. Fabrication and properties of composites of poly(ethylene oxide) and functionalized carbon nanotubes. *Advanced Materials* 2002; 14 : 1387–1390.
36. Velasco-Santos C, Martinez-Hernandez AL, Fisher FT, Ruoff R, Castano V. M. Improvement of thermal and mechanical properties of carbon nanotube composites through chemical functionalization. *Chem Mat* 2003; 23: 4470–4475.
37. Callister WD. *Materials Science and Engineering*: An Introduction. New York: Wiley, 2003.
38. Tucker CL, Liang E. Stiffness predictions for unidirectional short-fiber composites: review and evaluation. *Comp Sci Tech* 1999; 59:655

39. Ashton JE, Halpin JC, Petit PH. *Primer on composite materials: analysis.* Stamford Conn: Technomic, 1969.
40. Halpin JC, Kardos JL. The Halpin-Tsai equations: a review. *Polymer Engineering and Science* 1976;16:344–352.
41. Halpin JC. Stiffness and expansion estimates for oriented short fiber composites. *J Comp Mat* 1969;3: 732–41
42. Mori T, Tanaka K. Average stress in matrix and average elastic energy of materials with misfitting dislocation. *Acta Metall Mater* 1973; 21:571–4
43. Tandon GP, Weng GJ. The effect of aspect ratio of inclusions on the elastic properties of unidirectionally aligned composites. *Pol Comp* 1984;5: 327–33
44. Kearns JC, Shambaugh RL. Polypropylene fibers reinforced with carbon nanotubes. *J Appl Polym Sci* 2002; 86:2079–84.
45. Putz KW, Mitchell CA, Krishnamoorti R, Green PF. Elastic modulus of single-walled carbon nanotube/poly(methyl methacry- late) nanocomposites. *J Polym Sci Part B: Polym Phys* 2004; 42: 2286–93.
46. Deng H, Skipa T, Zhang R, Lellinger D, Bilotti E, Alig I, *et al.* Effect of melting and crystallization on the conductive network in conductive polymer composites. *Polymer* 2009; 50:3747–54
47. Kim YJ, Shin TS, Choi HD, Kwon JH, Chung YC, Yoon HG. Electrical conductivity of chemically modified multiwalled carbon nanotube/epoxy composites. *Carbon* 2005; 43. 23–30.
48. Barrau S, Demont P, Peigney A, Laurent C, Lacabanne C. DC and AC conductivity of carbon nanotubes-polyepoxy composites. *Macromolecules* 2003; 36: 5187–94
49. Kilbride BE, Coleman JN, Fraysse J, Fournet P, Cadek M, Drury A, *et al.* Experimental observation of scaling laws for alternating current and direct current conductivity in polymer-carbon nanotube composite thin films. J App Phys 2002; 92: 4024–30.
50. Pötschke P, Dudkin SM, Alig I. Dielectric spectroscopy on melt processed polycarbonate- multiwalled carbon nanotube composites. *Polymer* 2003; 44:5023–30.
51. Carmona F, Canet R, Delhaes P. Piezoresistivity of heterogeneous solids. *J Appl Phys* 1986; 61:2550–7.
52. Aneli JN, Zaikov GE, Khananashvili LM. Effects of mechanical deformations on the structurization and electric conductivity of electric conducting polymer composites. *J Appl Polym Sci* 1999; 74: 601–621.
53. Kovacs G. *Micromachined Transducers Sourcebook.* First edition. WCB/McGraw-Hill: Boston.;1998
54. Beken S, Vanhaecke T, De Smet K, Pauwels M, Vercruysse A, Rogiers V, *Methods Mol. Biol.* 1998 107, 303–9.
55. Vroman L. Effect of absorbed proteins on the wettability of hydrophilic and hydrophobic solids. *Nature.* 1962 Nov 3;196:476–7.
56. Lo CM, Keese CR, Giaever I. Impedance analysis of MDCK cells measured by electric cell-substrate impedance sensing. *Biophys J.* 1995 Dec;69(6):2800–7.

57. Huang X, Greve DW, Nguyen DD, Domach MM. Impedance based biosensor array for monitoring mammalian cell behavior. *Proc. IEEE Sensors* 2003, vol. 1, pp. 304–9
58. Vinci B, Cavallone D, Vozzi G, Mazzei D, Domenici C, Brunetto M, Ahluwalia A. In vitro liver model using microfabricated scaffolds in a modular bioreactor. *Biotechnol J*. 2010 Feb;5(2):232–41.
59. Vozzi G, Previti A, De Rossi D, Ahluwalia A. Microsyringe-based deposition of two-dimensional and three-dimensional polymer scaffolds with a well-defined geometry for application to *Tissue engineering*. *Tissue Eng*. 2002 Dec;8(6):1089–98.
60. Mariani M, Rosatini F, Vozzi G, Previti A, Ahluwalia A. Characterization of tissue-engineered scaffolds microfabricated with PAM. *Tissue Eng*. 2006 Mar;12(3):547–57.
61. Tirella A, Orsini A, Vozzi G, Ahluwalia A. A phase diagram for microfabrication of geometrically controlled hydrogel scaffolds. *Biofabrication*. 2009 Dec;1(4):045002.

13

Nanostructured Sensing Emulsion Droplets and Particles: Properties and Formulation by Membrane Emulsification

Emma Piacentini, Alessandra Imbrogno and Lidietta Giorno*

Institute on Membrane Technology, CNR-ITM, C/o University of Calabria, Rende (CS), Italy

Abstract
Responsive particles can adapt to surrounding environments regulating the transport of molecules with specific functional activity or convert chemical and biochemical signals into optical, electrical, thermal and mechanical signals, and vice versa. These materials are playing an increasingly important part in several ranges of applications, such as drug delivery, diagnostics, tissue engineering, biosensors, coatings and textiles. Microfabrication techniques which permit the creation of particles that possess a combination of structural and functional features have tremendous potential in the development of new products with improved quality for medical, food, cosmetic or chemical applications. Membrane emulsification is relatively new technology for the precise, selective and flexible manufacturing of micro- nano-particulates with target size, composition, structure and function. In this chapter, membrane emulsification technology and its use in the preparation of sensing particles are presented.

Keywords: Sensing emulsions, membrane emulsification, responsive materials, external stimuli, biochemical response

13.1 Introduction

Sensing particles represent a rapidly developing class of stimuli-responsive materials that find applications in drug delivery system, catalysis, and

**Corresponding author*: l.giorno@itm.cnr.it

Annarosa Gugliuzza (ed.) Smart Membranes and Sensors, (367–400) 2014 © Scrivener Publishing LLC

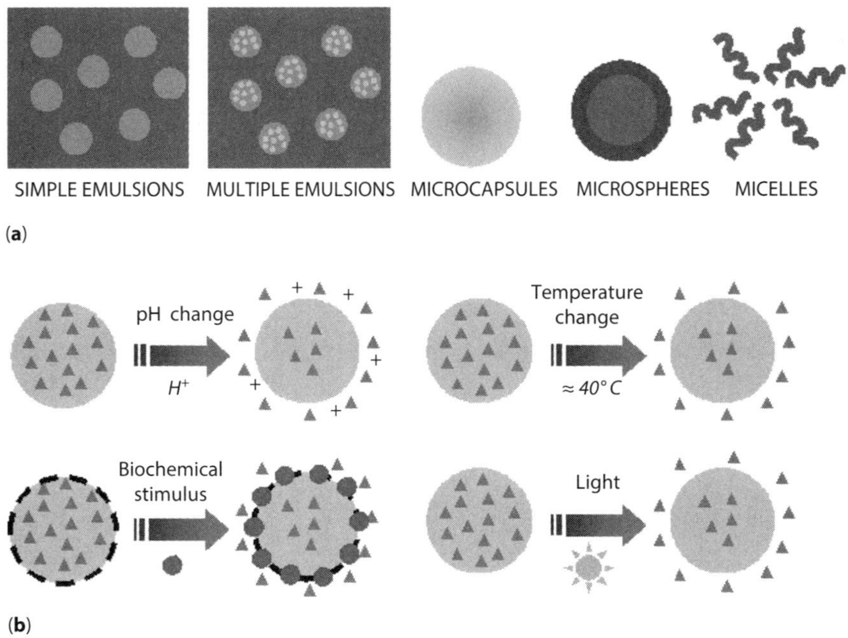

Figure 13.1 Micro-nanostructured stimuli-responsive systems a) and triggering release for chemical, thermal, biological and photo stimuli.

sensor. They include micro-nano- emulsions, capsules or spheres and micelles able to sense and respond directly to environmental conditions (Figure 13.1 a). They are of considerable interest in applications such as bioactive compounds preservation, fragrance release and drug delivery. The signal is derived from changes in the materials' environment due to chemical, thermal, biological, photo, electrical, and magnetic stimuli (Figure 13.1 b). The central operating principle of sensing particles lies in the fact that a specific stimulus of chemical, biochemical, or physical origin can modify the structural composition/conformation of the particles, thereby promoting release of the active species.

Micro- and nanostructure sensing systems can maximize the efficacy of therapeutic treatments because they have the ability to detect and respond to disease states directly at the site, without compromise physiologically healthy cells and tissues and thereby reducing the side effects. Functionalization of particles whit enzymes or the possibility of exposing or hiding functional groups on particles surface create new opportunities for bio- and chemical catalysis. In agricultural field, an increase in soil temperature can initiate delivery of nutrients or in personal-care industry, deodorant and antiperspirant materials can be released when a targeted temperature is reached.

Particle design is a critical determinant of the efficacy of particle-based products in food, chemical, medical and biotechnological applications. Both chemical and physical properties of particles have been recognized as critical in the success of sensing activity of these materials. Specific functional properties such as circulation times, reduction in secondary effects of bioactive compounds, accumulation in sites of interest, biocompatibility, and controlled degradation are strictly correlated whit the chemical and physical properties of the materials used in sensing particles preparation. The targeting and controlled delivery of sensing particles is closely related to their size and size distribution. Tailored particle size is required for specific applications such as in heterogeneous catalytic reactions where uniform small particles allow obtaining high interfacial area. Size plays an important role in biodistribution: particles with small size can easily pass through the fine capillary blood vessels and the lymphatic endothelium for applications as drug delivery systems. They have longer circulation time in the blood, higher binding capability and accumulation at the target sites, and give less inflammatory and immune response from the tissues and cells of the body than those with bigger size. Monodispersed particles give better control over the dose and release behavior of the encapsulated molecules of interest yields higher drug encapsulation efficiency and better biocompatibility that of polydispersed particles.

Micro and nanotechnologies can offer new opportunities towards the development of sensing particles Advancements in micro- and nano-fabrication methods are required to design hybrid materials with specified properties in terms of micro- or nano-fabricated structure and functional activity. In particular, the use of microporous architectures like membranes permits to meet these demanding requirements. The low local shear stress applied and the precisely controlled particle size and size distribution obtained make the membrane emulsification technique the most suitable method for sensitive formulations with controlled architecture containing labile molecules.

In this chapter, we examine the fabrication of stimuli-responsive particles by membrane methods. Technology for preparing monodispersed particles with a definite size has not been available until the membrane emulsification technique was developed. The combination of stimuli-responsive materials and membrane fabrication methods allow to design particles whit tailored smart and responsive properties for micro- and nanotechnologies applications.

In the first part of the chapter, the conventional and membrane-based methods to produce particles will be described while, in the second part, specific examples of the use of membrane emulsification in the production of sensing particles will be reported. In the third part, a summary of the stimulus-responsive particles prepared combining membrane emulsification method and one of the strategies described to introduce the stimulus responsive ability will be given.

13.2 Emulsions and Emulsification Methods

According to P. Becher [1] an emulsion is "a heterogeneous system", consisting of at least one immiscible liquid intimately dispersed in another in the form of droplets, whose diameters, in general, exceed 0.1 μm. The type of simple emulsion (water-in-oil or oil-in-water, commonly abbreviated as w/o or o/w) is decided mainly by the volume ratio of the two liquids, their order of addition and the nature of the emulsifier. An emulsion may also be an emulsion of an emulsion e.g. water-in-oil-in-water (W/O/W) and oil-in-water-in-oil (O/W/O) emulsions, also termed multiple or double emulsions. Due to their large interfacial area, emulsions are thermodynamically unstable and have a tendency to undergo coalescence, a process during which two droplets merge into a bigger one, minimizing the surface energy. Certain molecules or materials, such as surfactants, amphiphilic polymers or proteins, can be added to prevent coalescence, making emulsions kinetically stable. These emulsion stabilizers usually provide electrostatic repulsion, steric repulsion, and/or strength to the interfacial layer of the droplets, according to widely accepted theories. Except in special cases where spontaneous emulsification can occur, energy must be supplied to produce such metastable mixtures. Thermodynamically speaking, surfactants reduce the surface free energy required to increase any interfacial area by lowering the interfacial tension, and allow finely media to be created easily. In fact, a much higher amount of energy, with respect to this thermodynamic part, is necessary, since breaking of large droplets into smaller ones involves additional shear forces, so that the viscous resistance during agitation absorbs most of the energy [2]. Conventional mechanical methods used to provide the required energy to produce an emulsion are: a) Rotor-stator system; b) High-pressure system; c) Ultrasonic system.

13.2.1 Rotor-stator Systems

Rotor-stator systems can be operated in continuous or discontinuous mode. For the discontinuous or quasi-continuous production agitators, different geometry or gear-rim dispersion machine are usually used. In this system the dispersed phase droplets are broken- up to a large extent by forces of inertia and shearing in turbulent flow. The produced stress can be classified as medium to high depending on the number of revolutions and the geometry of the rotor-stator systems. Mean droplet diameter below 2 μm cannot be obtained with this system [3]

13.2.2 High-pressure Homogenizer

In a high-pressure homogenizer, the oil and water mixture is passed through a narrow orifice, or inject dispersion in which two jets of different components are made to collide head-on; pressures in the range of $5.0x10^6$–$3.5x10^7$ Pa are commonly used. In this system emulsion is subjected to intense turbulent and shear flow fields. Turbulence, the predominant mechanism, leads to the break-up of the dispersed phase into small droplets and laminar shear and cavitation can assist the process. With this system mean droplet diameter below 0.2 μm can be obtained with high product yields. However, the stress on the product is very high due to the high pressure gradients and flow rate [4].

13.2.3 Ultrasonication

In ultrasound emulsification, reported for the first time by Wood and Loomis, high frequency - vibrations applied to a diphasic liquid system provide a different breaking and dispersing of a bulk phase: large drops produced by the instability of interfacial waves are broken into smaller ones by acoustic cavitation. Due to the small product throughput this process is mainly applied in laboratories when mean droplets diameter of approximately 0.4 μm can be obtained [5].

13.2.4 Membrane Emulsification

The membrane emulsification (ME) technique was first proposed by Nakashima *et al.* [6] to prepare monodisperse emulsion using a particular glass membrane called Shirasu Porous Glass (SPG) membrane. The method has continued to attract attention due to its effectiveness in producing narrow droplets size distributions at low energy consumption. The use of membrane to manufacture emulsion follows two innovative approaches respect to conventional methods:

1. reduction of turbulent perturbations in the mixing processes that rupture the liquids;
2. manufacture of droplets individually (drop-by-drop) using micro-engineered structured systems;

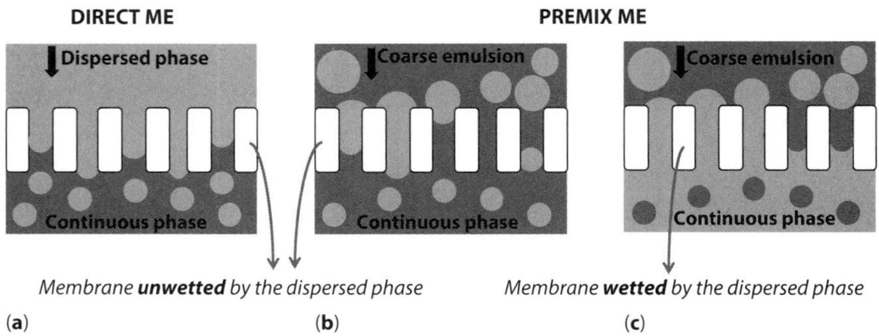

Figure 13.2 Direct and Premix membrane Emulsification.

The conventional direct ME involves the permeation of the dispersed phase through a porous membrane in order to form droplets at the opening pore successively detached from the membrane surface by the shear stress of the moving continuous phase (Figure 13.2 a). Suzuki *et al.* [7] introduced a modified form of the classic emulsification system, in which a coarse emulsion (prepared using a conventional stirrer mixer) is refined upon passage through a microporous membrane. The method is referred as premix ME (Figure 13.2 b, c). Usually, in premix ME method, the membrane is wetted by the continuous phase of the coarse emulsion and the emulsion is broken up into smaller droplets (Figure 13.2 b). Sometimes the membrane is wetted by the disperse phase of the coarse emulsion, and in this case phase inversion can take place, leading to very high disperse phase volume fractions (Figure 13.2 c). The energy costs for premix emulsification are relatively low, since no cross-flow is needed, the process is easier to control and operate than direct ME but the droplet polydispersity is higher [8].

In membrane dispersion process, droplet formation mainly consists of the following three processes: (I) droplet across the pore; (II) droplet growth; (III) droplet detachment.

The main forces that act on the forming droplets are:

1. the interfacial tension force, F_{Υ}, which represents the effect of dispersed phase adhesion around the edge of the opening pore:

$$F_{\Upsilon} = \pi d_p \gamma$$

2. the static pressure difference force, F_{sp}, due to the pressure difference between the dispersed phase and the continuous phase at the membrane surface:

$$F_{sp} = \frac{\gamma}{d_p} \pi d_d^2$$

3. the drag force, F_d, created by moving the continuous phase or using moving membranes:

$$F_d = \frac{3}{2} k_x \pi \tau_{c,s} d_d^2$$

4. the dynamic lift force, F_L, which results from the asymmetric velocity profile of the continuous phase near the droplet:

$$F_L = 0.761 \frac{\tau_{c,s}^{1.5} \rho_c^{0.5}}{\mu_c} d_d^3$$

5. the buoyancy force, F_B, due to the density difference between the continuous phase and the dispersed phase:

$$F_B = \frac{1}{6} \pi g \Delta\rho d_d^3$$

6. the inertial force, F_i, caused by the dispersed phase flow moving through the capillary as it inflates the droplet:

$$F_I = \rho_d \left(\frac{J_d}{\varepsilon} \right) A_N$$

with:

d_d: droplet diameter
A_N: cross-sectional area of the droplet neck,
k_x: equal to 1.7 and takes into account the wall correction factor for a single sphere touching an impermeable wall,
ε: membrane porosity,
$\tau_{c,s}$: wall shear stress,
γ: dynamic interfacial tension,
ρ_c: density of the continuous phase,
ρ_d: density of the dispersed phase,
μ_c: viscosity of the continuous phase,
Δρ: the difference between continuous and dispersed phase density.

These forces can be divided into holding and detaching forces, which hold the droplet on the sieve-like structure surface and detach the droplet

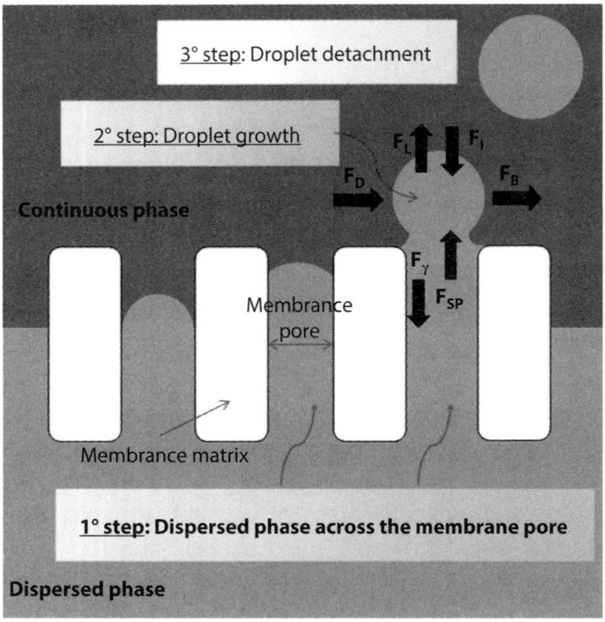

Figure 13.3 Forces acting on a droplet coming out of the membrane pores.

from the surface, respectively (Figure 13.3). Drag force, inertial force, static pressure force tend to detach the droplets whereas interfacial tension forces the drop to remain attached to the pore; when the detaching forces become higher than the holding forces, the formed droplet begins to go away from the surface.

The drag force, F_d, can be generated by the continuous phase flowing parallel to (cross flow ME) or stirring up (stirred ME) the membrane surface. Alternatively, vibrating or rotating membranes can be also used.

In the absence of any drag force, droplets can be spontaneously detached from the pore outlets. ME processes carried out in quiescent condition are referred as *static ME* while those carried out in moving conditions are referred as *dynamic ME*.

The main factors influencing membrane emulsification include:

- membrane parameters, including porosity, mean pore size, pore geometry, pore distance wettability;
- phase parameters, including interfacial tension, emulsifier type and concentration, viscosity and density of dispersed and continuous phases, phase composition, pH and ionic strength;

- process parameters, including wall shear stress, trans-membrane pressure, membrane module configuration, temperature;

13.2.5 Membrane Parameters

Shirasu porous glassy (SPG) membranes are among the first membranes specifically developed for emulsion preparation. SPG membranes are characterized by interconnected micro-pores, a wide spectrum of available mean pore size (0.05–30 μm) and high porosity (50–60%). The surface wettability can be changed by reaction with organic silanes [9]. Other commercial microfiltration membranes are attractive because of their availability in very large surface area, and their high flux through the membrane pores: ceramic aluminium oxide (α-Al2O3) membranes (Membraflow, Germany)[10], α-alumina and zirconia coated membranes (SCT, France) and polytetrafluoroethylene (PTFE) membranes [11].

Micropore Technologies (United Kingdom) developed flat-sheet metallic membranes characterized by cilyndrical pores, uniform and in a regular array with a distance between each pore of 200 μm. It has been generally observed that droplet size (d_d) of an emulsion can be related to the pore size (d_p) of the membrane by a linear relationship for given operating condition:

$$d_d = xd_p$$

where x can range typically from 2–10 for SPG membranes. For membranes other than SPG the values reported for x are typically from 3–50 [12]. To allow optimal production of mono-dispersed emulsions, the affinity between membrane surface, dispersed and continuous phase and electrical charge of the emulsifier, functional groups must be considered. For a preparation of O/W emulsions, hydrophilic membranes must be used in order to avoid the wetting and spreading of the oil on the membrane surface; on the contrary hydrophobic membrane must be used for the preparation of W/O emulsion. Similar principles may also apply to the choice of the emulsifiers; in fact, its functional groups must not carry opposite charge respect to the membrane surface in order to preserve hydrophilicity of the membrane. For example, an untreated SPG membrane has a negative surface charge within the pH range of 2 to 8, due to the dissociation of acid silanol groups, therefore cationic emulsifier must be avoided for the preparation of O/W emulsion [13].

Membrane purchased in a hydrophilic form can be made hydrophobic by chemical surface modification or by pre-soaking it in the oil phase. Pre-sonication in solutions of surfactants has also been shown to improve dispersed phase flux and the decreased interfacial tension in the pores leads to better filling.

The porosity of a membrane is also an important parameter because it determines the distance between adjacent pores, this distance increases as the porosity decreases. The porosity is critical to ensure that two adjacent droplets do not come sufficiently close to allow contact with each other and to coalesce. Therefore, the pores are preferred to be uniformly located on the surface to ensure maximum distance between any two adjacent pores for a given porosity [9].

13.2.6 Phase Parameters

The interfacial tension force, which represents the effects of dispersed phase adhesion around the edge of the opening pore, forces the drop to remain attached to the pore; therefore, the drop grows until a diameter is reached at which the shear force is higher than the interfacial tension force. Emulsifiers have two main roles in the formation of an emulsion. Firstly, it lowers the interfacial tension between oil and water; secondly, it stabilizes the droplets against coalescence and /or aggregation. Shröder *et al.* [14] studied the effect of dynamic interfacial tension on droplet formation. They found that the adsorption kinetics of the emulsifier determine the time needed to stabilize the droplets against coalescence in all stages of formation. The faster an emulsifier adsorbs at the newly formed interfaces, the smaller the droplets produced. Fast adsorbing emulsifiers diminish the influence of coalescence and allow smaller droplet to detach from the pore because of faster decrease of retaining forces [15]. Geerken *et al.* [16] found that using fast emulsifier, like SDS, the lag time, which is the period of inactivity of the pore after the detachment of a drop, can be avoided; after droplet detachment, a sufficient amount of emulsifier have to adsorb at the hemispherical interface before the critical Laplace pressure is reached again. For faster adsorbing emulsifiers the lag time reduces because the required critical interfacial tension is reached at earlier time. The viscosities of dispersed and continuous phase have also an important effect on the membrane emulsification process. According to Darcy's law, the dispersed flux is inversely proportional to the dispersed phase viscosity, therefore, if it is high the dispersed flux will be low, and as a consequence the droplet diameter will be small compared to the mean pore diameter [9].

13.2.7 Process Parameters in Dynamic Membrane Emulsification

The transmembrane pressure, ΔP_{tm}, is defined as the difference between the pressure of the dispersed phase, Pd, (feed in dead-end mode) and the mean pressure of the continuous phase:

$$\Delta P_{tm} = P_d - \frac{\left(P_{c,in} + P_{c,out}\right)}{2}$$

where $P_{c,in}$ and $P_{c,out}$ are the pressure of the flowing continuous phase at the inlet and outlet of the membrane module respectively. The minimum emulsification pressure required to allow the dispersed phase to pass through the pores into the other side of the membrane can be estimated by the capillary pressure, P_c:

$$P_c = \frac{4\Upsilon cos\delta}{d_p}$$

where Υ is the interfacial tension, δ is the contact angle of the droplets against the membrane surface wetted with the continuous phase and d_p is the average pore diameter.

The flux of the dispersed phase is correlated with the transmembrane pressure by the Darcy's law:

$$J_d = \frac{K\Delta P_{tm}}{\mu L}$$

where K is the membrane permeability, L the membrane thickness, and μ the dispersed phase viscosity. The mean droplet size decreases initially with the increasing of the transmembrane pressure, due to gradual activation of smaller pores, and then increases with further pressure increase because of increased droplet coalescence at the membrane surface.

The generation of a surface shear is the most conventional way to control droplet detachment in membrane emulsification. In general, the mean droplets size decreases exponentially whit increasing the wall shear stress. In cross-flow and stirred ME, this result is achieved by increasing:

- the continuous phase velocity in cross-flow and stirred ME
- the frequency of continuous phase pulsed flow in pulsed ME
- the rotation speed of the membrane in rotating ME
- the frequency vibration of the membrane in vibrating ME

13.2.8 Membrane Emulsifications Devices

The stirred ME device (Figure 13.4) is a dispersion emulsifying system in which the inside of tube-shaped membrane is filled with dispersed phase liquid and, the outside, with continuous phase liquid; then pressure (internal pressure) is given to the inside of the membrane by use of nitrogen gas while stirring the continuous phase with a rotator. Another configuration could be used in which the outside of tube-shaped is filled with dispersed phase liquid, and the inside, with continuous phase liquid; then pressure (external pressure) is given from outside to inside of SPG membrane by use of nitrogen gas while stirring the continuous phase liquid with a rotator. Alternatively, a flat-sheet membrane can be used and the dispersed phase is pressed through the membrane from the bottom to the top while a paddle blade stirrer, positioned on the top of the membrane, is used to stir the continuous phase. The stirred ME is useful for laboratory test to study the effect of different experimental conditions on the preparation characteristics but suitable for small scale production [17].

A typical cross-flow ME apparatus (Figure 13.4) includes a tubular or flat-sheet microfiltration membrane, a pump for the recirculation of the continuous phase along the lumen side of the membrane, a feed vessel, and a pressured (N_2) container for the disperse phase [18]. The dispersed phase, under a gas pressure, is forced through the pores of the membrane while the continuous phase flows along the membrane surface in order to detach the droplets at the opening pore after they reached a certain size. Cross-flow ME is suitable for large scale production and continuous or semi-continuous operation, however there are still problems in obtain narrow droplet size distribution at high dispersed flux and droplets break-up can occurs using high cross-flow velocities.

In rotating ME (Figure 13.4) the dispersed phase is introduced into the centre of a rotating tubular porous membrane, and dispatched through the pores in the membrane wall in radial direction into the stationary continuous phase [19]. The pressure is given to the inside of the membrane by use of nitrogen gas or a pump while the membrane was rotated using stirring the continuous phase with a overhead stirrer. In vibrating ME (Figure 13.4) the membrane is immersed into a beaker containing the continuous phase and filled inside with the dispersed phase [20]. The pressure is given to the inside of the membrane by use a pump while frequency and amplitude of the oscillation of the membrane was generated by an electrically driven vertical oscillator. Rotating and vibrating ME methods are suitable for fragile and structured particles, in which the droplets and/or particles could be subject to breakage during the pump circulation or continuous

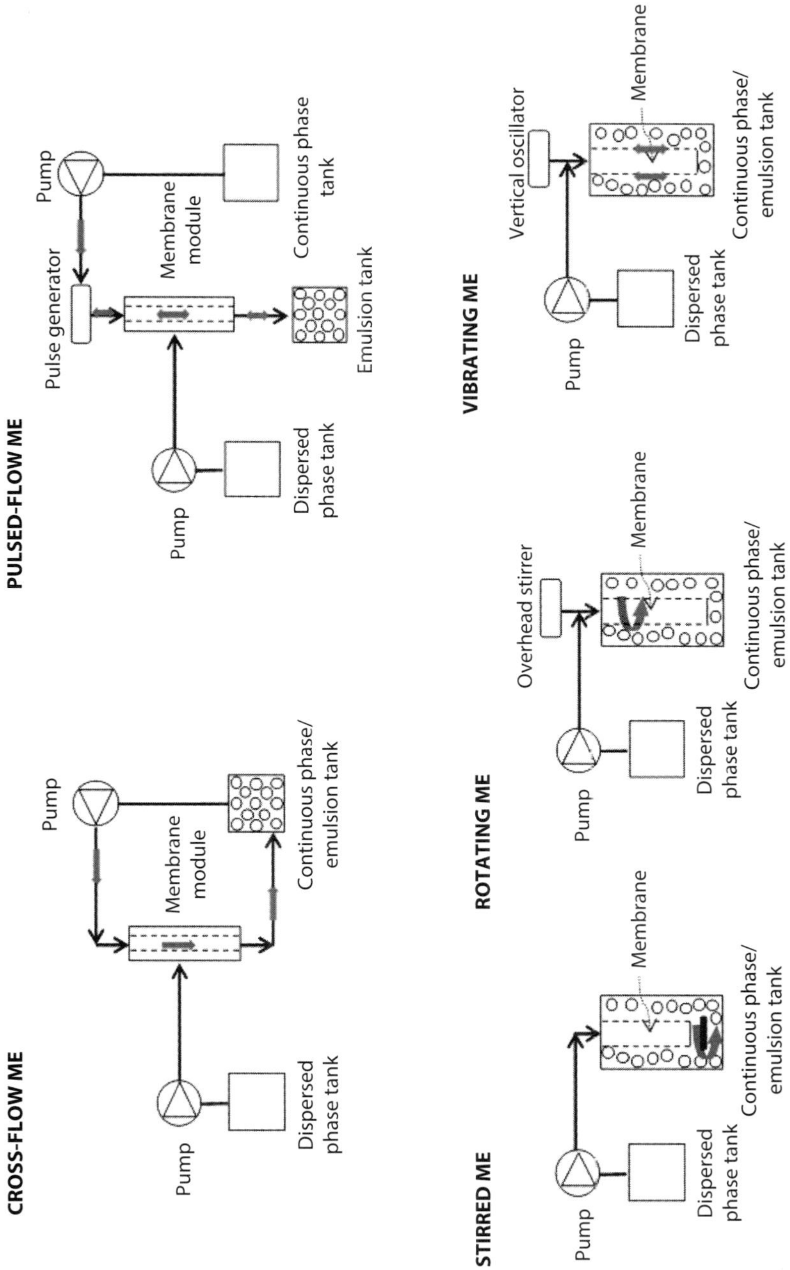

Figure 13.4 Membrane emulsification methods and devices.

phase stirring. They can be used for large scale production and continuous operation however they have a complicated and more expensive design.

Pulsed flow ME (Figure 13.4) is an alternative method of producing emulsions with a high dispersed to continuous phase ratio continuously in a single- pass operation of the continuous phase [21]. The dispersed and the continuous phase are injected using a pump while the pulsed flow is generated by a frequency generator. The potential benefits of the system are that it does not require a special module design, it is suitable for large scale production and it can be connected in series with a baffled reactor, to achieve simultaneous drop generation and chemical/physicochemical reaction in the produced emulsion.

13.2.9 Material Nature and Sensing Properties

Materials used in particles preparation play the most critical role for stimuli-responsive modality. The stimulus responsive ability can be obtained by using in particles manufacturing environmentally sensitive materials, such as polymers, or introducing specific "receptors" such as proteins or solid or micro-nanogels particles adsorbed to the interface of two immiscible liquids as oil and water (Figure 13.5). In the current literature emulsions stabilized by adsorbing particles to the interface are called "Pickering emulsions". The surface properties of the particle can be manipulated chemically, and thus the interfacial properties can be tailored. In recent years there has been increasing interest in the use of stimuli responsive or switchable particles able to control the breakage of such Pickering emulsions.

In Table 13.1 some of the most common polymes or proteins or solid/microgels particles used in sensing particles construction are reported.

13.2.10 Temperature and ph Responsive-Materials

Temperature changes can cause the melting of polymeric particles or can result in a phase change transition, transforming a swollen, hydrated state to a shrunken, dehydrated state or polymer decomposition. Temperature-responsive polymers may have a lower critical solution temperature (LCST), and become hydrophobic upon heating or an upper critical solution temperature (UCST) and become hydrophilic upon heating. The change of temperature causes a change in the solvation state of the material and a consequence volume phase transition by swelling-deswelling process. Typical LCST polymers are based on N-isopropylacrylamide (NIPAM), N,N-diethylacrylamide (DEAM), methylvinylether (MVE)

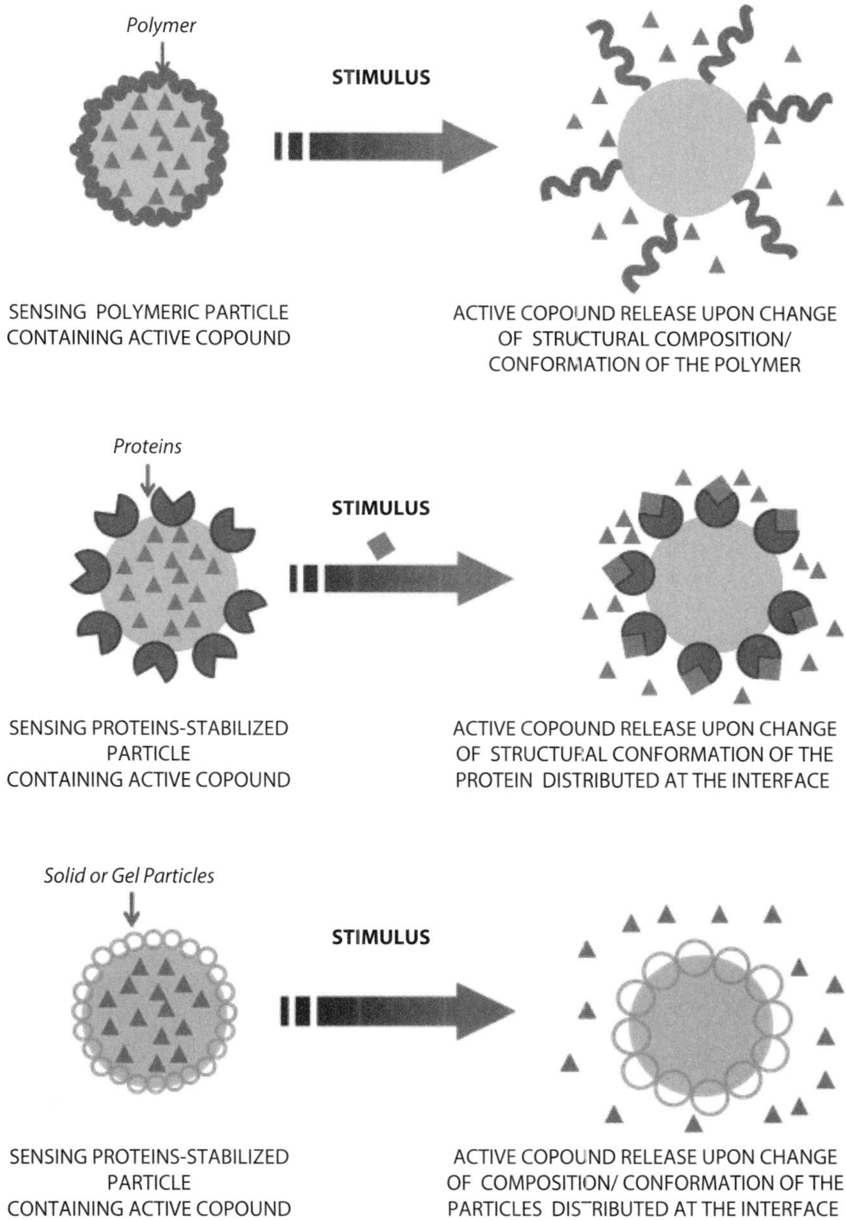

Figure 13.5 Stimulus responsive ability of particles manufactured using environmentally sensitive materials, such as polymers, proteins or solid/micro-nanogels particles adsorbed to the interface.

Table 13.1 Some of the most common polymers or proteins or solid/microgels particles used in sensing particles production.

pH and Temperature responsive polymers		
Poly(acrylamide) (PAAm) 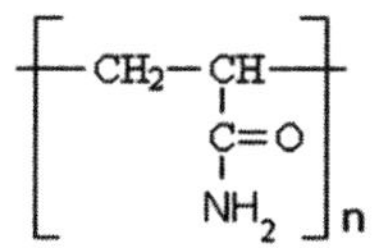	Poly(acrylic acid) (PAAc) 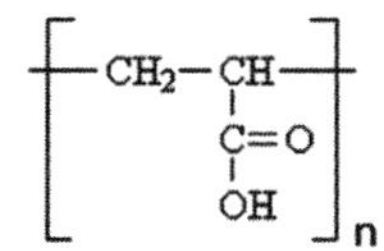	Poly(methacrylic acid) (PMAAc) 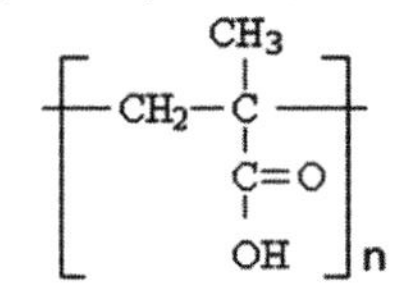
Poly(diethylaminoethyl methacrylate) (PDEAEMA) 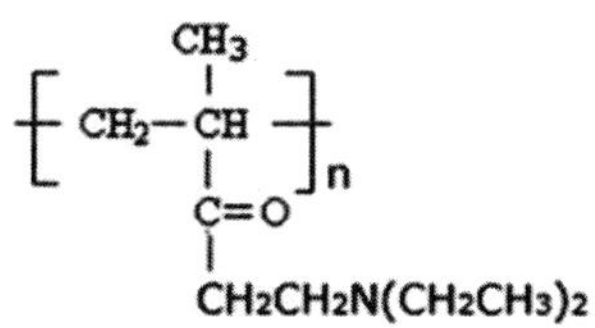	Poly(diethylaminoethyl methacrylate) (PDMAEMA) 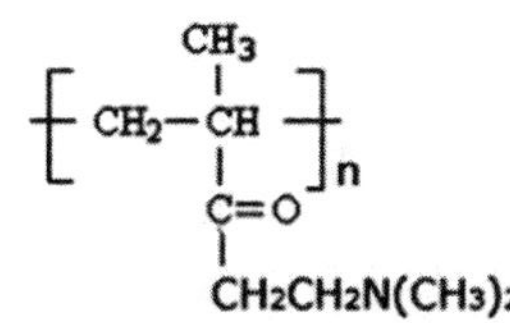	Poly (vinyl alcohol) (PVA) 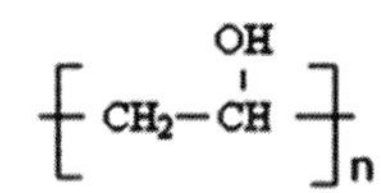
Poly(N-isopropylacrylamide) (PNIPAM) 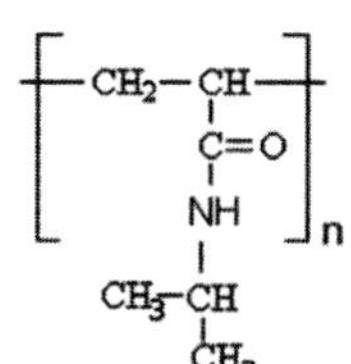	Poly(N-isopropylacrylamide-co--methacrylic acid) (PNIPAM-co-MAA) 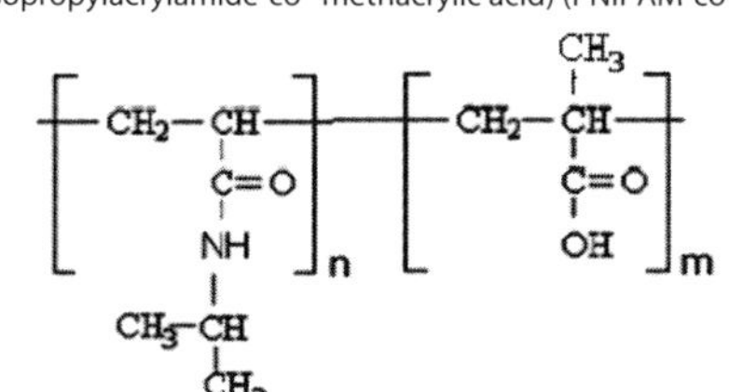	Poly(methylvinylether)(PMVE) 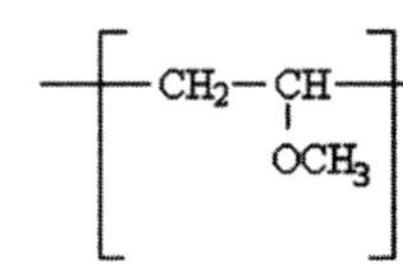

Table 13.1 Some of the most common polymers or proteins or solid/microgels particles used in sensing particles production.

Poly(N,N-diethylacrylamide) (PDEAAM)

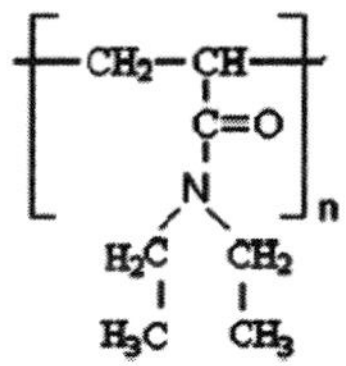

Poly(ethylene glycol) (PEG)

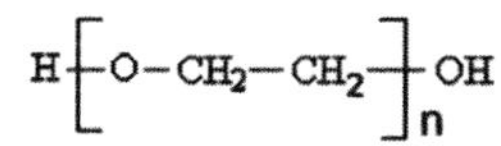

Polystyrene (PS)

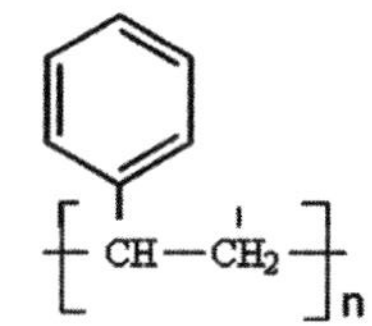

Poly(N-vinylcaprolactam)(PNVCL)

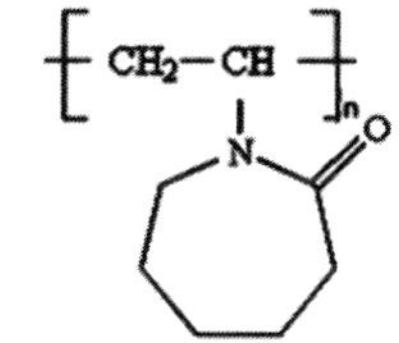

Alginate

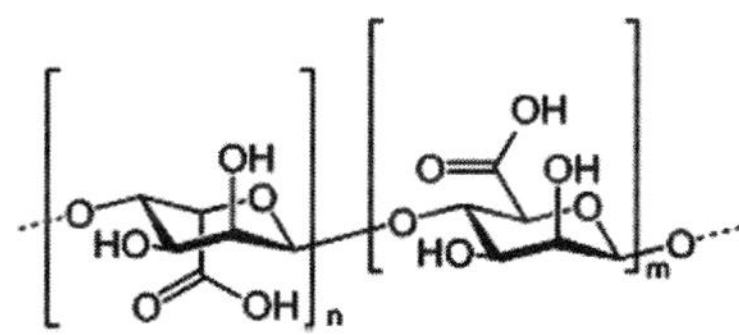

Chitosan

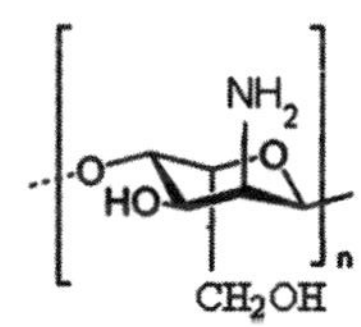

hyaluronic acid

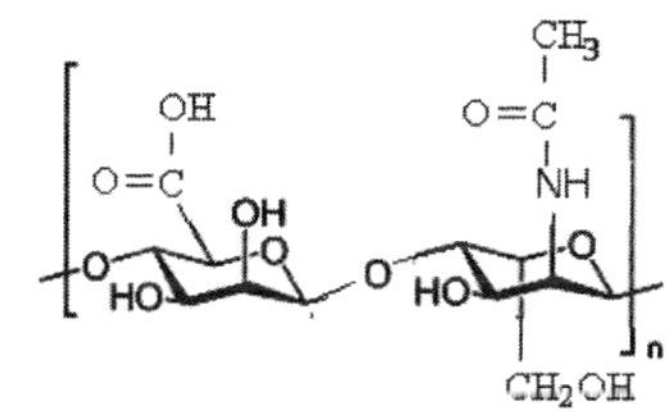

pH and Temperature responsive solid particles	Biochemical material
TiO_2	Glucose Oxidases (GOD)
P4VP/SiO_2 (4-vynilpridine)	Glucose Catalase (CAT)
8-HQ/SiO_2 (8-hydroxyquinoline)	Concanavalin (Con A)
KHP/SiO_2 (Potassium hydrogen phthalate) 8-HQ/SiO_2 (8-hydroxyquinoline)	Ovalbumin

and N-vinylcaprolactam (NVCL) as monomers. A typical UCST system is based on a combination of acrylamide (AAm) and acrylic acid (AAc) [22]. PNIPAAm is the most used termo-responsive hydrogels for the preparation of drug delivery systems because it is characterized by a LCST close to body temperature (32°C) , therefore it may change its conformation in local infections or diseased tissues (where the temperature is above the physiological one) releasing the drug. Above the LCST the amphiphilic chains hide the hydrophilic amide groups in a compact globule conformation; during this transition the hydrogen bonding formed between water molecules and the polymeric chains are disrupted resulting in a changed swelling capability of the material. The LCST of PNIPAM is independent of the molecular weight and the concentration, but it can be changed by shifting the hydrophilic/hydrophobic balance by copolymerization with a second monomer. Hydrophobic co-monomers increase the LCST, whereas hydrophilic co-monomers have the opposite effect [23]

Another class of thermal-responsive gels is the "shape memory gel" (SMPs) that can memorize their structure after being deformed and the initial geometry may be returned upon temperature change. Generally, in the original shape, internal stress is zero or negligible. If the SMP is subject to deformation, the deformation stress is stored in the cross-linking structure after cooling the polymer below its switch transition temperature. When the SMP is heated to a temperature above the switch transition temperature, it recovers its original shape with the release of internal stress stored in the cross-linking structure.

From the application point of view, the prerequisite for thermal-active polymers as shape memory materials is that their switch transition temperature needs to be greater than the temperature of the environment in which they are applied. Among these SMPs, shape memory polyurethanes (SMPUs) have been widely studied and commercially applied. These SMPs usually have a physical cross-linking structure, crystalline/amorphous hard phase, or chemical cross-linking structure [24].

In the case of pH-sensitive particles, the pH-tunable moieties that are generally incorporated into the particle structure include carboxyl and/or tertiary amino groups, which function as pH sensors; hydrophobicity or the swelling behavior of the polymers can be altered by protonation or deprotonation, when the ionisable groups are linked to the polymer structure. Classical monomers are acrylic acid (AAc), methacrylic acid (MAAc), maleic anhydride (MA) and N,N-dimethylaminoethyl methacrylate (DMAEMA). In the case of polyacids (such as poly-acrylic acid) an increase in their hydrodynamic volume is observed when their carboxylic groups pass from ionized to a deionized state, whereas the opposite effect is observed for polybases (such as N,N-dimethylaminoethyl methacrylate).

Besides synthetic non-biodegradable polymers, natural based polyelectrolytes, especially polysaccharides have been tested in the development of pH-responsive materials. Those biopolymers are especially attractive in biomedicine, including tissue engineering, due to their biocompatibility and biodegradability. Examples of anionic natural origin polymers are alginate and hyaluronic acid whereas chitosan is the only cationic polysaccharide found in nature.

Proteins are polyampholytes, and their charges and hydrophobicity/hydrophilicity character is pH-dependent, therefore, food protein hydrogels with nanoscale dimensions are ideal candidates as sensitive particles to load and release active molecules. Food protein-polysaccharide interactions have been studied by many scientists. For example, Yu *et al.* [25] produced nanogels of chitosan-ovalbumin taking advantages of gelation properties of ovalbumin to entrap chitosan chains in to the core and shell of the nanogel such as a cross-linker agent. Ovalbumin, the most abundant protein in the egg white, forms a transparent solution or transparent gel if the pH of the solution is far from the isoelectric point and the ionic strength is low; turbidity appears and the protein aggregates or gels when the pH is near the isoelectric point and/or the ionic strength.

pH sensitive nanocomposite material can be obtained by polymerizing 4-vinylpyridine (P4VP) in presence of silica sols. The size of the microgels depends on the protonation of P4VP that is pH dependent resulting in pH-sensitive emulsion stabilized only at high pH values [26]. Haase *et al.* [27] prepared stable Pickering emulsion under slightly acidic conditions by adsorption of 8-hydroxyquinoline (8-HQ) onto silica particles (Ludox TMA). With decreasing pH, protonation of silanol groups take place, resulting in the decay of the surface charge and consequently the adsorption of 8-HQ on the surface of silica nanoparticles promotes their interfacial adsorption by increasing their hydrophobicity. Another strategy to produce pH responsive emulsions with silica nanoparticles is the binding with potassium hydrogen phthalate (KHP). Dissociation of phthalate from the particle surface upon the addition of either base or acid leads to hydrophilic particles that are unable to stabilize the emulsions. Therefore, at either high or low pH, emulsions are demulsified, whereas they are stable in the pH range of 3.5–5.5. Thus, this system provides two reversible transitions using only one stimulus [28].

Temperature and pH sensitive microgels (such as PNIPAM and MAA) or copolymer of PNIPAM with epoxy groups such as glycidyl methacrylate (GMA) (thermosensitive poly(N-isopropylacrylamide-co-glycidyl methacrylate) microgels for controlled drug release] have also been used as emulsifiers for the production of sensitive Pickering emulsions. The emulsion becomes unstable and ripening of the oil droplets occurs if the temperature of the emulsion was above the LCST of PNIPAM or the emulsion was destabilized by acid addition [29].

Marchal *et al.* [30] synthesized an amphiphilic copolymer composed of polystyrene as hydropohobic polymer and 2-(dimethylamino)ethyl methacrylate (DMAEMA) as hydrophilic polymer, to produce a new pH and thermo-responsive surfactant able to stabilize invertible emulsions. It was possible to produce both O/W and W/O emulsions reversibly by changing the environmental conditions. At high pH and temperature this copolymer was mostly hydrophobic and formed W/O emulsions, opposite conditions promoted the formation of O/W emulsions.

13.2.11 Physical Sensitive Material (Light, Magnetic and Electrical Field)

A wide range of polymeric materials have been found to alter certain properties in response to light such as shape changes (contraction, bending, volume changes). Polymers or proteins used in the preparation of photoresponsive particles have to be equipped with photosensitive functional groups or fillers. Some of the most studied photosensitive molecules can be classified into three groups: i) photo-isomerizable molecules such as azobenzene that shows reversible trans-cis isomerization after exposure to the light; when azobenzene groups are linked to macromolecules, the interconversion between the two photoisomers can induce macroscopic changes in the polymeric material. ii) Triphenylmethane leuco derivatives that undergo photoinduced ionic dissociation; the back reaction by recombining the ion pair occurs thermally in the dark. If triphenylmethane leuco derivatives are incorporated in hydrogels, the reversible variation of electrostatic repulsion between photogenerated charges will generate the photoinduced expansion and shrinkage of hydrogels. iii) Photoreactive molecules such as cinnamates that is able to form photoreversible covalent cross-links in polymers when exposed to UV illuminations.

Light-active polymers have been proposed to be used in bio-separation, aviation, drug delivery, cardiovascular therapeutic device, medical stents and optical and micro-robot medicine.

Magnetic-active polymers show magnetic-active effect with the change in magnetic fields. They are composites of elastomers or gels filled with small magnetic particles. Based on the polymer matrix used, magnetic-active polymers are categorized into magnetic-active elastomers that change their shape upon a magnetic field and ferrogels, a chemically cross-linked polymer swollen by a ferrofluid. A variety of electrically sensitive materials have also been incorporated in the particles structures and, depending on the material, the responses included in this section range from alignment in electric fields, electrical conductivity, and redox reactions. For example hydrogels based on acrylic acid/acrylamide blended with conductive polypyrrole-carbon black composites were found to have a fast and reversibly electroactuation [31].

13.2.12 Biochemical Responsive Materials

The incorporation of highly specific, high affinity binding proteins or highly specific enzymatic cleavable proteins and peptide is used to produce particles able to respond to different biological conditions. In addition, polymeric materials that can be digested by specific enzymes are regarded as promising candidates for fabricating biochemical- sensitive particles. Designing synthetic polymers with the ability to adapt their properties in response to specific interactions with bio-macromolecules and small molecules may facilitate the application of smart polymers in drug delivery, diagnostics, sensing, separations, etc. The conjugation of a protein (enzyme, antibody, and receptor) to a polymer may offer several advantages:

- the polymer can promote or attenuate the accessibility of the active binding-site of the protein by changing the environmental conditions;
- the polymer can mask a toxic protein or avoid adverse immune response before the protein reaches its target tissue;
- a polymer- protein conjugate can be easily separate and purified when the polymer goes through the phase transition and become insoluble [32];

Examples of biochemical responsive systems are:

- *Glucose-responsive polymeric systems* are typically based on enzymatic oxidation of glucose by glucose oxidase (GOx), binding of glucose with concanavalin A (Con A), or reversible covalent bond formation between glucose and boronic acids.
- *Enzyme-responsive systems* are typically composed by an enzyme-sensitive substrate and another component that

directs or controls interactions which lead to macroscopic transitions. Catalytic action of the enzyme on the substrate can lead to change in supramolecular architectures, swelling/collapse of gels, or the transformation of surface properties. Another approach to convey enzyme-sensitivity to a material is the incorporation of functional groups that react under enzymatic conditions. Exposure of the groups to a specific enzyme can lead to the creation of new covalent linkages that cause a change in macroscopic properties.
- *Immuno-responsive polymeric systems* are typically hydrogels with antigens or antibodies entrapped in the network by physically entrapping, chemical conjugation or with antigen antibody pair used as reversible cross-linker within the network. Antigen–antibody interactions are highly specific and are associated with complex immune responses that help recognize and neutralize foreign infection-causing objects in the body [33].

13.2.13 Phase Change Material (PCM)

PCM which can convert from solid to liquid or from liquid to solid state is the most frequently used latent heat storage material, suitable for the manufacturing of heat-storage and thermo-regulated textiles, packaging, medical applications and for the use of solar energy in buildings. Phase change materials have the ability to change their state with a certain temperature range; They absorb energy during the heating process as phase change takes place and successively this energy can be transferred to the environment in the phase change range during a reverse cooling process. The insulation effect reached by the PCM is defined as "dynamic thermal insulation" because it takes place only during the phase change (in the temperature range of the phase change) and terminates when the phase change in all of the PCMs is complete. The advantage of PCM storage compared to sensible heat-storage systems is its potential to store large amounts of heat with only a small temperature swing. However, the disadvantage is the low heat-conductivity of the material that can be enhanced by using metal filler, carbon nanofiber/fiber fillers. Textiles containing micro-PCM capsules react immediately with changes in environmental temperatures, and the temperatures in different areas of the body. When an increase in temperature occurs, the PCM microcapsules react by absorbing heat and storing this energy in the liquefied phase change materials. When the temperature decreases, the microcapsules release this stored heat energy and the phase change materials solidify again [34]. A wide spectrum of phase change material is available with different heat storage capacity and phase change temperature. A classical example of phase change material is the paraffin wax, which can be microencapsulated and then either integrated into fiber or used as a coating. Polyethylene glycol (PEG) is other important PCMs for textile applications. Hydrated inorganic salt with 'n' water molecules, can also be used in the manufacturing of heat storage and thermo-regulated textiles and clothing which usually, has a heat absorbing and releasing temperature range of about 20–40 °C. Hydrated salts are attractive materials for use in thermal energy storage due to their high volumetric storage density and thermal conductivity, low cost compared to paraffin wax [35].

Others attractive materials recently studied for thermal energy storage are fatty acids (capric, lauric, palmitic and stearic acids) and their binary mixtures, because they melt at temperatures useful for thermal energy storage [36].

13.3 Senging Particles Produced by Membrane-Based Process

13.3.1 Temperature and pH Responsive-Materials

Monodisperse hollow poly(N-isopropylacrylamide) (PNIPAM) microcapsules were prepared combining membrane emulsification and UV-initiated polymerization at the interface of W/O single emulsions at a temperature below the lower critical solution temperature of PNIPAM [37]. Before and after the polymerization, the span of the particles distribution was maintained at about 0.27. PNIPAM is the most widely studied thermo-sensitive polymer and the main advantage demonstrated by Cheng and co-workers was that the membrane method allowed the production of monodisperse hollow polymeric microcapsules.

pH-sensitive particles were also synthesized using Acrylic polymers (Eudragit® L100 or Eudragit® E100) dissolved in water at a set pH [38]. The polymer solution passed through the pores of a membrane into a continuous phase of a different pH and the contact with the continuous phase causes the pH-sensitive polymer to precipitate. pH-sensitive quaternized chitosan microspheres were prepared by combining membrane emulsification technique and thermal-gelation method [39]. Quaternized chitosan is obtained introducing quaternary amino groups into chitosan chains natural polysaccharide in order to increase water-solubility, cationic activity and bioadhesive properties. The water phase was a mixture of quaternized chitosan solution and α-β- glycerophosphate (α-β-GP used as crosslinking reagent) while the oil phase was a mixture of liquid paraffin and petroleum ether (v/v = 7/5) containing PO-500 4.0 wt % as emulsifier. A uniform W/O emulsion was produced using the SPG membrane and the droplets solidified into microspheres at 37 °C by thermal-gelation method. The influence of process conditions on the property of prepared microspheres was investigated and the optimized preparation condition was obtained. As a result, the coefficient of variation (C.V.) of obtained microspheres diameters was below 15%. The microspheres had porous structure and showed apparent pH-sensitivity by dissolving rapidly in acid solution (pH 5) and kept stable in neutral solution (pH 7.4). The pH-sensitivity of microspheres was demonstrated to influence the release of bovine serum albumin (BSA) used as a model drug and BSA was released rapidly in acid solution and slowly in neutral medium. Other two methods for the preparation of monodisperse chitosan microcapsules with hollow structures using the SPG membrane emulsification technique were proposed by Akamatsu *et al.* [40]. SPG membranes were employed for the preparation of two different W/O emulsions: one prepared using alginate dissolved in the dispersed phase and the second one using chitosan. The first emulsion was used to produce Ca-alginate particles by using calcium chloride as cross-linking agent. Ca-alginate particles were successively mixed with the second emulsion and monodisperse alginate/chitosan core-shell microspheres were obtained since electrostatic interaction between cationic chitosan and anionic alginate chains. Chitosan cross-linking shell was obtained by two methods (Figure 13.6):

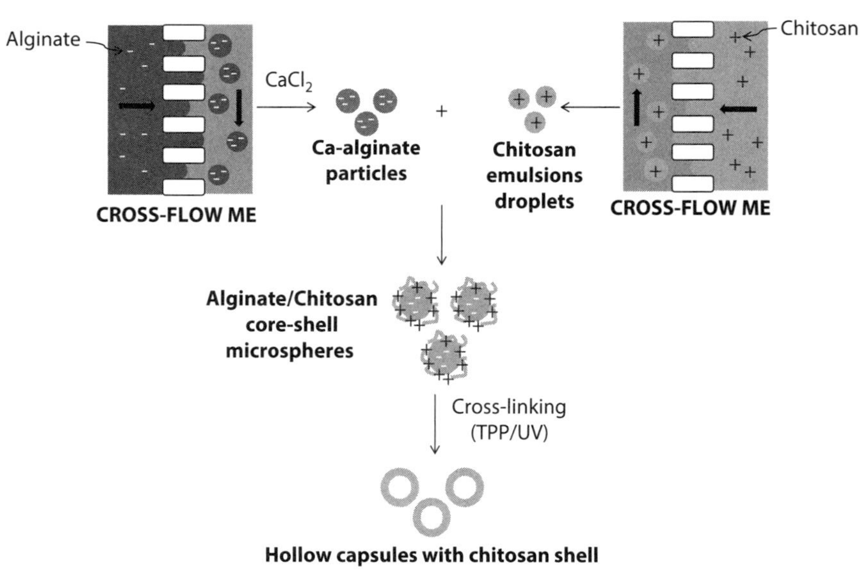

Figure 13.6 Alginate/chitosan core-shell microspheres produced by membrane emulsification.

the first method consisted in a physical cross-linking of chitosan by tripolyphosphate treatment that at the same time acts as capture agent for calcium alginate resulting in hollow structured particles. The second procedure was a chemical cross-linking method consisting in the use of UV irradiation of photo-cross-linkable chitosan containing azido groups and tripolyphosphate treatment was used to solubilize and remove the core particles of calcium alginate. The use of membrane emulsification for the production of the emulsions permitted to provide a uniform reaction environment and obtain monodisperse microparticles with average diameters of 4.4 μm by using 4.9 μm pore size SPG membrane.

Chitosan has been also used by Vladisavljević *et al.* [41] in the electrostatic deposition method to modify interfacial characteristics of uniformly sized droplets produced by premix membrane emulsification. The interfacial properties of emulsion droplets can be changed by using surfactant-displacement and/or electrostatic deposition methods. The surfactant-displacement method simply involves mixing a preformed emulsion with an emulsifier solution so that the original emulsifier is partially or fully displaced from the droplet surfaces by the new emulsifier. The electrostatic deposition technique involves depositing polyelectrolytes onto oppositely charged droplet surfaces. This technique offers a promising way to improve the stability of emulsions against environmental stresses such as pH, ionic strength, freezing, and heating. Lipophilic active components can be encapsulated within the oil droplets, whereas charged hydrophilic components can be incorporated within the interfacial layers. Potentially, active ingredients can be released at the site of action in response to a specific environmental stimulus. In this work the displacement of surfactant was carried out by adding to the O/W emulsion stabilized by Tween 20 a displacing surfactant such as SDS, DTAB or β-Lg in acetic acid buffer. The original non-ionic surfactant molecules were fully or partly displaced from the droplets surface by the charged surfactant and the successively mixing with a polyelectrolytes solution permitted to obtain a secondary emulsion stabilized by Tween 20-SDS-chitosan interfacial layers.

J. Ma *et al.* [42], used membrane emulsification coupled with internal gelation for the preparation of Ca-alginate gel beads with small and uniform size. The disperse phase consisted in a mixture of calcium carbonate and sodium alginate extruded through the microporous membrane under N2 pressure in a continuous phase composed of paraffin oil and surfactant. The droplets were detached by the flowing of the continuous phase along the lumen of the membrane. Gelation step started after the adding of glacial acetic acid to the W/O emulsion in order to dissolve $CaCO_3$ and to release Ca_2, the cross-linker. A study of the morphology and size distribution of

Ca-alginate gel beads as a function of mean pore size of the membrane, trans-membrane pressure, polymer concentration and surfactant concentration was carried out. Alginate has several unique properties that have enabled it to be used as a matrix for the entrapment and /or delivery of a variety of biological agents. Thanks to its bioadhesive properties and non-immunogenicity it can be advantageous for the site specific delivery to mucosal tissues and as protein delivery system. Moreover, the presence of anionic charged groups make it sensible to pH changes.

Membrane emulsification technique has been also used for the production of poly-acrylamide-co-acrylic acid microspheres. Microspheres size can be finely controlled by choosing SPG membranes with different pore size and the relationship between the electric properties of the surfaces of polyacrylamide-co-acrylic acid microspheres and their sizes as well as the size dependence of the microsphere structure were studied by Nagashima *et al.* [43]. An aqueous monomer solution containing different concentration of acrylamide and acrylic acid was dispersed in a continuous phase of cyclohexane, SUNSOFT818H and 2,2-azobis-isobutyronitrile (photoinitiator) to prepare a W/O emulsion by using the micro porous glass apparatus. Hydrophobic membranes with different pore size and pre-treated with octadecyltrichlorosilane and trimethylchlorosilane were used for the emulsification. The polymerization was conducted by stirring the emulsion prepared for 3 h at 70 °C under nitrogen atmosphere. The electrophoretic mobility of poly-acrylamide-co-acrylic acid microspheres was negative at pH 7.4 and more negative mobility values were obtained with smaller poly-acrylamide-co-acrylic acid microspheres than with the larger ones. The electrokinetic study revealed that copolymerization of acrylamide monomers and acrylic acid monomers does not proceed homogeneously within a microsphere: acrylic acid monomers and charged oligomers move toward the interior of microemulsions while polymerization proceeds. Therefore, acrylic acid concentration is higher in the core region than in the surface layer of the microspheres and the surface charge density in the surface layer increases as the size of microspheres decreases. Electrical surface properties and size of microspheres are primarily important when microspheres are used as drug devices; in fact the device interacts with various kinds of biological cell surfaces carrying different electrical charges.

13.3.2 Biochemical Responsive Materials

W/O/W emulsions able to respond to a biochemical stimulus have been produced using two-step membrane emulsification method [44] (Figure 13.7). A multiple emulsion containing a bio-receptor (Con A) that

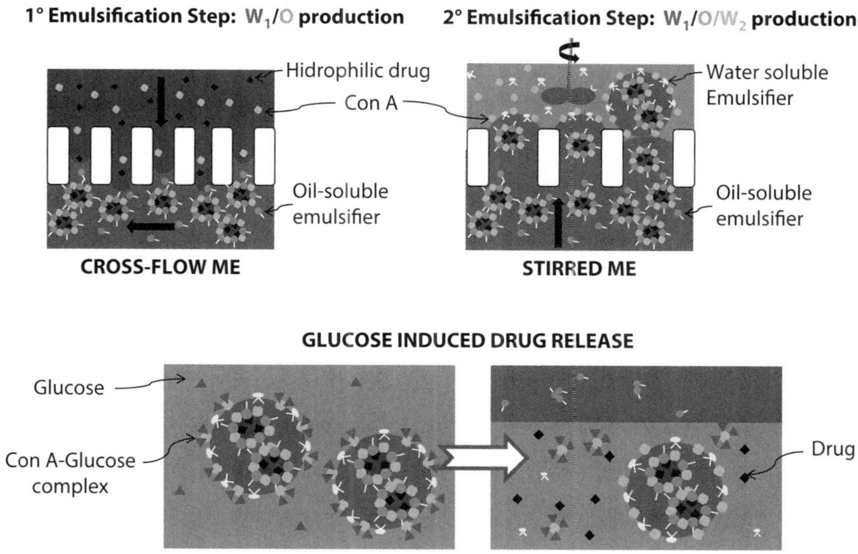

Figure 13.7 Glucose-sensor multiple emulsions preparation by membrane emulsification and schematic representation of stimulus-induced drug release.

specifically recognizes and interacts with an artificial ligand (Glucose) was manufactured by the membrane process and used as a model system. Cross-flow membrane emulsification and stirred membrane emulsification were used to produce a W/O emulsion and a multiple W/O/W emulsion, respectively. The best operative conditions able to control droplets emulsions size without causing loss of functional properties of specific ingredients were identified. The multiple emulsions produced demonstrate the ability to modulate the drug release as a function of emulsion interface composition. It was supposed that the Con A-glucose interaction at emulsion interface determines changes in the properties of Con A interfacial film. This promoted the release of bioactive molecules added in the inner water phase observed in the presence of effective stimulus concentration.

A simple emulsion with a heterogenyzed enzyme was produced by membrane emulsification [45] (Figure 13.8). The lipase was used as model enzyme. Lipase is an enantioselective phase transfer biocatalyst frequently used in esterifications, transesterifications and hydrolysis reactions accepting a broad range of even hydrophobic substrates with vast industrial importance. The enantioselective hydrolysis of racemic naproxen methyl ester was used as reaction model to produce (S)-naproxen, a member of the arylacetic acid group of nonsteroidal anti-inflammatory drugs. When the emulsion was prepared by membrane emulsification, lipase is distributed

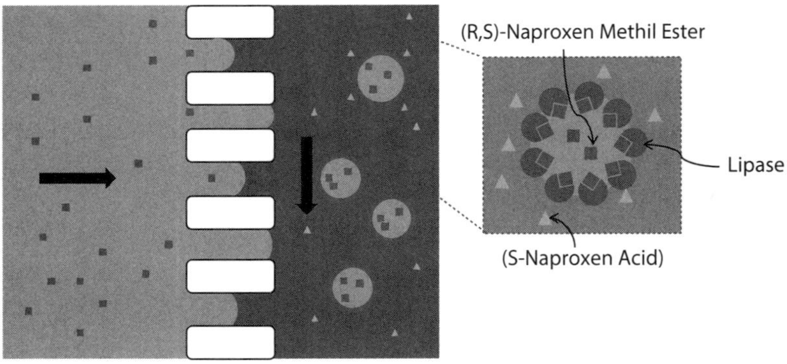

Figure 13.8 O/W emulsion with heterogenyzed enzyme (lipase) produced by membrane emulsification. Lipase at emulsion interface catalyzes the enantioselective hydrolysis of racemic naproxen methyl ester (dissolved in the dispersed phase) to produce (S)-naproxen acid (extracted in the continuous phase).

at the O/W interface while drops grow at the membrane pore opening and the substrate of the reaction was dissolved in the dispersed phase. Both droplets micromanufacturing and biocatalysts immobilization can be performed simultaneously and continuously. The interface with controlled and uniform size provided a constant reaction interface at the steady-state. The method permitted the controlled fabrication of monodispersed microstructured biocatalytic droplets for a highly efficient enzymatic reaction. Lipase demonstrated 100 % enantioselectivity converting 90% of (S)-isomer in less than 24 h. This performance was achieved thanks to the enzyme optimal distribution at the interface of stable, uniform and small droplets obtained produced in mild conditions able to prevent enzyme inacativation.

Size-controlled and monodisperse enzyme-encapsulated chitosan microspheres have been developed by membrane emulsification technique [46]. An aqueous solution containing a mixture of chitosan polymer and lysozymes used as enzyme was emulsified to form W/O emulsions. Microspheres were generated by a cross-linking reaction with glutaraldehyde. The encapsulated lysozyme was demonstrated to maintain about half of its activity during the preparation procedure by conducting the activity tests using the subtstrate ethyleneglycolchitin.

Omi S. *et al.* [47] prepared enzyme-responsive systems consisting in hydrophobic crosslinked polystyrene (PS) and hydrophilic polymethyl methacrylate (PMMA) spheres with Glucoamylase (GluA) immobilized. For the preparation of the spheres a certain volume of monomer –solvent mixture, containing also an oil-soluble initiator was permeated trough the micropores of the SPG membrane in an aqueous solution of PVA, and polymerized under nitrogen atmosphere with gentle stirring. The strong hydrophilic character of SPG membrane favors the formation of PS particles; a wider size dispersion was obtained when hydrophilic methyil methacrylate was used as monomer and the problem has been overcome by using the swelling technique, in which the primary hydrophobic droplets were swollen with the secondary emulsion droplets of MAA and other hydrophilic components and the subsequent polymerization yielded PMMA microspheres. Cross-linking agents such as divynilbenzene for PS and ethyleneglycol dimetacrylate for PMAA were used. GluA immobilization was carried out through two methods: covalent bonding by using a small amount of acrylic acid or glycidyl methacrylate (GMA) and physical adsorption thanks to porous structure and high specific area of the microspheres. Although GluA activity has been difficult retained after the immobilization process, the porous structure of the carriers favored the enzyme immobilization respect to covalent binding and a maximum 55% activity was obtained by the physical adsorption to PMMA spheres.

13.3.3 Physical Sensitive Material

Fluorescent chitosan microspheres were produce combining membrane emulsification method and cross-linking reaction [48]. Chitosan microspheres were found to exhibit fluorescent properties without conjugation to any fluorescent agent. The fluorescence color varied with different crosslinkers and can be modulated by further chemical reduction, whereas the fluorescence intensity can be controlled by tuning the particle size and degree of crosslinking. The chitosan microspheres were studied as photostable tracers to study the phagocytosis of Human hepatocellular carcinoma cells (HepG2).

Floating photocatalytic composite particles were created by injecting an oil phase into an aqueous suspension of TiO_2 nanoparticles using the process of membrane emulsification to control the (Pickering) emulsion size [49]. The composite particles were in the range between 80 and 300 μm. They possessed photocatalytic activity, which was further enhanced by chemically incorporating silver particles into the TiO_2 shell. The floating catalytic particles were ever to become a viable alternative to conventional advanced oxidation.

Another application is the synthesis and characterization of polymeric microcapsules for ultrasound-triggered delivery of lipophilic drugs [50]. Microcapsules were made through pre-mix membrane emulsification.

Fluorinated end-capped polymer poly(L-lactic acid), (pLA-pF) and hexadecane oil were chosen as the shell material and drug-carrier reservoir. Hexadecane and cyclodecane were added at a ratio of 0:1 for the production of completely gas-filled microcapsule, at a ratio 1:1 for the production of half oil-filled microcapsules and at a ratio 1:0 for the production of almost completely oil-filled microcapsules. Using diagnostic ultrasound, the ability to crack the oil-filled microcapsules, thereby releasing the encapsulated gas and drug has been demonstrated. In addition, non-destructive imaging of the microcapsules was demonstrated indicating that guidance and monitoring of therapy will be possible. Among the microcapsules studied, the oil-filled microcapsules had the highest lipophilic drug-reservoir whilst diagnostic ultrasound could still be used to trigger drug release. These microcapsules therefore have great potential for ultrasound triggered local delivery of lipophilic drugs.

Nanoparticles [51] and polymeric microspheres containing magnetite nanoparticles [52] have been produced using membrane emulsification method to create the reaction spaces in which the metal or metal oxide precursors were reacted to form the final product. Kakazu *et al.* produced W/O microemulsions containing silver aqueous solutions using the SPG membranes. Then, after adding hydrazine, silver nanoparticles were formed in each W/O microdroplet. The obtained nanoparticles were around 10 nm in diameters, with 15–20% CVs, and crystalline sphere-like shape. Yang *et al.* fabricated uniform magnetic poly(styrene-co-2-hydroxyethyl methacrylate) (PST-HEMA) microspheres combining premix membrane emulsification of double emulsion and an in situ magnetization technique. The method involved, first, the preparation of uniform double W/O/W emulsions with ferrous ferric solution as the internal water phase and dichloromethane dissolved with PST-HEMA as the oil phase by premix SPG membrane emulsification. Then, the inner ferrous ferric solution was in situ magnetized into nanoparticles by adding the ammonia solution into the external water phase. This W/O/S emulsion was finally solidified into microspores by solvent evaporation. In situ magnetization technique determined overcoming the encapsulation difficulties in traditional synthesis routes due to poorer affinity between iron oxide nanoparticles and polymer. A new method, that does not require any further reactions after the emulsification, was developed combining membrane emulsification and solvent pervaporation process [53]. A hexane phase containing primary magnetite nanoparticles (10nm) is dispersed through the SPG membrane of a fixed pore size into an aqueous phase containing a minimal amount of surfactant. Droplets of the hexane phase with narrow size distributions were formed, and dense clusters of magnetite nanoparticles remained after the solvent was removed. The clusters can be easily coated with a layer of silica using the Stöber method, providing them with a protective shell and functionalizable surface. The possibility to integrate membrane emulsification method and membrane pervaporation is an attractive strategy for the development of a continuous process for the synthesis of various polymeric and hybrid beads.

13.3.4 Molecular Imprinting

Molecularly imprinted nanospheres have been prepared by premix membrane emulsification technique [54]. Chloramphenicol (CAP) was used as template molecule and methacrylicacid (MAA) was used as functional monomer. Reproducible detection or extraction methods are required to control and monitor CAP residues in food of animal origin because CAP is a broad-spectrum antibiotic with demonstrated serious side-effects on humans and it has been banned for use in food-producing animals in many countries. The template molecule (CAP), the functional monomer MAA, the cross-linker ethylene glycol dimethacrylate (EDMA) and the initiator 2,2-azobis (2-isobutyroni-trile) (AIBN) were dissolved in ethylacetate as oil phase and mixed whit the PVA solution using a homogenizer. The coarse emulsions were extruded through the uniform pores of the SPG membrane to obtain uniform smaller droplets, which were further polymerized to achieve the nanospheres. The resulted nanospheres were relatively uniform with a diameter in the range between 300 and 800 nm and demonstrated good selectivity.

13.4 Conclusions

There is an increasing use of membrane emulsification technology in the preparation of sensing particles as demonstrated by literature data during the last years. Results demonstrated that sensing particles produced by membrane-based methods are able to combine the tuned structural properties such as size, size distribution and chemical compositions with the functional activity properly designed for specific applications. The flexibility of the method allowed the application in the production of particles from nano- to micro-scale using (bio)-polymeric materials, conventional emulsifiers, solvents of different chemical nature. Due to the low shear stress required in particles production, the method has been successful used in the production of formulations containing biofunctional proteins such as sensor or enzyme. The convergence of membrane micromanufacturing and biology and chemistry will lead to new approaches in the productions of advanced particles-based products and may provide advantages over existing technologies. The development of new methodologies such as pulsed flow membrane emulsification is expected to introduce new concepts in the manufacturing of particles in combination with chemical/ physicochemical reaction also at large scale production. The operational flexibility, reproducibility, straightforward up-scaling and reduction in the ratio of equipment size to production capacity of droplets production methods membrane based may lead to novel process intensification in particles production.

References

1. P. Becher, *Emulsions, Theory and Practice*, The New York, 1965.
2. P. Walstra, *Chemical Engineering Science*, Vol.48, p. 333–349, 1993.
3. A. Mersmann, H. Grobmann, *Chemie Ingenieur Technik*, Vol.52, p. 621, 1980 .
4. J. Floury, A. Desrumaux , J. Lardieres , *Innovative Food Science & Emerging Technology*, Vol.1, p. 127–134, 2000.
5. W. T. Richards, A. L. Loomis, *Journal of the American Chemical Society*, Vol.49, p. 3086–3100, 1927 .
6. T. Nakashima, M. Shimizu, M. Kukizaki, *Key Engineering Material*, Vol.61–62, p. 513–516, 1991.
7. K. Suzuki , I. Shuto , Y. Hagura, *Food Science and Technology International*, Vol.2, p.43, 1996.
8. A. Nazir, K. Schroën , R. Boom, *Journal of Membrane Science*, Vol. 362, p.1–11, 2010.
9. C. Charcosset, *Journal of Food Engineering*, Vol. 92, p. 241–249, 2009 .
10. V. Schroder, H. Schubert, *Colloids and Surfaces A: Physicochemical and Engineering Asp*ects, Vol.152,p. 103–109, 1999.
11. K. Suzuki, I Fujiki, Y. Hagura, *Food Science and Technology International*, Vol. 4, p.164–167, 1998.
12. S.M. Joscelyne, G. Trägårdh, *Journal of Membrane Science*, Vol. 169, p.107–117, 2000.
13. T. Nakashima , M. Shimizu, M. Kikizaki, *Advanced Drug Delivery Reviews*, Vol. 45, P. 47–56, 2000.
14. V. Schroder, O. Behrend, and H. Schubert, *Journal of colloid and interface science*, Vol. 202, p. 334–340, 1998.
15. GT. Vladisavljević, H. Schubert, *Journal of Membrane Science*, Vol. 225, p. 15–23, 2003.
16. M. J. Geerken, R. G.H. Lammertink, M. Wessling, *Journal of Colloid and Interface Sci*ence, Vol. 312, p. 460–469, 2007.
17. M. T. Stillwell, R. G. Holdich, S. R. Kosvintsev, G. Gasparini, and W. Cumming, *Industrial & Engineering Chemistry Research*, Vol. 46, p. 965–972, 2007.

18. R. A. Williams, S.J. Peng, D.A. Wheeler, N.C. Morley, D. Taylor, M. Whalley, D.W. Houldsworth, *Chemical Engineering Research and Design,* Vol. 76, p. 902–910, 1998.
19. G. T. Vladisavljević, R. A. Williams, *Journal of Colloid and Interface Science,* Vol. 299, p. 396–402, 2006.
20. R. G. Holdich, M. M. Dragosavac, G. T. Vladisavljevic´, and S. R. Kosvintsev, *Industrial & Engineering Chemistry Research,* Vol. 49, p. 3810–3817, 2010.
21. R. G. Holdich, M. M. Dragosavac, G. T. Vladisavljević, E. Piacentini, *Industrial & Engineering Chemistry Research* , Vol.52, p. 507-515, 2013.
22. D. Schmaljohann, *Advanced Drug Delivery Reviews,* Vol. 58, p. 1655–1670, 2006.
23. J. F. Mano, *Advanced Engineering Materials,* Vol. 10, p. 515–527, 2008.
24. H. Meng, J. Hu, *Journal of Intelligent Material Systems and Structures,* Vol. 21, p. 859, 2010 .
25. S. Yu , J. Hu , X. Pan , P. Yao , and M. Jiang, *Langmuir,* Vol. 22, p. 2754–2759, 2006.
26. S. Fujii, E.S. Read, B.P. Binks, S.P. Armes, *Advanced Materials,* Vol. 17, p. 1014–1018, 2005.
27. M. F. Haase, D. Grigoriev , H. Moehwald, B. Tiersch, and D. G. Shchukin, *Journal of Physical Chemistry.* Vol. 114, p 17304–17310, 2010.
28. J. Li and H. D. H. Stöver, *Langmuir,* Vol. 24, p. 13237–13240, 2008.
29. B. Brugger , B. A. Rosen, and W. Richtering, *Langmuir,* Vol. 24, p. 12202–12208, 2008.
30. F. Marchal , A. Roudot , N. Pantoustier , P. Perrin , J. Daillant , and P. Guenoun, *Journal of Physical and Chemistry,*Vol. 111, p 13151–13155, 2007.
31. M. Zrínyi, *Colloid and Polymer Science,* Vol. 278, p. 98–103, 2000.
32. D. Schmaljohann, *Advanced Drug Delivery Reviews,* Vol. 58, p. 1655–1670, 2006.
33. D. Roy, J. N. Cambre, B. S. Sumerlin, *Progress in Polymer Science,* Vol. 35, p. 278–301, 2010.
34. S. Mondal, *Applied Thermal Engineering,* Vol. 28, p. 1536–1550, 2008 .
35. A. Saito, S. Okawa, T. Shintani, R. Iwamoto, *International Journal of Heat and Mass Transfer,* Vol. 44, p. 4693–4701, 2001 .
36. D. Feldman, M.M. Shapiro, *Solar Energy Material,* Vol. 18, p. 201–216, 1989.
37. C. J. Cheng, L. Y. Chu, P. W. Ren, J. Zhang, L. Hu, *Journal of Colloidal Interface Science,* Vol.313 , p. 383–388, 2007.
38. N. Sheibat-Othman, T. Burne, C. Charcosset, H. Fessi, *Colloids and Surfaces A,* Vol. 315, p. 13–22, 2008.
39. J. Wu, W. Wei, L.Y. Wang., Z. G. Sua, G.H. Ma, *Colloids and Surfaces B: Biointerfaces,* Vol. 63, p.164–175, 2008.
40. K. Akamatsu, W. Chen, Y. Suzuki, T. Ito, A. Nakao, T. Sugawara, R. Kikuchi, S. Nakao, *Langmuir,* Vol. 26, p. 14854–14860, 2010.
41. G. T. Vladisavljević, D. J. McClementsc, *Colloids and Surfaces A: Physicochem. Eng. Aspects,* Vol. 364, p. 123–131, 2010.

42. J. Ma, Q. Yuan, *Journal of Applied Polymer Science*, Vol. 87, p. 848–852, 2003.
43. S. Nagashima, M. Koide, S. Ando, K. Makino, T. Tsukamoto, H. Ohshima, *Colloids and Surfaces A: Physiochemical and Engineering Aspects,* Vol. 153, p. 221–227, 1999.
44. E. Piacentini, E. Drioli, L. Giorno, *Biotechnology and Bioengineering*, Vol. 10, p. 913–923, 2011.
45. L. Giorno, E. Piacentini, R. Mazzei, E. Drioli. *Journal of Membrane Science*, Vol. 317, p. 19–25, 2008.
46. K. Akamatsu, Y. Ikeuchi, A. Nakao, S. Nakao, *Journal of Colloid and Interface Science*, Vol. 371, p. 46–51, 2012.
47. S. Omi, K. Kaneko, A. Nakayama, K. Katami, T. Taguchi, M. Iso, M. Nagai, G. H. Ma, *Journal of applied polymer science*, Vol. 65, p. 2655–2664, 1997.
48. W. Wei, L.Y. Wang, L. Yuan, Q. Wei, X.D. Yang, Z.G. Su, G.H. Ma, *Adv. Funct. Mater*, Vol. 17, p. 3153–3158, 2007.
49. R. G. Holdich, I. Y. Ipek, M. Lazrigh, G. Shama, *Industrial & Engineering Chemistry Research*, Vol. 51, p. 12509–12516, 2012.
50. K. Kooiman, M.R. Böhmer, M. Emmer, H.J. Vos, C. Chlon, W.T. Shi, C.S. Hall, S.H.P.M. de Winter, K. Schroën, M. Versluis, N. de Jong, A. van Wamel, *Journal of Controlloed Release,* Vol. 133, p. 109–118, 2009.
51. E. Kakazu, T. Murakami, K. Akamatsu, T. Sugawara, R. Kikuchi, S. Nakao, *Journal of Membrane Science,* Vol. 354, p. 1–5, 2010.
52. J. Yang, D. Hao, C. Bi, Z. Su, L. Wang, G. Ma, *Industrial & Engineering Chemistry Research*, Vol. 49,p. 6047–6053, 2010.
53. E. P. Chang, T.A. Hatton, *Langmuir*, Vol. 28, p. 9748–9758, 2012.
54. X. Kou, Q.Li, J. Lei, L. Geng, H. Deng, G. Zhang, G. Ma, Z. Su, Q. Jiang, *Journal of Membrane Science*, Vol. 417, p. 87–95, 2012

14

Membranes for Ultra-Smart Textiles

Annarosa Gugliuzza[1] and Enrico Drioli[1,2,3]

[1]*Institute on Membrane Technology ITM-CNR, University of Calabria, Rende (CS), Italy*
[2]*Department of Chemical Engineering and Materials, University of Calabria, Rende (CS), Italy*
[3]*WCU Energy Engineering Department, College of Engineering, Hanyang University, Seoul, Korea*

Abstract
Textiles can benefit from using membranes capable to work as multi-compartment arrays, where different and complementary functions can be allocated. Today, textiles markets face great competition but supply does not meet demand fully. Membrane science along with sensor technology is expected to bring innovation within the textiles manufacturing industry. The design of intelligent materials for the creation of ultra-smart textiles, which can provide self-maintenance, adaptability, auto-adjustment and long-distance communication, represent the future. This chapter is dedicated to deal with smart materials for the fabrication of sensing membranes showing adjustable functions, including breathability, heat exchange, self-cleaning, odor capture, drug delivery, power storage and communication. Every membrane function is discussed in relation to specific work activities and needs. Present and prospects of intelligent membranes for the creation of ultra-smart textiles are discussed as well.

Keywords: Adaptive membranes, ultra-smart textiles, clothing, breathability, power storage, long-distance communication, self-cleaning, perspiration, barrier membranes, fabrics

14.1 Introduction

Membranes find large application in different textiles fields; they are used to enhance fabrics comfort qualities, waterproofness, moisture and static

**Corresponding author*: a.gugliuzza@itm.cnr.it

Annarosa Gugliuzza (ed.) Smart Membranes and Sensors, (401–420) 2014 © Scrivener Publishing LLC

resistance along with self-cleaning, adaptability and interactive functions [1]. The opportunity to combine different materials in different structures makes the membranes attractive for textile applications, the molecular manipulation being a fundamental step to tailor desired properties on different length scale. Indeed, membranes take the advantage of incorporating antibacterical and antiallergenic compounds, scents and adsorbing capsules along with thermo, and photocromic chemicals. They can be made from polymers/materials having memory-shape, adaptive, conductive, piezoelectric and thermoelectric functions. Membranes can support textiles working as heat exchangers and extra barrier against environmental hazards, chemical and biological agents as well as mechanical forces, preserving together esthetics, thermo-physiological, and sensorial comfort [2].

Membranes can be regarded as a further alternative to traditional natural and synthetic raw materials currently used in various areas of the textiles industry, which includes clothing, medical, marine, architectures, agricultural, environmental, transport and aerospace applications (Figure 14.1).

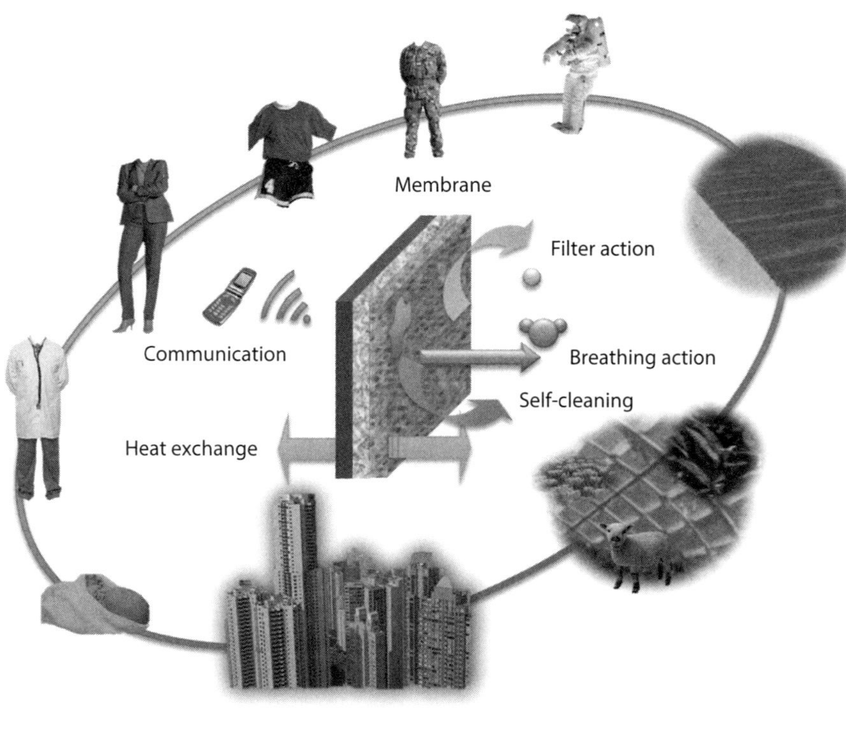

Figure 14.1 Represented multifunctional membranes coupled with fabrics for multi-applications.

Indeed, the dramatic decline of the demand for natural fibers over the last years – less than 20–50% - in addition to the inclination of many small-scale farmers to switch from cotton to maize, groundnut and soybean, which are regarded as important sources of bio-energy against greenhouse gas emissions [3], is forcing to check new practices to produce high-tech and functional textiles looking towards advanced nanotechnologies, including membrane science.

This chapter covers ultra-smart concepts related membranes, describing classes of materials, functions and 'sense-to react' mechanisms involved in microclimate regulation, heat exchange, defense against harmful agents, self-cleaning procedures, integrated wearable electronics, power storage and long-distance communication. Prospects and application of membrane science in the creation of ultra-smart textile marketplaces are also discussed.

14.2 Membranes and Comfort

Membranes work as semi-permeable interfaces, allowing the passage of some species while others are rejected. Different types of mechanisms can control the mass and/or energy transfer through membranes, depending on structure, porosity, and chemical composition of the films along with process engineering (Figure 14.2):

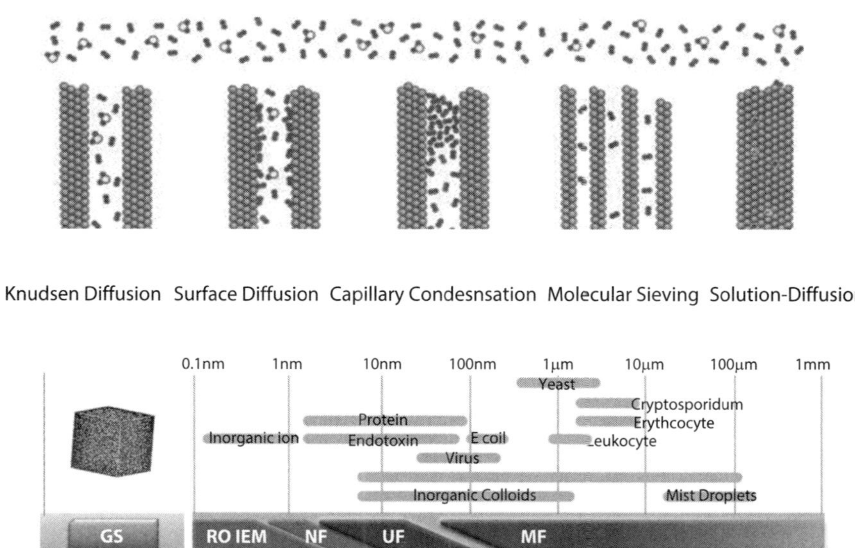

Figure 14.2 Membrane separation processes depending on membrane pore size (a) and dimension of the species to separate.

14.2.1 Breathable Membranes

Concerning the perspiration, membranes would keep the wearer dry and comfortable [4]. This means that water vapor and heat exchange can be regulated, while liquid entry has to be disallowed. Water vapor transmission rate (*WVTR*) can be calculated as a flux, energy or resistance according to various testing approaches listed in Table 14.1:

Because there is no a statistical correlation, one rate value cannot easily be converted into another and *vice versa,* thereby making difficult a comparison between the various artifacts.

Figure 14.3 shows how the breathability values estimated for some kinds of membranes change with selected testing methodology [5–6].

Considering that the water vapor mass transfer through membranes changes with temperature and external humidity, larger amounts of vapor are expected when large difference of concentration are obtained. This can be achieved by increasing the difference of temperature between feed and permeate streams or removing quickly the permeating vapors in order to maintain constant the driving force that control the overall process [7–8].

The breathability is usually calculated under steady state conditions and expressed as:

$$P = 22.414\frac{\left(W_f - W_o\right)}{18At}\frac{l_m}{\Delta P} \tag{14.1}$$

where P is the permeability; W_f and W_o the final and initial weight of a cup hermetically sealed by the membrane, A the effective membrane area, t the length of time of a measurement; ΔP the trans-membrane water vapor pressure. Permeability and flux are related by the following equation:

$$J = P\frac{\left(p_h - p_l\right)}{l_m} \tag{14.2}$$

where l_m is the membrane thickness, p_h the upstream (*high pressure*) and p_l the downstream pressure (*low pressure*). This equation is used when there is no dependence on concentration and the Fick's first law is obeyed. When there is diffusion coefficient dependence, the flux is, instead, expressed as:

$$J = \frac{D_0}{\beta l_m}\left(e^{\beta C_0} - 1\right) \tag{14.3}$$

Table 14.1 Description of usual testing approaches used to estimate WVTR.

Test method	***Description***	***Regulation***	***WVTR Units***
Upright Cup	A cup of water is sealed with a fabric and an air gap is left between liquid and fabric. The cup is weighed before and after to determine how much vapor has been lost.	*ASTM E96*	$g \cdot m^{-2} \cdot day^{-1}$
Inverted Cup	A cup of water is sealed with a fabric and inverted, keeping the water in direct contact with fabric. This causes absorption of large amount of moisture, increasing up to 5 times the breathability.	*ASTM E96*	$g \cdot m^{-2} \cdot day^{-1}$
The Desiccant Inverted Cup	A liquid desiccant is used to move water through the fabric. The cup is weighed before and after to quantify the permeability that can be 20 times higher than that produced by the Upright Cup.	*JIS 1099*	$g \cdot m^{-2} \cdot day^{-1}$
Dynamic Moisture Permeation Cell	A dry and water-saturated nitrogen streams passing over each side of the fabric are used. The control of flow rates, temperature and humidity of the two flows, and the measurement of the humidity afterwards, give indication about the water vapor transmission rate or the fabrics resistance. The breathability rate can be 4 to 6 times those of the Upright Cup.	*ASTM F 2298*	$g \cdot m^{-2} \cdot day^{-1}$
Evaporative Resistance	A "sweating hot plate", electrically powered and controlled by a thermostat, is used to maintain a steady temperature. The amount of power necessary to maintain that temperature defines how much energy has been lost through the fabric covering it by evaporation.	*ISO 11092*	m^2PaW^{-1}

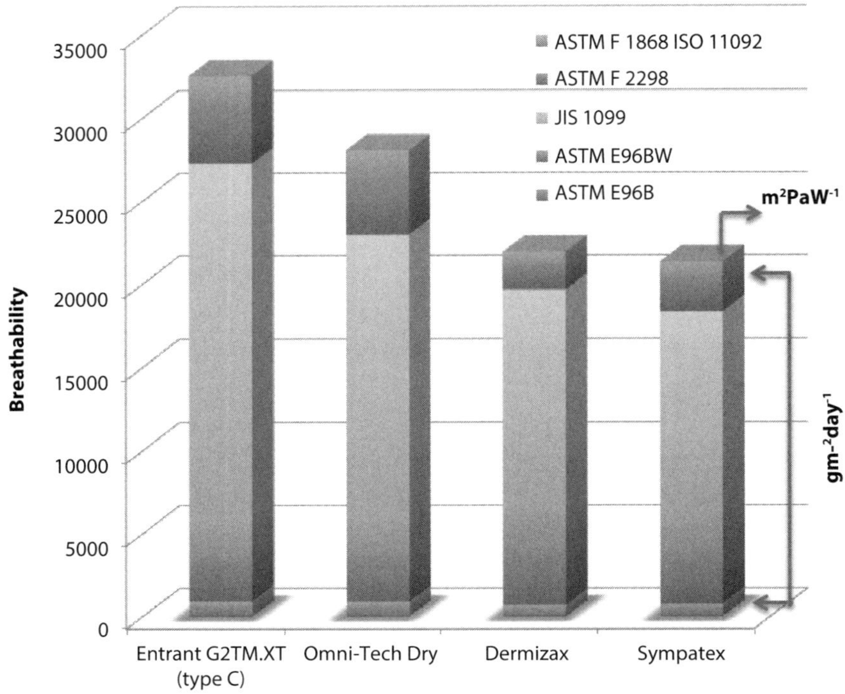

Figure 14.3 Comparison of water vapor mass transfer estimated for various types of trade membranes according to different testing approaches.

where β is a constant and D_0 the relationship between the diffusion coefficient as the concentration is close to zero [1].

14.2.2 Membranes as Heat Exchangers

It is instructive to consider the ability of membranes to work as a heat exchanger as well [1]. Based on the principle that regulates water distillation plants, the heat transfer can be mathematically described by adding three contributions, including (4) convective heat transfer from feed side to membrane surface (Q_f); (5) heat transfer across the membrane (Q_m); (6) convective heat transfer from membrane surface to permeate side (Q_p):

$$Q_f = h_f\left(T_f - T_1\right) \tag{14.4}$$

where h_f is the heat transfer coefficient in feed boundary layer, T_f the temperature of bulk feed stream, T_1 the temperature at feed membrane surface.

$$Q_m = N\Delta H_v + \frac{k_m}{l_m}(T_1 - T_2) \tag{14.5}$$

where N is permeation flux, ΔH_v the heat of vaporization, k_m the thermal conductivity of the membrane, T_2 the temperature at permeate membrane surface.

$$Q_p = h_p(T_2 - T_p) \tag{14.6}$$

where h_p is the heat transfer coefficient in permeate boundary layer, T_p the temperature of bulk permeate stream.

The temperatures T_1 and T_2 can be calculated at steady state according to the following equations.

$$T_1 = \frac{h_m\left(T_p + (h_f / h_p)T_f\right) + h_f T_f - N\Delta H_v}{h_m + h_f(1 + h_m / h_p)} \tag{14.7}$$

$$T_2 = \frac{h_m\left(T_f + (h_p / h_f)T_p\right) + h_p T_p - N\Delta H_v}{h_m + h_p(1 + h_m / h_f)} \tag{14.8}$$

Another important parameter to consider is the temperature polarization (τ), which occurs when the temperature near the membrane surface is lower than the bulk region, i.e. body, due to absorption from sweat:

$$\tau = \frac{T_1 - T_2}{T_f - T_p} \tag{14.9}$$

Ideal working conditions imply values of temperature polarization near unity, whereas values approaching the zero are considered poor enough.

14.3 Adaptive Membranes for Smart Textiles

In the last years, different kinds of hydrophobic microporous membranes along with breathable monolithic films have been proposed for maintaining regulated microclimate conditions [1]; polyurethanes showing sponge-like structure as well as monolithic co-polyester and polyamide membranes are

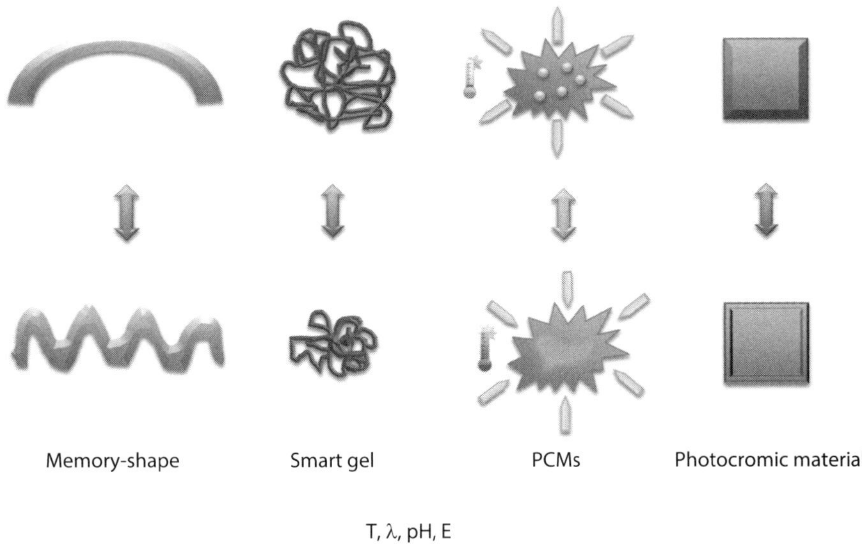

Figure 14.4 Representative scheme of stimuli-responsive materials for the creation of ultra-smart membranes.

used for breathing fabrics [9–10]. Polyvinyl chloride (PVC) and polyvinyl fluoride (PVDF) in combination with other perfluoropolymers such as Hylar® and Hyflon AD® have also resulted to be attractive materials for the fabrication of waterproof and breathable clothing solutions [11–14].

Nowadays, the major ambition remains the fabrication of smart membranes enabling to keep the wearer comfortable independently of external climatic environment. With this concern, some classes of adaptive materials have been proposed to design ultra-smart textiles for auto-adjusting indoor/outdoor and wearable clothing solutions [15] (Figure 14.4):

14.3.1 Shape Memory-Based Membranes

Membranes with shape memory functions are frequently proposed to regulate water vapor mass transfer in fabrics [16–17]. These materials have a particular ability to deform into a temporary shape and recover their memorized original shape under the action of a specific external trigger. The commonest shape memory polymers, including mainly polyurethanes along with poly(ether/amide), polyethers, polyamides, polyacrylates, polysiloxanes, polyethers esters and urethane/butadiene copolymers, exhibit a thermally induced deformation, which causes the segment chains to move from randomly coiled to constrained conformations via temporary hydrogen bonding formation. This morphological change can be used to regulate the water vapor rate through membranes. This transposition occurs frequently at melting or glass transition temperature of the material used [18–21]. So, the major efforts are directed towards the fabrication of materials showing transitions in the region of 10–40 °C at least. Recently, polyurethanes-based membranes have been proposed for obtaining thermally induced deformation in desired range by using light, infrared irradiation and electro-activation [22–23].

14.3.2 Responsive Gel-Based Membranes

The category of gels consists of 3D cross-linked networks, which are between a solid and liquid phase. Depending on the solvent, gels can be classified as:

- Aerogels or xerogels when the solvent is air;
- Hydrogels, when the solvent is water;
- Organic gels, when the solvent is oil;

The gels smartness is due to their capability of undergoing abrupt and reversible volume transitions. This modification implies a dynamic transposition from extended and swollen polymer chains to collapsed and shrunk segments. Reversible hydrogen bonding and/or electrostatic repulsion/attraction regulate changes in polymer conformation for more than 1000 times under the action of external stimuli such as temperature, pH, solvent, light, ionic strength, magnetic or electrical field and pressure, thereby making smart gel-membranes interesting materials for auto-adjusting textiles.

Gels showing a critical solution temperature (LCST) such as poly(acrylamide) have a volume phase transition regulated by unbalanced hydrophilic/hydrophobic interactions, which cause the chains to swell when polymer-solvent interactions predominate - under LCST -, while induce shrinking when polymer-polymer interactions are over polymer-solvent attractions - above LCST -. Membranes based on temperature-sensitive gels with LCST of 32 to 34 °C are promising materials for the regulation of WVTR through mechanisms of hydration/dehydration or swelling/deswelling [24–25]. As an example, PNIPPAm has been photo-grafted onto carboxylmethyl cellulose (CMC) coupled with monomers of acid acrylic (AA); large adsorptions of water below LCST and important shrinking above the transition temperature have been observed [26]. Similarly, poly-NiPAAm/chitosan (PNCS) microgels have been incorporated into fabrics, causing swelling events to close the membrane pores with a reduction of the mass transport [27]. On the contrary, collapse of the structure with opening of the pores has been obtained bringing the systems above LCST [28]. It is also relevant to examine the possibility to use pH-responsive gels, enabling to induce conformational transitions dependently of pH conditions used. Poly(acrylic)acid has capability to swell at high pH, whilst it shrinks under acid conditions. On the contrary, chitosan interacts with water when the amino groups are fully protonated, while mutual repulsions are significantly reduced at higher pH. Another interesting class of thermally adaptive gels is that of polyethylene glycols, enabling to move from an order to disorder conformational state and *vice versa* by formation/breaking of hydrogen bonding with adsorption/release of heat [29].

14.3.3 Phase Changing Materials (PCMs) in Membranes

Thermo-regulated membranes are greatly attractive for the fabrication of ultra-smart gloves for skiers, helmets to keep the wearer cool, sleeping bags, smart jackets for astronauts or for instruments to keep protect against extreme fluctuation of temperature in space, but also garments for people working under extreme environmental conditions or to keep warm longer than conventional insulation.

Thermal regulation depends on environmental conditions, including temperature and humidity, but also on physical activities. Dynamic phase change materials, enabling to store and release heat by changing the phase, represent an interesting class of materials for textiles applications [30]. Moving from solid to liquid phase and back, these materials can absorb or release thermal energy at a constant temperature, resulting suitable to interact with body temperature. Considering that the human body temperature is around 37°C, the possibility to incorporate in membranes compounds with ability to regulate the wearer comfort by buffering or reducing overheating is strongly attractive for the fabrication of textiles needed to the control of perspiration through thermo-regulation mechanisms. PCMs are usually encapsulated in micro-spheres having dimension of a few micrometers to a few millimeters by using different approaches [32]. They have high mechanical, chemical and heat resistance. When the surrounding temperature increases the microcapsules react by absorbing heat. Then, the PCMs melt pulling heat from the external environment and storing the surplus energy. As a result, a cooling effect is achieved. The stored energy is again released as thermal heat when the temperature decreases, providing heating effect. These microcapsules can be easily incorporated in membranes during lamination, coupling, melt spinning, in-situ polymerization, or extrusion processes. Hydrated inorganic salt with *n* water molecules, hydrophobic linear hydrocarbon, polyethylene glycol (PEG), fatty capric, lauric, palmitic and stearic acids and many other compounds are usually incorporated in capsules and used for drawing heat storage and thermo-regulated textiles.

Membranes containing PCMs microcapsules with melting point around 28–32 °C are successfully used in commercial multifunctional textiles such as the Dull® where the ability to control absorption/release of heat is associated with a large mass transfer of water vapor (>70000 g/m^2/day JIS L1099 B1) and a high resistance to the liquid entry (>5000 mm/H_2O JIS L1092) [33].

Microencapsulated PCM slurry containing paraffin waxes and having an average diameter of 4.5 mm has been recently used in melamine resin-based membranes, demonstrating a very quick absorption of large amounts of heat with little or no temperature changes [34]. Carbon nanofibers are often used as enhancer for improving the thermal conductivity of PMCs [35–36].

14.3.4 Photochromic Compounds for Smart Membranes

Photochromic dyes are frequently used in textiles to alert for over exposition to harmful ultra and violet light [37]. Their ability to change color when in direct contact with light makes the compounds useful for the fabrication of protective and safety textiles, especially for military purposes or skin barrier against sunlight [38].

These compounds could be easily encapsulated in liquid or solid bubbles and then dispersed in breathable membranes by using traditional manufacturing approaches.

14.4 Barrier Functions of Membranes

Breathable fabrics allow keeping comfortable conditions, while thermo-regulated garments work as heat exchangers. However, these two functions are not enough to keep high hygiene conditions or give protection against harmful and dangerous agents [39]. Natural or synthetic polymers are often used for the fabrication of surgical apparels, towels, gloves, diapers, napkins, incontinence products, wipes or gauzes as well as socks, sportswear and common *t*-shirts. However, microencapsulation and dispersion of anti-bacterial agents, aromatic scents and chemical inhibitors in membranes is becoming a frequent practice to make biologically and chemically safe textiles.

14.4.1 Waterproof Function

Depending on the final application, various barrier functions can be requested. Waterproof function is frequently obtained with superhydrophobic porous membranes or laminated dense films [1]. However, chemical and biological contaminants dissolved in water can dangerously reduce the surface free tension of the liquid, causing penetration inside the membranes with the risk of contact with human skin. In order to prevent these events, hydrophobic and oleophobic membranes based on PVDF, PTFE, polyesters and polyamides and having porous and dense structure have been developed to prevent water, oil and organic liquid contaminations [40–52].

14.4.2 Antibacterial Action

Polymers such as chitosan with antibacterial and antifungal properties are usefully used for the fabrication of membranes with biocide activity. This kind of polymers has been combined with PNIPAAm/PU/cellulose and silk fibroin with the intent to fabricate laminates having barrier properties against *Staphylococcus Auereus and Escherichia coli* [53–54]. But, the additional use of microcapsules containing substances enables membranes to prevent bacteria growth, skin irritation and stimulate healing process through drug, vitamins, nutrients and antiseptics release.

PNIPAAm/PU films encapsulating vitamin C have been grafted to non-woven and its release has been successfully investigated as a function of LCST [26]. Similarly, the release of vitamin E from poly(2-ethoxyethyl vinyl ether) (EOVE200) and poly(hydroxyethyl vinyl ether) (HOVE00) has been evaluated in relation to LCST, according to the hydrophilic/hydrophobic ratio of the polymer segments [24], yielding useful indication about its possible application for skin care.

Silver nanoparticles, ammonium salts along with zinc oxide (ZnO), copper oxide (CuO), magnesium oxide (MgO), and titanium dioxide (TiO_2) are frequently used for their ability to interact with and denature bacterial cell walls via redox and/or photo-oxidation mechanisms [55–57].

As an example, poly-L-Lactic acid (Ag/PLLA) fibrous membranes containing silver particles have been recently proposed for the fabrication of protective layers enabling to respond to infections [58].

14.4.3 Scents Release and Superabsorbent Action

The use of nanoparticles with antimicrobial action is a practical route for contrasting bacteria and fungi growth and preventing development of unpleasant odors. Tiny bubbles and capsules containing scents, smell sequesters and maskers are usefully incorporated in membranes when healthcare and safe textiles are requested, i.e. underwear, sportswear, socks and medical garments.

In this respect, β-cyclodextrins are regarded as an attractive class of compounds for the production of scented membrane-based textiles [59]. These cyclic oligosaccharides have been used as reservoirs vanillin [60] as well as β-citronellol, menthol geraniol and so others [61]. The fragrance release has been controlled with time by exploring the role of host–guest complexation and non-specific interactions between substrates and fiber surfaces.

Guar gum, chitin, or polyacrylates and polyacrylamides have been, instead, interestingly used as materials for absorbing liquid over 100% of uptake [62].

14.4.4 Warfare Agent Defense

Military actions, terrorist attacks and accidents represent hard situations wherein the risk to come in contact with hazardous substances is real. In these cases, advanced clothing solutions enabling to offer protection against toxic chemical and biological agents, keeping the wearer dry and comfortable, are needed. Membranes prepared from PVA, PEI have been proposed to decontaminate fabrics from pinacolyl methyl phosphonofluoridate, known as the Soman (GD) [63, 64]. Chitosan membranes have been demonstrated to block the permeation of nerve agents such as dimethyl methyl phosphonate (DMMP) and GB [65], whilst sulphonated polymers with a suitable degree of sulfonation are claimed to preserve a good WVTR combined with a low permeability to bis(chloroethyl) sulfide, which is classified as a nerve agent [64]. Membranes containing activated carbon against air-borne nanoparticles have been also proposed [66].

14.5 Membrane Materials for Self-cleaning Function

Considering that dirty and fouling events can compromise the performance of membranes causing flux decline and loss of surface properties, there is an increasing need to make membranes and coatings with ability to destroy and remove organic and inorganic contaminates. Depending on the materials used, two different self-cleaning mechanisms can be preferably induced: (a) rolling water droplets; (b) photo-catalysis. The first one exploits the membrane rough surface, whilst the second one is based on the membrane catalytic function.

Poly(N-isopropylacrylamide)-nanoclay (synthetic hectorite, $[Mg_{5.34}Li_{0.66}Si_8O_{20}(OH)_4]Na_{0.66}$) hydrogels have been used to realize hierarchical textures with superoleophobic character (154–159°) [67]. Similarly, silica nanoparticles have been dispersed into pH-responsive block copolymers achieving switchable superoleophilicity and superoleophobicity in aqueous media [68]. Carbon nanotubes (CNTs) have been assembled onto cotton surface in order to get lotus-like structures; as a result, super-hydrophobic and electrically conductive fabrics have been achieved.

Another common practice to promote self-decontamination through oxidation and reduction mechanisms is the incorporation of oxide nanoparticles such as Al_2O_3, TiO_2 and MgO in membranes [69]. As an example, TiO_2-treated fabrics have been realized to decompose coffee and red wine stains under UV radiation after multiple staining and washing steps [70]. Also, TiO_2 nanoparticles have been used to prevent formation of bacteria films by photocatalysis [71, 72].

For coming, the major ambition will be, however, to get super-amphiphobic textiles able to resist spills, repel and release.

14.6 Interactive Membranes for Wearable Electronics

Health monitoring, aided communication and detected locations are some dominant challenges for the feature electronic market intended for sportswear, sick, soldier and/or astronauts life-suites. Indeed, interactive textiles, fabrics and materials with integrated electrical circuits or conductive fibres can offer concrete solutions to long-distance communication and work [1]. Great benefits are expected from using electronic textiles due to the concrete possibility to monitor vital functions, people position (*Global Positioning Systems* (GPS) incorporated into walking shoes or clothing) and protection as well as to give more comfortable facilities such as mobile phone connectors in the form of Bluetooth technology and MP3 players with wireless control panels. Therefore, miniaturization and integration of computers and wearable electronics into garments and textiles represent an important target, which have also the attractive ambition to use forms of power such as solar, battery and human power.

Considering that the activation of a mobile phone would require around the 10% of the energy that materials such as carbon, steel, and silver can provide, the incorporation of electro-active materials (EAMs) is expected to be a reliable route for converting electrical energy into mechanical one [73, 74]. In this respect, different classes of materials can be used, including carbon nanotubes (CNTs), ionic polymer–metal composites (IPMCs), conductive polymers (CPs) as well as ferroelectronic polymers and dielectric elastomers.

Carbon materials are regarded as an interesting and promising category of compounds enabling textiles to get new interactive functions such as sensing, actuation, rechargeable battery, storage energy, and semiconductors. The actuation power of CNTs has been studied confirming the ability to make textiles and membranes highly conductive when very low voltages are applied [75–79]. Membranes working as semiconductors have been tailored by building up a CNTs network on honeycomb interfaces with passage of electrical current of the order of 10^{-3} A at low voltage [80]. Interesting results have been also obtained with ion exchange membranes containing metals such as Pt, Au, Ag, or Pd [81]. An actuation of membranes can be induced by applying an electrical potential, which cause the membrane to flex depending on the ion flux direction [82]. Among the polymers, polypyrrole (PPy), polyaniline (PANi), poly(p-phenylene vinylene) (PPV) and polythiophen (PT) have been studied as potential conductive actuators [83], whereas the polyvinylidene fluoride (PVDF) in its polar β-phase has been demonstrated to work as a mechanical-electrical energy converter [84].

14.7 Conclusions and Prospects

The creation of ultra-smart textiles is expected to benefit from using exiting materials assembled in new devices. In this respect, membranes represent a powerful tool for realizing multi-compartment arrays where different and complementary functions can be accommodated [85–86]. Polymers, including perfluoropolymers, conductive macromolecules and elastomers, can be easily combined with nanoparticles, carbon nanotubes, strong-solid particles, nanowires, ionic liquid crystals, thermal and light-driven compounds as well as core–shell and valved structures, enzymes, dyes, actuators and gels. Advanced adaptive membranes can be created directly onto fabrics or inside the wefts, wherein desired function of breathing can be integrated with ability to self-clean, actuate, power and communicate. These targets can be achieved with a multidisciplinary approach, which include also computational studies enabling to give provisional models for the creation of new devices with advanced functions [87–91].

Many areas of textiles include healthcare, comfortable fabrics, telemedicine, automotive, architecture and house indoor for energy reduction, along with earth, marine, agriculture, outdoor plants, canal liners, road and pavements, environment protection, and communication.

However, many issues are still unsolved such as biodegradability, recycling, durability and biocompatibility. The necessity to reduce the impact on the environment makes the research of biodegradable materials urgent matter as well as materials enabling the textiles to offer protection against radioactive heat, liquids, allergens, chemical and biological contaminants are in great demand. Another concern is the creation of materials with durable performance and resistance to mechanical stress, laundry-washing cycles and fouling events.

The prospect is, hence, innovation with choice and combination of intelligent materials for making advanced textiles capable to interact with body and environment, provide self-maintenance and adaptability as well as keep the wearer constantly powered and connected to remote panels.

The achievement of these targets is expected to open new frontiers into textiles markets, including water, space and military industry.

References

1. A. Gugliuzza, E. Drioli, *J. Membr. Sci.*, Vol. 446, p. 350, 2013.
2. A. Gugliuzza, *Membranes* Vol. 3, p.151, 2013.
3. A. Gugliuzza. A. Basile, *Membranes for clean and renewable power applications,* Cambridge, Woodhead Publishing Limited, 2014.
4. S.A. Brewer, *Recent Patents on Materials Science* Vol. 4, p. 1, 2011.
5. E.A. McCullough, M. Kwon, H. Shim, *Meas. Sci. Technol.*, Vol. 14, p.1402, 2003.
6. P.W. Gibson, D. Rivin, C. Kendrick, Int. J. Cloth. Sci. Technol. Vol.12, p. 90, 2000
7. P.W. Gibson, *Journal of Polymer Testing*, Vol.19, p.673, 2000.
8. P.W. Gibson, H.L. Schreuder-Gibson, *J Eng Fiber Fabr*, Vol.4, p.11, 2000.
9. B.R. Bohringer, H.J.M.Van de Ven, J.C.W. Spijkers, US6706413, 2004.
10. R. Biju, C.P.R. Nair, *High Performance Polymers*, p.1, 2013.
11. A. Gugliuzza, M.C. Aceto, E. Drioli, *Polym Int*, Vol.58, p.1452, 2009.
12. A. Gugliuzza, E. Drioli, *Desalination* Vol. 240, p.14, 2009.
13. A. Gugliuzza, F. Ricca, E. Drioli, *Desalination*, Vol. 200 p. 26, 2006.
14. A. Gugliuzza, E. Drioli, *J. Membr. Sci.*, Vol.300, p.51, 2007.
15. https://www.smitherssapex.com
16. H.R. Han, S.E. Chung, C.H. Park, *Textile Research Journal,* p.17, 2012.
17. T. Pretsch, *Polymer*, Vol. 2, p.120, 2010.
18. R. Biju, C.P.R. Nair, *High Performance Polymers*, p.1, 2013.
19. T. Xie, I.A. Rousseau, Polymer Vol. 50, p.1852, 2009.
20. H.J. Lee, J.J. Lee, *Smart Mater. Struct.*, Vol. 9, p.817, 2000.
21. S.M. Kang, M.G. Kim, S.H. Kwon, H. Park, H.M. Jeong, B.K. Kim, *J. Mater. Res.* Vol. 27, p. 2837, 2012.

22. A. Lendlein, H.Y. Jiang, O. Junger, R. Langer, *Nature*, Vol. 434 (2005), p. 879, 2005.
23. A. Lendlein, *Shape-memory polymers*, Berlin Herdelberg, Springer-Verlag, 2010.
24. Y.Y. Liu, X.D. Fan, *Polymer*, Vol.43 p. 4997, 2002.
25. M. Serra, *Smart Materials Bulletin*, Vol.8, p.7, 2002
26. J. Hu, *Environmentally sensitive polymer gel and its application in the textiles field in: Shape Memory polymers and textiles*, Cambridge, Whoodhead Publishing Ltd, , 2007.
27. P.K. Lavrič, M.M.C.G. Warmoeskerken, D.Jocic, *Cellulose* , Vol.19, p.257, 2012.
28. P.K. Lavrič, B. Tomsic, B. Simovic, M.M.C.G. Warmoeskerken, D. Jocic, *Cellulose*, Vol. 19, p.273, 2013.
29. T.L. Vigo, A.F. Turbak, *High-tech Fibrous materials: Composites, Biomedical Materials, Protective Clothing, and Geotextiles,* Washington, DC: ACS, 1991.
30. S. Mondal, *Appl. Therm. Eng.*, Vol. 28, p.1536, 2008.
31. V.V. Tyagia, S.C. Kaushik, S.K. Tyagi, T. Akiyama. Renewable and Sustainable Energy Reviews Vol. 15, p. 1373, 2011.
32. V.V. Tyagia, S.C. Kaushik, S.K. Tyagi, T. Akiyama. Renewable and Sustainable Energy Reviews, Vol. 15, p.1373, 2011.
33. http://www.everest.com.tw
34. Zhang Yanlai, *Appl. Mechanics and Materials* Vol.110, p.571, 2012.
35. A. Elgafy, K. Lafdi, *Carbon*, Vol. 43 p. 3067, 2005.
36. I.M. Bugaje, *Inter. J. Energy Research* Vol. 21, p.759, 1997.
37. F. P. Nicoletta, D. Cupelli, P. Formoso, G. De Filpo, V. Colella, A. Gugliuzza, *Membranes*, Vol. 2, p. 134, 2012.
38. _http://www.screenweb.com/index.php/channel/6/id/1425/
39. W. Zhong, *Healthcare protective textiles in: An Introduction to Healthcare and Medical Textiles*, DEStech Publications, Inc, 2012.
40. S.B. Little, R.B. Quincy, J.A. Rotella, P.A. Schorr, S.R.Stopper, *US7381666*, 2008.
41. D.C Potts, G.A. Young, D.S. Everhart, J.G. McDonald, R.S. Nohr, *US5145727*, 1992.
42. R.R Sargent, J.R. Alender, *US5459188*, 1995
43. R.B. Quincy, A. Yahiaoui, J.L. McManus, *US6203889*, 2001.
44. J.S. Mrozinski, C.P. Jariwala, *WO03104310 2003* and *US6858290, 2005.*
45. W.K. Weimer, G.E. Keenan, R.J. Kinney, J.S. Mrozinski, P.D. Radovanovic, *US5690949*, 1997.
46. J.S. Mrozinski, H.J. Seppala, *WO9308019*, 1993.
47. RW. Gore, *US 4187390*, 1980.
48. C. Srinivas, S. Gerts, D. Lee, C. Rogers, R.W. Pekela, *WO2009065092*, 2009.
49. B. Gustafson, *US20040170852*, 2004.
50. G. Becke, D.J. Carmody, M.J. Dobosy, *US20080131676*, 2008.
51. L.S. White, *US5824405*, 1998.

52. D.Y. Cheng, S.J. Wiersma, *US4419242*, 1983 and *US4316772*, 1982.
53. F.K. Tavaria, J.C. Soares, I.L. Reis, M.H. Paulo, F.X. Malcata, M.E. Pintado, *J Appl Microbiol.*, Vol.112, p.1034, 2012.
54. Z.X. Cai, X.M. Mo, K.H. Zhang, L.P. Fan, A.L. Yin, C.I. He, H.S. Wang, *Int. J. Mol. Sci.* Vol.1, p. 3529, 2010.
55. Y.A. Son, B.S. Kim, K. Ravikumar, S.G. Lee, *Eur Polym J*, Vol. 42, p.3059, 2006.
56. Z. Chen, Y. Sun, *Ind Eng Chem Res.* Vol. 45 p. 2634, 2006.
57. K. M. Reddy, K. Feris, J. Bell, D.G. Wingett, C. Hanley, A. Punnoose, *Appl. Phys. Lett.* Vol. 90, p. 213902, 2007.
58. L. Li, Y. Li, J. Li, L. Yao, A,F,T, Mak, F.Ko, L. Qin, J. Nanomater., p.1, 2009.
59. H.J. Buchman, U. Denter, D. Knitter, E. Schollmeyer, J. Textile Institute, Vol. 89, p.554, 1998.
60. M.H. Lee, K.J. Yoon, S.W. Ko, *J. Appl. Polym. Sci.*, Vol.78, p.1986, 2000.
61. B. Martel, M. Morcellet, D. Ruffin, F. Vinet, M. Weltrowski, *J. Incl. Phenom. Macrocyclic Chem.*, Vol.44, p.439, 2002.
62. N. Sau-Fun, H. Chi-Leung, W. Lai-Fan, *Textile Progress*, Vol. 43, p. 235, 2011.
63. C.R. Willis, S.A. Brewer, M.R. Dennis, C.A. Stone, *WO2008113996*, 2008.
64. J.P. Critchley, G.J. Knight, W.W. Wright, *Heat resistant polymers technologically useful materials*, New York, Plenum Press, 1983.
65. R.D. Giglia, E.A. Battistelli, Non-woven activated carbon fabric, US456572,7, 1985.
66. P. Gibson, H. Schreuder-Gibson, D. Rivin, *Colloids and Surfaces A: Physicochemical and Engineering Aspects, Vol.* 187, p. 469, 2001.
67. L. Lin, M.J. Liu, L. Chen, P.P. Chen, J. Ma, *et al.*, *Adv. Mater.*, Vol. 22, p.4826, 2010.
68. L. Zhang, Z. Zhang, P. Wang, *NPG Asia Materials*, Vol.4, p.1, 2012.
69. A.L. Costa, *Proceedings of NANOTECHITALY*, 31-Nanomat, 2012.
70. N.M. Bedford, A.J. Steckl, ACS Appl. Mater. Interf., Vol. 2, p.2448, 2010.
71. D. Wu, M. Long, J. Zhou, W. Cai, X. Zhu, C. Chen, Y. Wu, *Surface & Coatings Technology*, Vol.203, p.3728, 2009.
72. K. Qi, X. Wang, J.H. Xin, *Textile Res. J.* Vol.81, p.101, 2011.
73. K. Laxminarayana, N. Jalili, *Textile Res J* , Vol.75, p.670, 2005.
74. B. Safadi, R. Andrews, E.A. Grulke, J. Appl. Polym. Sci. Vol. 84, p.2660,2002.
75. R.H. Baughman, C. Cui, A.A. Zakhidov, Z. Iqbal, J.N. Barisci, G.M. Spinks, G.G- Wallace, A. Mazzoldi, D. De Rossi, A.G. Rinzler, O. Jaschinski, S. Roth, M.Kertesz, *Science*, Vol. 284, p.1340, 1999.
76. S. Hara, T. Zama, W. Takashima, K. Kaneto, *Synth. Met.* Vol.156, p. 351, 2006.
77. T.J. Kang, A. Choi, D.H. Kim, K. Jin, D.K. Seo, D.H. Jeong, S.H. Hong, Y.W. Park, Y.H. Kim, *Smart Mater. Struct.* Vol. 20, p.1, 2010.
78. X.H. Zhong, Y.L. Li, J.M. Feng, Y.R. Kang, S.S. Han, *Nanoscale* Vol.4, p.5614, 2010.
79. T. Mirfakhrai, J. Oh, M.E. Kozlov, S. Fang, M. Zhang, R.H. Baughman, J.D.W. Madden, Ieee/Asme Transactions on Mechatronics, Vol.16, p.90, 2011.
80. V. Pingitore, A. Gugliuzza, J. Chem. Phys. C, Vol.117, p.26562, 2013.

81. B.J. Akle, M.D. Bennett, D.J. Leo, Sens. Actuators A Vol.126, p.173, 2006.
82. J.D.W. Madden, NA Vandesteeg, PA Anquetil, PGA Madden, A Taksh, RZ Pytel, SR Lafontaine, PA Wieringa, IW. Hunter, *IEEE J. Oceanic Eng.*, Vol.3, p.706, 2004.
83. Y. Bar-Cohen, Bellingham, SPIE Press, 2004.
84. S.O. Catarino, J.M. Miranda, S. Lanceros-Mendez, G Minas, *J. Phys. Conf. Ser.*, Vol. 395, p.01209.1–8, 2012.
85. V Speranza, F. Trotta, E. Drioli, A. Gugliuzza, *ACS Appl. Mater. Interf.*, Vol.2, p.459, 2010.
86. A. Gugliuzza, M.C. Aceto, F. Macedonio, E. Drioli, *J. Phys. Chem. B*, Vol.112, p.10483, 2008.
87. A. Gugliuzza, G. De Luca, E. Tocci, L. De Lorenzo, E. Drioli, *J. Phys. Chem. B*, Vol.111 p.8868, 2007.
88. G. De Luca, A. Gugliuzza, E. Drioli, *J. Phys. Chem. B*, Vol. 113, p.5473, 2009.
89. L. De Lorenzo, E. Tocci, A. Gugliuzza, E. Drioli, *Membrane*, Vol.2, p.346, 2012.
90. L. De Lorenzo, E. Tocci, A. Gugliuzza, E. Drioli, J. *Membr. Sci.*, Vol.421–422, p.75, 2012.
91. E. Tocci, A. Gugliuzza, L. De Lorenzo, M. Macchione, G. De Luca, E. Drioli, *J. Membr. Sci.*, Vol.323, p.316, 2008.

INDEX

Also of Interest

Check out these published and forthcoming related titles from Scrivener Publishing

Advanced Biomaterials and Biodevices
Edited by Ashutosh Tiwari and Anis N. Nordin
Published 2014. ISBN 978-1-118-77363-5

Biosensors Nanotechnology
Edited by Ashutosh Tiwari and Anthony P. F. Turner
Published 2014 ISBN 978-1-118-77351-2

Advanced Sensor and Detection Materials
Edited by Ashutosh Tiwari and Mustafa M. Demir
Published 2014. ISBN 978-1-118-77348-2

Advanced Healthcare Materials
Edited by Ashutosh Tiwari
Published 2014 ISBN 978-1-118-77359-8

Advanced Carbon Materials and Technology
Edited by Ashutosh Tiwari and S.K. Shukla
Published 2014. ISBN 978-1-118-68623-2

Responsive Materials and Methods
State-of-the-Art Stimuli-Responsive Materials and Their Applications
Edited by Ashutosh Tiwari and Hisatoshi Kobayashi
Published 2013. ISBN 978-1-118-68622-5

Nanomaterials in Drug Delivery, Imaging, and Tissue Engineering
Edited by Ashutosh Tiwari and Atul Tiwari
Published 2013. ISBN 978-1-118-29032-3

Biomimetics Advancing Nanobiomaterials and Tissue Engineering
Edited by Murugan Ramalingam, Xiumei Wang, Guoping Chen, Peter Ma, and Fu-Zhai Cui
Published 2013. ISBN 9781118469620

Encapsulation Nanotechnologies
Edited by Vikas Mittal
Published 2013. ISBN 978-1-118-34455-2

Advances in Contact Angle, Wettability and Adhesion
Volume 1
Edited by K.L. Mittal
Published 2013. ISBN 978-1-118-47292-7

Biomedical Materials and Diagnostic Devices Devices
Edited by Ashutosh Tiwari, Murugan Ramalingam, Hisatoshi Kobayashi and Anthony P.F. Turner
Published 2012. ISBN 978-1-118-03014-1

Intelligent Nanomaterials
Processes, Properties, and Applications
Edited by Ashutosh Tiwari Ajay K. Mishra, Hisatoshi Kobayashi and Anthony P.F. Turner
Published 2012. ISBN 978-0-470-93879-9

Integrated Biomaterials for Biomedical Technology
Edited by Murugan Ramalingam, Ashutosh Tiwari, Seeram Ramakrishna and Hisatoshi Kobayashi
Published 2012. ISBN 978-1-118-42385-1

Integrated Biomaterials in Tissue Engineering
Edited by Murugan Ramalingam, Ziyad Haidar, Seeram Ramakrishna, Hisatoshi Kobayashi, and Youssef Haikel
Published 2012. ISBN 978-1-118-31198-1

The Physics of Micropdroplets
Jean Berthier and Kenneth Brakke
Published 2012. ISBN 978-0-470-93880-0

Introduction to Surface Engineering and Functionally Engineered Materials
By Peter Martin
Published 2011 ISBN 978-0-470-63927-6